Optimization Algorithms on
Matrix Manifolds

Optimization Algorithms on Matrix Manifolds

P.-A. Absil
R. Mahony
R. Sepulchre

PRINCETON UNIVERSITY PRESS

PRINCETON AND OXFORD

Published by Princeton University Press
41 William Street, Princeton, New Jersey 08540

In the United Kingdom: Princeton University Press
3 Market Place, Woodstock, Oxfordshire OX20 1SY

Library of Congress Control Number: 2007927538
ISBN: 978-0-691-13298-3

British Library Cataloging-in-Publication Data is
available

This book has been composed in Computer Modern in
LaTeX

The publisher would like to acknowledge the authors
of this volume for providing the camera-ready copy
from which this book was printed.

Printed on acid-free paper. ∞

press.princeton.edu

Printed in the United States of America

10 9 8 7 6 5 4 3 2 1

To our parents

Contents

List of Algorithms

Foreword

Constrained optimization is quite well established as an area of research, and there exist several powerful techniques that address general problems in that area. In this book a special class of constraints is considered, called geometric constraints, which express that the solution of the optimization problem lies on a manifold. This is a recent area of research that provides powerful alternatives to the more general constrained optimization methods. Classical constrained optimization techniques work in an embedded space that can be of a much larger dimension than that of the manifold. Optimization algorithms that work on the manifold have therefore a lower complexity and quite often also have better numerical properties (see, e.g., the numerical integration schemes that preserve invariants such as energy). The authors refer to this as unconstrained optimization in a constrained search space.

The idea that one can describe difference or differential equations whose solution lies on a manifold originated in the work of Brockett, Flaschka, and Rutishauser. They described, for example, isospectral flows that yield time-varying matrices which are all similar to each other and eventually converge to diagonal matrices of ordered eigenvalues. These ideas did not get as much attention in the numerical linear algebra community as in the area of dynamical systems because the resulting difference and differential equations did not lead immediately to efficient algorithmic implementations.

An important book synthesizing several of these ideas is *Optimization and Dynamical Systems* (Springer, 1994), by Helmke and Moore, which focuses on dynamical systems related to gradient flows that converge exponentially to a stationary point that is the solution of some optimization problem. The corresponding discrete-time version of this algorithm would then have linear convergence, which seldom compares favorably with state-of-the-art eigenvalue solvers.

The formulation of higher-order optimization methods on manifolds grew out of these ideas. Some of the people that applied these techniques to basic linear algebra problems include Absil, Arias, Chu, Dehaene, Edelman, Eldén, Gallivan, Helmke, Hüper, Lippert, Mahony, Manton, Moore, Sepulchre, Smith, and Van Dooren. It is interesting to see, on the other hand, that several basic ideas in this area were also proposed by Luenberger and Gabay in the optimization literature in the early 1980s, and this without any use of dynamical systems.

In the present book the authors focus on higher-order methods and include Newton-type algorithms for optimization on manifolds. This requires a lot more machinery, which cannot currently be found in textbooks. The main focus of this book is on optimization problems related to invariant subspaces of matrices, but this is sufficiently general to encompass well the two main aspects of optimization on manifolds: the conceptual algorithm and its convergence analysis based on ideas of differential geometry, and the efficient numerical implementation using state-of-the-art numerical linear algebra techniques.

The book is quite deep in the presentation of the machinery of differential geometry needed to develop higher-order optimization techniques, but it nevertheless succeeds in explaining complicated concepts with simple ideas. These ideas are then used to develop Newton-type methods as well as other superlinear methods such as trust-region methods and inexact and quasi-Newton methods, which precisely put more emphasis on the efficient numerical implementation of the conceptual algorithms.

This is a research monograph in a field that is quickly gaining momentum. The techniques are also being applied to areas of engineering and robotics, as indicated in the book, and it sheds new light on methods such as the Jacobi-Davidson method, which originally came from computational chemistry. The book makes a lot of interesting connections and can be expected to generate several new results in the future.

Paul Van Dooren   January 2007

Notation Conventions

$\mathcal{M}, \mathcal{N}$	manifolds
x, y	points on a manifold
ξ, η, ζ, χ	tangent vectors or vector fields
$\xi_x, \eta_x, \zeta_x, \chi_x$	tangent vectors at x
φ, ψ	coordinate charts
A, B	square matrices
W, X, Y, Z	matrices
$\mathcal{W}, \mathcal{X}, \mathcal{Y}, \mathcal{Z}$	linear subspaces

Conventions related to the definition of functions are stated in Section A.3.

Chapter One

Introduction

This book is about the design of numerical algorithms for computational problems posed on smooth search spaces. The work is motivated by matrix optimization problems characterized by symmetry or invariance properties in the cost function or constraints. Such problems abound in algorithmic questions pertaining to linear algebra, signal processing, data mining, and statistical analysis. The approach taken here is to exploit the special structure of these problems to develop efficient numerical procedures.

An illustrative example is the eigenvalue problem. Because of their scale invariance, eigenvectors are not isolated in vector spaces. Instead, each eigendirection defines a linear subspace of eigenvectors. For numerical computation, however, it is desirable that the solution set consist only of isolated points in the search space. An obvious remedy is to impose a norm equality constraint on iterates of the algorithm. The resulting spherical search space is an *embedded submanifold* of the original vector space. An alternative approach is to "factor" the vector space by the scale-invariant symmetry operation such that any subspace becomes a single point. The resulting search space is a *quotient manifold* of the original vector space. These two approaches provide prototype structures for the problems considered in this book.

Scale invariance is just one of several symmetry properties regularly encountered in computational problems. In many cases, the underlying symmetry property can be exploited to reformulate the problem as a nondegenerate optimization problem on an embedded or quotient manifold associated with the original matrix representation of the search space. These constraint sets carry the structure of nonlinear matrix manifolds. This book provides the tools to exploit such structure in order to develop efficient matrix algorithms in the underlying total vector space.

Working with a search space that carries the structure of a nonlinear manifold introduces certain challenges in the algorithm implementation. In their classical formulation, iterative optimization algorithms rely heavily on the Euclidean vector space structure of the search space; a new iterate is generated by adding an update increment to the previous iterate in order to reduce the cost function. The update direction and step size are generally computed using a local model of the cost function, typically based on (approximate) first and second derivatives of the cost function, at each step. In order to define algorithms on manifolds, these operations must be translated into the language of differential geometry. This process is a significant research program that builds upon solid mathematical foundations. Advances in that direction have been dramatic over the last two decades and have led to a solid conceptual framework. However, generalizing a given optimization algorithm on an abstract manifold is only the first step towards the objective of this book. Turning the algorithm into an efficient numerical procedure is a second step that ultimately justifies or invalidates the first part of the effort. At the time of publishing this book, the second step is more an art than a theory.

Good algorithms result from the combination of insight from differential geometry, optimization, and numerical analysis. A distinctive feature of this book is that as much attention is paid to the practical implementation of the algorithm as to its geometric formulation. In particular, the concrete aspects of algorithm design are formalized with the help of the concepts of *retraction* and *vector transport*, which are relaxations of the classical geometric concepts of motion along geodesics and parallel transport. The proposed approach provides a framework to optimize the efficiency of the numerical algorithms while retaining the convergence properties of their abstract geometric counterparts.

The geometric material in the book is mostly confined to Chapters 3 and 5. Chapter 3 presents an introduction to Riemannian manifolds and tangent spaces that provides the necessary tools to tackle simple gradient-descent optimization algorithms on matrix manifolds. Chapter 5 covers the advanced material needed to define higher-order derivatives on manifolds and to build the analog of first- and second-order local models required in most optimization algorithms. The development provided in these chapters ranges from the foundations of differential geometry to advanced material relevant to our applications. The selected material focuses on those geometric concepts that are particular to the development of numerical algorithms on embedded and quotient manifolds. Not all aspects of classical differential geometry are covered, and some emphasis is placed on material that is nonstandard or difficult to find in the established literature. A newcomer to the field of differential geometry may wish to supplement this material with a classical text. Suggestions for excellent texts are provided in the references.

A fundamental, but deliberate, omission in the book is a treatment of the geometric structure of Lie groups and homogeneous spaces. Lie theory is derived from the concepts of symmetry and seems to be a natural part of a treatise such as this. However, with the purpose of reaching a community without an extensive background in geometry, we have omitted this material in the present book. Occasionally the Lie-theoretic approach provides an elegant shortcut or interpretation for the problems considered. An effort is made throughout the book to refer the reader to the relevant literature whenever appropriate.

The algorithmic material of the book is interlaced with the geometric material. Chapter 4 considers gradient-descent line-search algorithms. These simple optimization algorithms provide an excellent framework within which to study the important issues associated with the implementation of practical algorithms. The concept of retraction is introduced in Chapter 4 as a key step in developing efficient numerical algorithms on matrix manifolds. The later chapters on algorithms provide the core results of the book: the development of Newton-based methods in Chapter 6 and of trust-region methods in Chapter 7, and a survey of other superlinear methods such as conjugate gradients in Chapter 8. We attempt to provide a generic development of each of these methods, building upon the material of the geometric chapters. The methodology is then developed into concrete numerical algorithms on specific examples. In the analysis of superlinear and second-order methods, the concept of vector transport (introduced in Chapter 8) is used to provide an efficient implementation of methods such as conjugate gradient and other quasi-Newton methods. The algorithms obtained in these sections of the book are competitive with state-of-the-art numerical linear algebra algorithms for certain problems.

The running example used throughout the book is the calculation of invariant subspaces of a matrix (and the many variants of this problem). This example is by far, for variants of algorithms developed within the proposed framework, the problem with the broadest scope of applications and the highest degree of achievement to date. Numerical algorithms, based on a geometric formulation, have been developed that compete with the best available algorithms for certain classes of invariant subspace problems. These algorithms are explicitly described in the later chapters of the book and, in part, motivate the whole project. Because of the important role of this class of problems within the book, the first part of Chapter 2 provides a detailed description of the invariant subspace problem, explaining why and how this problem leads naturally to an optimization problem on a matrix manifold. The second part of Chapter 2 presents other applications that can be recast as problems of the same nature. These problems are the subject of ongoing research, and the brief exposition given is primarily an invitation for interested researchers to join with us in investigating these problems and expanding the range of applications considered.

The book should primarily be considered a research monograph, as it reports on recently published results in an active research area that is expected to develop significantly beyond the material presented here. At the same time, every possible effort has been made to make the book accessible to the broadest audience, including applied mathematicians, engineers, and computer scientists with little or no background in differential geometry. It could equally well qualify as a graduate textbook for a one-semester course in advanced optimization. More advanced sections that can be readily skipped at a first reading are indicated with a star. Moreover, readers are encouraged to visit the book home page[1] where supplementary material is available.

The book is an extension of the first author's Ph.D. thesis [Abs03], itself a project that drew heavily on the material of the second author's Ph.D. thesis [Mah94]. It would not have been possible without the many contributions of a quickly expanding research community that has been working in the area

---

[1]http://press.princeton.edu/titles/8586.html

over the last decade. The Notes and References section at the end of each chapter is an attempt to give proper credit to the many contributors, even though this task becomes increasingly difficult for recent contributions. The authors apologize for any omission or error in these notes. In addition, we wish to conclude this introductory chapter with special acknowledgements to people without whom this project would have been impossible. The 1994 monograph [HM94] by Uwe Helmke and John Moore is a milestone in the formulation of computational problems as optimization algorithms on manifolds and has had a profound influence on the authors. On the numerical side, the constant encouragement of Paul Van Dooren and Kyle Gallivan has provided tremendous support to our efforts to reconcile the perspectives of differential geometry and numerical linear algebra. We are also grateful to all our colleagues and friends over the last ten years who have crossed paths as coauthors, reviewers, and critics of our work. Special thanks to Ben Andrews, Chris Baker, Alan Edelman, Michiel Hochstenbach, Knut Hüper, Jonathan Manton, Robert Orsi, and Jochen Trumpf. Finally, we acknowledge the useful feedback of many students on preliminary versions of the book, in particular, Mariya Ishteva, Michel Journée, and Alain Sarlette.

Chapter Two

Motivation and Applications

The problem of optimizing a real-valued function on a matrix manifold appears in a wide variety of computational problems in science and engineering. In this chapter we discuss several examples that provide motivation for the material presented in later chapters. In the first part of the chapter, we focus on the eigenvalue problem. This application receives special treatment because it serves as a running example throughout the book. It is a problem of unquestionable importance that has been, and still is, extensively researched. It falls naturally into the geometric framework proposed in this book as an optimization problem whose natural domain is a matrix manifold—the underlying symmetry is related to the fact that the notion of an eigenvector is scale-invariant. Moreover, there are a wide range of related problems (eigenvalue decompositions, principal component analysis, generalized eigenvalue problems, etc.) that provide a rich collection of illustrative examples that we will use to demonstrate and compare the techniques proposed in later chapters.

Later in this chapter, we describe several research problems exhibiting promising symmetry to which the techniques proposed in this book have not yet been applied in a systematic way. The list is far from exhaustive and is very much the subject of ongoing research. It is meant as an invitation to the reader to consider the broad scope of computational problems that can be cast as optimization problems on manifolds.

2.1 A CASE STUDY: THE EIGENVALUE PROBLEM

The problem of computing eigenspaces and eigenvalues of matrices is ubiquitous in engineering and physical sciences. The general principle of computing an eigenspace is to reduce the complexity of a problem by focusing on a few relevant quantities and dismissing the others. Eigenspace computation is involved in areas as diverse as structural dynamics [GR97], control theory [PLV94], signal processing [CG90], and data mining [BDJ99]. Considering the importance of the eigenproblem in so many engineering applications, it is not surprising that it has been, and still is, a very active field of research.

Let $\mathbb{F}$ stand for the field of real or complex numbers. Let A be an $n \times n$ matrix with entries in $\mathbb{F}$. Any nonvanishing vector $v \in \mathbb{C}^n$ that satisfies

$$Av = \lambda v$$

for some $\lambda \in \mathbb{C}$ is called an *eigenvector* of A; λ is the associated *eigen-*

value, and the couple (λ, v) is called an *eigenpair*. The set of eigenvalues of A is called the *spectrum* of A. The eigenvalues of A are the zeros of the *characteristic polynomial* of A,

$$\mathcal{P}_A(z) \equiv \det(A - zI),$$

and their *algebraic multiplicity* is their multiplicity as zeros of $\mathcal{P}_A$. If T is an invertible matrix and (λ, v) is an eigenpair of A, then (λ, Tv) is an eigenpair of TAT^{-1}. The transformation $A \mapsto TAT^{-1}$ is called a *similarity transformation* of A.

A *(linear) subspace* $\mathcal{S}$ of $\mathbb{F}^n$ is a subset of $\mathbb{F}^n$ that is closed under linear combinations, i.e.,

$$\forall x, y \in \mathcal{S}, \ \forall a, b \in \mathbb{F} : (ax + by) \in \mathcal{S}.$$

A set $\{y_1, \ldots, y_p\}$ of elements of $\mathcal{S}$ such that every element of $\mathcal{S}$ can be written as a linear combination of $y_1, \ldots, y_p$ is called a *spanning set* of $\mathcal{S}$; we say that $\mathcal{S}$ is the *column space* or simply the *span* of the $n \times p$ matrix $Y = [y_1, \ldots, y_p]$ and that Y *spans* $\mathcal{S}$. This is written as

$$\mathcal{S} = \mathrm{span}(Y) = \{Yx : x \in \mathbb{F}^p\} = Y\mathbb{F}^p.$$

The matrix Y is said to have full (column) rank when the columns of Y are linearly independent, i.e., $Yx = 0$ implies $x = 0$. If Y spans $\mathcal{S}$ and has full rank, then the columns of Y form a *basis* of $\mathcal{S}$. Any two bases of $\mathcal{S}$ have the same number of elements, called the *dimension* of $\mathcal{S}$. The set of all p-dimensional subspaces of $\mathbb{F}^n$, denoted by $\mathrm{Grass}(p, n)$, plays an important role in this book. We will see in Section 3.4 that $\mathrm{Grass}(p, n)$ admits a structure of manifold called the *Grassmann manifold*.

The kernel $\ker(B)$ of a matrix B is the subspace formed by the vectors x such that $Bx = 0$. A scalar λ is an eigenvalue of a matrix A if and only if the dimension of the kernel of $(A - \lambda I)$ is greater than zero, in which case $\ker(A - \lambda I)$ is called the *eigenspace of A related to λ*.

An $n \times n$ matrix A naturally induces a mapping on $\mathrm{Grass}(p, n)$ defined by

$$\mathcal{S} \in \mathrm{Grass}(p, n) \ \mapsto \ A\mathcal{S} := \{Ay : y \in \mathcal{S}\}.$$

A subspace $\mathcal{S}$ is said to be an *invariant subspace* or *eigenspace* of A if $A\mathcal{S} \subseteq \mathcal{S}$. The *restriction* $A|_\mathcal{S}$ of A to an invariant subspace $\mathcal{S}$ is the operator $x \mapsto Ax$ whose domain is $\mathcal{S}$. An invariant subspace $\mathcal{S}$ of A is called *spectral* if, for every eigenvalue λ of $A|_\mathcal{S}$, the multiplicities of λ as an eigenvalue of $A|_\mathcal{S}$ and as an eigenvalue of A are identical; equivalently, $X^T A X$ and $X_\perp^T A X_\perp$ have no eigenvalue in common when $[X|X_\perp]$ satisfies $[X|X_\perp]^T[X|X_\perp] = I_n$ and $\mathrm{span}(X) = \mathcal{S}$.

In many (arguably the majority of) eigenproblems of interest, the matrix A is real and symmetric ($A = A^T$). The eigenvalues of an $n \times n$ symmetric matrix A are reals $\lambda_1 \leq \cdots \leq \lambda_n$, and the associated eigenvectors $v_1, \ldots, v_n$ are real and can be chosen *orthonormal*, i.e.,

$$v_i^T v_j = \begin{cases} 1 \text{ if } i = j, \\ 0 \text{ if } i \neq j. \end{cases}$$

Equivalently, for every symmetric matrix A, there is an orthonormal matrix V (whose columns are eigenvectors of A) and a diagonal matrix Λ such that $A = V\Lambda V^T$. The eigenvalue λ_1 is called the *leftmost eigenvalue* of A, and an eigenpair (λ_1, v_1) is called a *leftmost eigenpair*. A p-dimensional *leftmost invariant subspace* is an invariant subspace associated with $\lambda_1, \ldots, \lambda_p$. Similarly, a p-dimensional *rightmost invariant subspace* is an invariant subspace associated with $\lambda_{n-p+1}, \ldots, \lambda_n$. Finally, *extreme eigenspaces* refer collectively to leftmost and rightmost eigenspaces.

Given two $n \times n$ matrices A and B, we say that (λ, v) is an eigenpair of the *pencil* (A, B) if

$$Av = \lambda Bv.$$

Finding eigenpairs of a matrix pencil is known as the *generalized eigenvalue problem*. The generalized eigenvalue problem is said to be *symmetric / positive-definite* when A is symmetric and B is symmetric positive-definite (i.e., $x^T Bx > 0$ for all nonvanishing x). In this case, the eigenvalues of the pencil are all real and the eigenvectors can be chosen to form a B-orthonormal basis. A subspace $\mathcal{Y}$ is called a *(generalized) invariant subspace* (or a *deflating subspace*) of the symmetric / positive-definite pencil (A, B) if $B^{-1}Ay \in \mathcal{Y}$ for all $y \in \mathcal{Y}$, which can also be written $B^{-1}A\mathcal{Y} \subseteq \mathcal{Y}$ or $A\mathcal{Y} \subseteq B\mathcal{Y}$. The simplest example is when $\mathcal{Y}$ is spanned by a single eigenvector of (A, B), i.e., a nonvanishing vector y such that $Ay = \lambda By$ for some eigenvalue λ. More generally, every eigenspace of a symmetric / positive-definite pencil is spanned by eigenvectors of (A, B). Obviously, the generalized eigenvalue problem reduces to the standard eigenvalue problem when $B = I$.

2.1.1 The eigenvalue problem as an optimization problem

The following result is instrumental in formulating extreme eigenspace computation as an optimization problem. (Recall that $\mathrm{tr}(A)$, the *trace* of A, denotes the sum of the diagonal elements of A.)

Proposition 2.1.1 *Let A and B be symmetric $n \times n$ matrices and let B be positive-definite. Let $\lambda_1 \leq \cdots \leq \lambda_n$ be the eigenvalues of the pencil (A, B). Consider the* generalized Rayleigh quotient

$$f(Y) = \mathrm{tr}(Y^T AY(Y^T BY)^{-1}) \tag{2.1}$$

defined on the set of all $n \times p$ full-rank matrices. Then the following statements are equivalent:

(i) span(Y_*) *is a leftmost invariant subspace of (A, B);*
(ii) Y_* *is a global minimizer of (2.1) over all $n \times p$ full-rank matrices;*
(iii) $f(Y_*) = \sum_{i=1}^{p} \lambda_i$.

Proof. For simplicity of the development we will assume that $\lambda_p < \lambda_{p+1}$, but the result also holds without this hypothesis. Let V be an $n \times n$ matrix for which $V^T BV = I_n$ and $V^T AV = \mathrm{diag}(\lambda_1, \ldots, \lambda_n)$, where $\lambda_1 \leq \cdots \leq \lambda_n$.

Such a V always exists. Let $Y \in \mathbb{R}^{n \times p}$ and put $Y = VM$. Since $Y^T BY = I_p$, it follows that $M^T M = I_p$. Then

$$\operatorname{tr}(Y^T AY) = \operatorname{tr}(M^T \operatorname{diag}(\lambda_1, \ldots, \lambda_n)M)$$

$$= \sum_{i=1}^{n} \lambda_i \sum_{j=1}^{p} m_{ij}^2$$

$$= \sum_{j=1}^{p} \left(\lambda_p + \sum_{i=1}^{p} (\lambda_i - \lambda_p)m_{ij}^2 + \sum_{i=p+1}^{n} (\lambda_i - \lambda_p)m_{ij}^2 \right)$$

$$= \sum_{i=1}^{p} \lambda_i + \sum_{i=1}^{p} (\lambda_p - \lambda_i)\left(1 - \sum_{j=1}^{p} m_{ij}^2 \right) + \sum_{j=1}^{p} \sum_{i=p+1}^{n} (\lambda_i - \lambda_p)m_{ij}^2.$$

Since the second and last terms are nonnegative, it follows that $\operatorname{tr}(Y^T AY) \geq \sum_{i=1}^{p} \lambda_i$. Equality holds if and only if the second and last terms vanish. This happens if and only if the $(n - p) \times p$ lower part of M vanishes (and hence the $p \times p$ upper part of M is orthogonal), which means that $Y = VM$ spans a p-dimensional leftmost invariant subspace of (A, B). □

For the case $p = 1$ and $B = I$, and assuming that the leftmost eigenvalue λ_1 of A has multiplicity 1, Proposition 2.1.1 implies that the global minimizers of the cost function

$$f : \mathbb{R}_*^n \to \mathbb{R} : y \mapsto f(y) = \frac{y^T Ay}{y^T y} \tag{2.2}$$

are the points $v_1 r$, $r \in \mathbb{R}_*$, where $\mathbb{R}_*^n$ is $\mathbb{R}^n$ with the origin removed and v_1 is an eigenvector associated with λ_1. The cost function (2.2) is called the *Rayleigh quotient* of A. Minimizing the Rayleigh quotient can be viewed as an optimization problem on a manifold since, as we will see in Section 3.1.1, $\mathbb{R}_*^n$ admits a natural manifold structure. However, the manifold aspect is of little interest here, as the manifold is simply the classical linear space $\mathbb{R}^n$ with the origin excluded.

A less reassuring aspect of this minimization problem is that the minimizers are not isolated but come up as the continuum $v_1 \mathbb{R}_*$. Consequently, some important convergence results for optimization methods do not apply, and several important algorithms may fail, as illustrated by the following proposition.

Proposition 2.1.2 *Newton's method applied to the Rayleigh quotient (2.2) yields the iteration $y \mapsto 2y$ for every y such that $f(y)$ is not an eigenvalue of A.*

Proof. Routine manipulations yield $\operatorname{grad} f(y) = \frac{2}{y^T y}(Ay - f(y)y)$ and $\operatorname{Hess} f(y)[z] = \mathrm{D}(\operatorname{grad} f)(y)[z] = \frac{2}{y^T y}(Az - f(y)z) - \frac{4}{(y^T y)^2}(y^T Azy + y^T zAy - 2f(y)y^T zy) = H_y z$, where $H_y = \frac{2}{y^T y}(A - f(y)I - \frac{2}{y^T y}(yy^T A + Ayy^T - 2f(y)yy^T)) = \frac{2}{y^T y}(I - 2\frac{yy^T}{y^T y})(A - f(y)I)(I - 2\frac{yy^T}{y^T y})$. It follows that H_y is singular if and only if $f(y)$ is an eigenvalue of A. When $f(y)$ is not an eigenvalue of A, the Newton equation $H_y \eta = -\text{grad} f(y)$ admits one and only one solution, and it is easy to check that this solution is $\eta = y$. In conclusion, the Newton iteration maps y to $y + \eta = 2y$. □

This result is not particular to the Rayleigh quotient. It holds for any function f homogeneous of degree zero, i.e., $f(y\alpha) = f(y)$ for all real $\alpha \neq 0$.

A remedy is to restrain the domain of f to some subset $\mathcal{M}$ of $\mathbb{R}^n_*$ so that any ray $y\mathbb{R}_*$ contains at least one and at most finitely many points of $\mathcal{M}$. Notably, this guarantees that the minimizers are isolated. An elegant choice for $\mathcal{M}$ is the unit sphere

$$S^{n-1} := \{y \in \mathbb{R}^n : y^T y = 1\}.$$

Restricting the Rayleigh quotient (2.2) to S^{n-1} gives us a well-behaved cost function with isolated minimizers. What we lose, however, is the linear structure of the domain of the cost function. The goal of this book is to provide a toolbox of techniques to allow practical implementation of numerical optimization methods on nonlinear embedded (matrix) manifolds in order to address problems of exactly this nature.

Instead of restraining the domain of f to some subset of $\mathbb{R}^n$, another approach, which seems *a priori* more challenging but fits better with the geometry of the problem, is to work on a domain where all points on a ray $y\mathbb{R}_*$ are considered just one point. This viewpoint is especially well suited to eigenvector computation since the useful information of an eigenvector is fully contained in its direction. This leads us to consider the set

$$\mathcal{M} := \{y\mathbb{R}_* : y \in \mathbb{R}^n_*\}.$$

Since the Rayleigh quotient (2.2) satisfies $f(y\alpha) = f(y)$, it induces a well-defined function $\tilde{f}(y\mathbb{R}_*) := f(y)$ whose domain is $\mathcal{M}$. Notice that whereas the Rayleigh quotient restricted to S^{n-1} has two minimizers $\pm v_1$, the Rayleigh quotient $\tilde{f}$ has only one minimizer $v_1\mathbb{R}_*$ on $\mathcal{M}$. It is shown in Chapter 3 that the set $\mathcal{M}$, called the *real projective space*, admits a natural structure of *quotient manifold*. The material in later chapters provides techniques tailored to (matrix) quotient manifold structures that lead to practical implementation of numerical optimization methods. For the simple case of a single eigenvector, algorithms proposed on the sphere are numerically equivalent to those on the real-projective quotient space. However, when the problem is generalized to the computation of p-dimensional invariant subspaces, the quotient approach, which leads to the Grassmann manifold, is seen to be the better choice.

2.1.2 Some benefits of an optimization framework

We will illustrate throughout the book that optimization-based eigenvalue algorithms have a number of desirable properties.

An important feature of all optimization-based algorithms is that optimization theory provides a solid framework for the convergence analysis.

Many optimization-based eigenvalue algorithms exhibit almost global convergence properties. This means that convergence to a solution of the optimization problem is guaranteed for almost every initial condition. The property follows from general properties of the optimization scheme and does not need to be established as a specific property of a particular algorithm.

The speed of convergence of the algorithm is also an intrinsic property of optimization-based algorithms. Gradient-based algorithms converge *linearly*; i.e., the contraction rate of the error between successive iterates is asymptotically bounded by a constant $c < 1$. In contrast, Newton-like algorithms have *superlinear* convergence; i.e., the contraction rate asymptotically converges to zero. (We refer the reader to Section 4.3 for details.)

Characterizing the global behavior and the (local) convergence rate of a given algorithm is an important performance measure of the algorithm. In most situations, this analysis is a free by-product of the optimization framework.

Another challenge of eigenvalue algorithms is to deal efficiently with *large-scale* problems. Current applications in data mining or structural analysis easily involve matrices of dimension $10^5 - 10^6$ [AHLT05]. In those applications, the matrix is typically sparse; i.e., the number of nonzero elements is $O(n)$ or even less, where n is the dimension of the matrix. The goal in such applications is to compute a few eigenvectors corresponding to a small relevant portion of the spectrum. Algorithms are needed that require a small storage space and produce their iterates in $O(n)$ operations. Such algorithms permit matrix-vector products $x \mapsto Ax$, which require $O(n)$ operations if A is sparse, but they forbid matrix factorizations, such as QR and LU, that destroy the sparse structure of A. Algorithms that make use of A only in the form of the operator $x \mapsto Ax$ are called *matrix-free*.

All the algorithms in this book, designed and analyzed using a differential geometric optimization approach, satisfy at least some of these requirements. The trust-region approach presented in Chapter 7 satisfies all the requirements. Such strong convergence analysis is rarely encountered in available eigenvalue methods.

2.2 RESEARCH PROBLEMS

This section is devoted to briefly presenting several general computational problems that can be tackled by a manifold-based optimization approach. Research on the problems presented is mostly at a preliminary stage and the discussion provided here is necessarily at the level of an overview. The interested reader is encouraged to consult the references provided.

2.2.1 Singular value problem

The singular value decomposition is one of the most useful tasks in numerical computations [HJ85, GVL96], in particular when it is used in dimension reduction problems such as principal component analysis [JW92].

Matrices U, Σ, and V form a *singular value decomposition* (SVD) of an arbitrary matrix $A \in \mathbb{R}^{m \times n}$ (to simplify the discussion, we assume that $m \geq n$) if

$$A = U\Sigma V^T, \tag{2.3}$$

with $U \in \mathbb{R}^{m \times m}$, $U^T U = I_m$, $V \in \mathbb{R}^{n \times n}$, $V^T V = I_n$, $\Sigma \in \mathbb{R}^{m \times n}$, Σ diagonal with diagonal entries $\sigma_1 \geq \cdots \geq \sigma_n \geq 0$. Every matrix A admits an SVD. The diagonal entries σ_i of Σ are called the singular values of A, and the corresponding columns u_i and v_i of U and V are called the left and right singular vectors of A. The triplets (σ_i, u_i, v_i) are then called singular triplets of A. Note that an SVD expresses the matrix A as a sum of rank-1 matrices,

$$A = \sum_{i=1}^{n} \sigma_i u_i v_i^T.$$

The SVD is involved in several least-squares problems. An important example is the best low-rank approximation of an $m \times n$ matrix A in the least-squares sense, i.e.,

$$\arg \min_{X \in \mathcal{R}_p} \|A - X\|_F^2,$$

where $\mathcal{R}_p$ denotes the set of all $m \times n$ matrices with rank p and $\|\cdot\|_F^2$ denotes the Frobenius norm, i.e., the sum of the squares of the elements of its argument. The solution of this problem is given by a truncated SVD

$$X = \sum_{i=1}^{p} \sigma_i u_i v_i^T,$$

where (σ_i, u_i, v_i) are singular triplets of A (ordered by decreasing value of σ). This result is known as the Eckart-Young-Mirsky theorem; see Eckart and Young [EY36] or, e.g., Golub and Van Loan [GVL96].

The singular value problem is closely related to the eigenvalue problem. It follows from (2.3) that $A^T A = V \Sigma^2 V^T$, hence the squares of the singular values of A are the eigenvalues of $A^T A$ and the corresponding right singular vectors are the corresponding eigenvectors of $A^T A$. Similarly, $AA^T = U\Sigma^2 U^T$, hence the left singular vectors of A are the eigenvectors of AA^T. One approach to the singular value decomposition problem is to rely on eigenvalue algorithms applied to the matrices $A^T A$ and AA^T. Alternatively, it is possible to compute simultaneously a few dominant singular triplets (i.e., those corresponding to the largest singular values) by maximizing the cost function

$$f(U, V) = \text{tr}(U^T A V N)$$

subject to $U^T U = I_p$ and $V^T V = I_p$, where $N = \text{diag}(\mu_1, \ldots, \mu_p)$, with $\mu_1 > \cdots > \mu_p > 0$ arbitrary. If (U, V) is a solution of this maximization problem, then the columns u_i of U and v_i of V are the ith dominant left and right singular vectors of A. This is an optimization problem on a manifold; indeed, constraint sets of the form $\{U \in \mathbb{R}^{n \times p} : U^T U = I_p\}$ have the structure of an embedded submanifold of $\mathbb{R}^{n \times p}$ called the *(orthogonal) Stiefel manifold* (Section 3.3), and the constraint set for (U, V) is then a product manifold (Section 3.1.6).

2.2.2 Matrix approximations

In the previous section, we saw that the truncated SVD solves a particular kind of matrix approximation problem, the best low-rank approximation in the least-squares sense. There are several other matrix approximation problems that can be written as minimizing a real-valued function on a manifold.

Within the matrix nearness framework

$$\min_{X \in \mathcal{M}} \|A - X\|_F^2,$$

we have, for example, the following symmetric positive-definite least-squares problem.

$$\begin{aligned}
\text{Find} \quad & C \in \mathbb{R}^{n \times n} \\
\text{to minimize} \quad & \|C - C_0\|^2 \\
\text{subject to} \quad & \operatorname{rank}(C) = p, \ C = C^T, \ C \succeq 0,
\end{aligned} \tag{2.4}$$

where $C \succeq 0$ denotes that C is positive-semidefinite; i.e., $x^T C x \geq 0$ for all $x \in \mathbb{R}^n$. We can rephrase this constrained problem as a problem on the set $\mathbb{R}_*^{n \times p}$ of all $n \times p$ full-rank matrices by setting $C = YY^T$, $Y \in \mathbb{R}_*^{n \times p}$. The new search space is simpler, but the new cost function

$$f : \mathbb{R}_*^{n \times p} \to \mathbb{R} : Y \mapsto \|YY^T - C_0\|^2$$

has the symmetry property $f(YQ) = f(Y)$ for all orthonormal $p \times p$ matrices Q, hence minimizers of f are not isolated and the problems mentioned in Section 2.1 for Rayleigh quotient minimization are likely to appear. This again points to a quotient manifold approach, where a set $\{YQ : Q^TQ = I\}$ is identified as one point of the quotient manifold.

A variation on the previous problem is the best low-rank approximation of a correlation matrix by another correlation matrix [BX05]:

$$\begin{aligned}
\text{Find} \quad & C \in \mathbb{R}^{n \times n} \\
\text{to minimize} \quad & \|C - C_0\|^2 \\
\text{subject to} \quad & \operatorname{rank}(C) = p, \ C_{ii} = 1 \ (i = 1, \ldots, n), \ C \succeq 0.
\end{aligned} \tag{2.5}$$

Again, setting $C = YY^T$, $Y \in \mathbb{R}_*^{n \times p}$, takes care of the rank constraint. Replacing this form in the constraint $C_{ii} = 1$, $i = 1, \ldots, n$, yields $\operatorname{diag}(YY^T) = I$. This constraint set can be shown to admit a manifold structure called an *oblique manifold*:

$$\mathcal{OB} := \{Y \in \mathbb{R}_*^{n \times p} : \operatorname{diag}(YY^T) = I_n\};$$

see, e.g., [Tre99, TL02, AG06]. This manifold-based approach is further developed in [GP07].

A more general class of matrix approximation problems is the *Procrustes problem* [GD04]

$$\min_{X \in \mathcal{M}} \|AX - B\|_F^2, \quad A \in \mathbb{R}^{l \times m}, B \in \mathbb{R}^{l \times n}, \tag{2.6}$$

where $\mathcal{M} \subseteq \mathbb{R}^{m \times n}$. Taking $\mathcal{M} = \mathbb{R}^{m \times n}$ yields a standard least-squares problem. The orthogonal case, $\mathcal{M} = O_n = \{X \in \mathbb{R}^{n \times n} : X^T X = I\}$, has a closed-form solution in terms of the polar decomposition of $B^T A$ [GVL96]. The case $\mathcal{M} = \{X \in \mathbb{R}^{m \times n} : X^T X = I\}$, where $\mathcal{M}$ is a Stiefel manifold, is known as the *unbalanced orthogonal Procrustes problem*; see [EP99] and references therein. The case $\mathcal{M} = \{X \in \mathbb{R}^{n \times n} : \text{diag}(X^T X) = I_n\}$, where $\mathcal{M}$ is an oblique manifold, is called the *oblique Procrustes problem* [Tre99, TL02].

2.2.3 Independent component analysis

Independent component analysis (ICA), also known as blind source separation (BSS), is a computational problem that has received much attention in recent years, particularly for its biomedical applications [JH05]. A typical application of ICA is the "cocktail party problem", where the task is to recover one or more signals, supposed to be statistically independent, from recordings where they appear as linear mixtures. Specifically, assume that n measured signals $x(t) = [x_1(t), \ldots, x_n(t)]^T$ are instantaneous linear mixtures of p underlying, statistically independent source signals $s(t) = [s_1(t), \ldots, s_p(t)]^T$. In matrix notation, we have

$$x(t) = As(t),$$

where the $n \times p$ matrix A is an unknown constant *mixing matrix* containing the mixture coefficients. The ICA problem is to identify the mixing matrix A or to recover the source signals $s(t)$ using only the observed signals $x(t)$.

This problem is usually translated into finding an $n \times p$ *separating matrix* (or *demixing matrix*) W such that the signals $y(t)$ given by

$$y(t) = W^T x(t)$$

are "as independent as possible". This approach entails defining a cost function $f(W)$ to measure the independence of the signals $y(t)$, which brings us to the realm of numerical optimization. This separation problem, however, has the structural symmetry property that the measure of independence of the components of $y(t)$ should not vary when different scaling factors are applied to the components of $y(t)$. In other words, the cost function f should satisfy the invariance property $f(WD) = f(W)$ for all nonsingular diagonal matrices D. A possible choice for the cost function f is the log likelihood criterion

$$f(W) := \sum_{k=1}^{K} n_k (\log \det \text{diag}(W^* C_k W) - \log \det(W^* C_k W)), \qquad (2.7)$$

where the C_k's are covariance-like matrices constructed from $x(t)$ and $\text{diag}(A)$ denotes the diagonal matrix whose diagonal is the diagonal of A; see, e.g., [Yer02] for the choice of the matrices C_k, and [Pha01] for more information on the cost function (2.7).

The invariance property $f(WD) = f(W)$, similarly to the homogeneity property observed for the Rayleigh quotient (2.2), produces a continuum of minimizers if W is allowed to vary on the whole space of $n \times p$ matrices. Much as in the case of the Rayleigh quotient, this can be addressed by restraining the domain of f to a constraint set that singles out finitely many points in each equivalence class $\{WD : D \text{ diagonal}\}$; a possible choice for the constraint set is the oblique manifold

$$\mathcal{OB} = \{W \in \mathbb{R}_*^{n \times p} : \text{diag}(WW^T) = I_n\}.$$

Another possibility is to identify all the matrices within an equivalence class $\{WD : D \text{ diagonal}\}$ as a single point, which leads to a quotient manifold approach.

Methods for ICA based on differential-geometric optimization have been proposed by, among others, Amari et al. [ACC00], Douglas [Dou00], Rahbar and Reilly [RR00], Pham [Pha01], Joho and Mathis [JM02], Joho and Rahbar [JR02], Nikpour et al. [NMH02], Afsari and Krishnaprasad [AK04], Nishimori and Akaho [NA05], Plumbley [Plu05], Absil and Gallivan [AG06], Shen et al. [SHS06], and Hüeper et al. [HSS06]; see also several other references therein.

2.2.4 Pose estimation and motion recovery

In the pose estimation problem, an object is known via a set of landmarks $\{m_i\}_{i=1,\dots,N}$, where $m_i := (x_i, y_i, z_i)^T \in \mathbb{R}^3$ are the three coordinates of the ith landmark in an object-centered frame. The coordinates m_i' of the landmarks in a camera-centered frame obey a rigid body displacement law

$$m_i' = Rm_i + t,$$

where $R \in SO_3$ (i.e., $R^T R = I$ and $\det(R) = 1$) represents a rotation and $t \in \mathbb{R}^3$ stands for a translation. Each landmark point produces a normalized image point in the image plane of the camera with coordinates

$$u_i = \frac{Rm_i + t}{e_3^T(Rm_i + t)}.$$

The pose estimation problem is to estimate the pose (R, t) in the manifold $SO_3 \times \mathbb{R}^3$ from a set of point correspondences $\{(u_i, m_i)\}_{i=1,\dots,N}$. A possible approach is to minimize the real-valued function

$$f : SO_3 \times \mathbb{R}^3 \to \mathbb{R} : (R, t) \mapsto \sum_{i=1}^{N} \|(I - u_i u_i^T)(Rm_i + t)\|^2,$$

which vanishes if and only if the points u_i and m_i' are collinear, i.e., u_i is indeed the coordinate vector of the projection of the ith landmark onto the image plane of the camera. This is an optimization problem on the manifold $SO_3 \times \mathbb{R}^3$. Since rigid body motions can be composed to obtain another rigid body motion, this manifold possesses a group structure called the *special Euclidean group* SE_3.

A related problem is motion and structure recovery from a sequence of images. Now the object is unknown, but two or more images are available from different angles. Assume that N landmarks have been selected on the object and, for simplicity, consider only two images of the object. The coordinates m'_i and m''_i of the ith landmark in the first and second camera frames are related by a rigid body motion

$$m''_i = Rm'_i + t.$$

Again without loss of generality, the coordinates of the projections of the ith landmark onto each camera image plane are given by $p_i = \frac{m'_i}{e_3^T m'_i}$ and $q_i = \frac{m''_i}{e_3^T m''_i}$. The motion and structure recovery problem is, from a set of corresponding image points $\{(p_i, q_i)\}_{i=1,\dots,N}$, to recover the camera motion (R, t) and the three-dimensional coordinates of the points that the images correspond to. It is a classical result in computer vision that corresponding coordinate vectors p and q satisfy the *epipolar constraint*

$$p^T R^T t^\wedge q = 0,$$

where $t^\wedge$ is the 3×3 skew-symmetric matrix

$$t^\wedge := \begin{bmatrix} 0 & -t_3 & t_2 \\ t_3 & 0 & -t_1 \\ -t_2 & t_1 & 0 \end{bmatrix}.$$

To recover the motion $(R, t) \in SO_3 \times \mathbb{R}^3$ from a given set of image correspondences $\{(p_i, q_i)\}_{i=1,\dots,N}$, it is thus natural to consider the cost function

$$f(R, t) := \sum_{i=1}^{N} (p_i^T R^T t^\wedge q_i)^2, \quad p_i, q_i \in \mathbb{R}^3, (R, t) \in SO_3 \times \mathbb{R}^3.$$

This function is homogeneous in t. As in the case of Rayleigh quotient minimization, this can be addressed by restricting t to the unit sphere S^2, which yields the problem of minimizing the cost function

$$f(R, t) := \sum_{i=1}^{N} (p_i^T R^T t^\wedge q_i)^2, \quad p_i, q_i \in \mathbb{R}^3, (R, t) \in SO_3 \times S^2.$$

Equivalently, this problem can be written as the minimization of the cost function

$$f(E) =: \sum_{i=1}^{N} (p_i^T E q_i)^2, \quad p_i, q_i \in \mathbb{R}^3, E \in \mathcal{E}_1,$$

where $\mathcal{E}_1$ is the *normalized essential manifold*

$$\mathcal{E}_1 := \{Rt^\wedge : R \in SO_3, \ t^\wedge \in \mathfrak{so}_3, \ \tfrac{1}{2}\mathrm{tr}((t^\wedge)^T t^\wedge) = 1\}.$$

($\mathfrak{so}_3 = \{\Omega \in \mathbb{R}^{3\times 3} : \Omega^T = -\Omega\}$ is the Lie algebra of SO_3, and the tr function returns the sum of the diagonal elements of its argument.)

For more details on multiple-view geometry, we refer the reader to Hartley and Zisserman [HZ03]. Applications of manifold optimization to computer vision problems can be found in the work of Ma et al. [MKS01], Lee and Moore [LM04], Liu et al. [LSG04], and Helmke et al. [HHLM07].

2.3 NOTES AND REFERENCES

Each chapter of this book (excepting the introduction) has a Notes and References section that contains pointers to the literature. In the following chapters, all the citations will appear in these dedicated sections.

Recent textbooks and surveys on the eigenvalue problem include Golub and van der Vorst [GvdV00], Stewart [Ste01], and Sorensen [Sor02]. An overview of applications can be found in Saad [Saa92]. A major reference for the symmetric eigenvalue problem is Parlett [Par80]. The characterization of eigenproblems as minimax problems goes back to the time of Poincaré. Early references are Fischer [Fis05] and Courant [Cou20], and the results are often referred to as the Courant-Fischer minimax formulation. The formulation is heavily exploited in perturbation analysis of Hermitian eigenstructure. Good overviews are available in Parlett [Par80, §10 and 11, especially §10.2], Horn and Johnson [HJ91, §4.2], and Wilkinson [Wil65, §2]. See also Bhatia [Bha87] and Golub and Van Loan [GVL96, §8.1].

Until recently, the differential-geometric approach to the eigenproblem had been scarcely exploited because of tough competition from some highly efficient mainstream algorithms combined with a lack of optimization algorithms on manifolds geared towards computational efficiency. However, thanks in particular to the seminal work of Helmke and Moore [HM94] and Edelman, Arias, and Smith [Smi93, Smi94, EAS98], and more recent work by Absil et al. [ABG04, ABG07], manifold-based algorithms have now appeared that are competitive with state-of-the-art methods and sometimes shed new light on their properties. Papers that apply differential-geometric concepts to the eigenvalue problem include those by Chen and Amari [CA01], Lundström and Eldén [LE02], Simoncinin and Eldén [SE02], Brandts [Bra03], Absil et al. [AMSV02, AMS04, ASVM04, ABGS05, ABG06b], and Baker et al. [BAG06]. One "mainstream" approach capable of satisfying all the requirements in Section 2.1.2 is the Jacobi-Davidson conjugate gradient (JDCG) method of Notay [Not02]. Interestingly, it is closely related to an algorithm derived from a manifold-based trust-region approach (see Chapter 7 or [ABG06b]).

The proof of Proposition 2.1.1 is adapted from [Fan49]. The fact that the classical Newton method fails for the Rayleigh quotient (Proposition 2.1.2) was pointed out in [ABG06b], and a proof was given in [Zho06].

Major references for Section 2.2 include Helmke and Moore [HM94], Edelman et al. [EAS98], and Lippert and Edelman [LE00]. The cost function suggested for the SVD (Section 2.2.1) comes from Helmke and Moore [HM94, Ch. 3]. Problems (2.4) and (2.5) are particular instances of the *least-squares covariance adjustment problem* recently defined by Boyd and Xiao [BX05]; see also Manton et al. [MMH03], Grubisic and Pietersz [GP07], and several references therein.

Chapter Three

Matrix Manifolds: First-Order Geometry

The constraint sets associated with the examples discussed in Chapter 2 have a particularly rich geometric structure that provides the motivation for this book. The constraint sets are *matrix manifolds* in the sense that they are manifolds in the meaning of classical differential geometry, for which there is a natural representation of elements in the form of matrix arrays.

The matrix representation of the elements is a key property that allows one to provide a natural development of differential geometry in a matrix algebra formulation. The goal of this chapter is to introduce the fundamental concepts in this direction: manifold structure, tangent spaces, cost functions, differentiation, Riemannian metrics, and gradient computation.

There are two classes of matrix manifolds that we consider in detail in this book: embedded submanifolds of $\mathbb{R}^{n \times p}$ and quotient manifolds of $\mathbb{R}^{n \times p}$ (for $1 \leq p \leq n$). Embedded submanifolds are the easiest to understand, as they have the natural form of an explicit constraint set in matrix space $\mathbb{R}^{n \times p}$. The case we will be mostly interested in is the set of orthonormal $n \times p$ matrices that, as will be shown, can be viewed as an embedded submanifold of $\mathbb{R}^{n \times p}$ called the Stiefel manifold $\mathrm{St}(p, n)$. In particular, for $p = 1$, the Stiefel manifold reduces to the unit sphere S^{n-1}, and for $p = n$, it reduces to the set of orthogonal matrices $O(n)$.

Quotient spaces are more difficult to visualize, as they are not defined as sets of matrices; rather, each point of the quotient space is an equivalence class of $n \times p$ matrices. In practice, an example $n \times p$ matrix from a given equivalence class is used to represent an element of matrix quotient space in computer memory and in our numerical development. The calculations related to the geometric structure of a matrix quotient manifold can be expressed directly using the tools of matrix algebra on these representative matrices.

The focus of this first geometric chapter is on the concepts from differential geometry that are required to generalize the steepest-descent method, arguably the simplest approach to unconstrained optimization. In $\mathbb{R}^n$, the steepest-descent algorithm updates a current iterate x in the direction where the first-order decrease of the cost function f is most negative. Formally, the update direction is chosen to be the unit norm vector η that minimizes the directional derivative

$$\mathrm{D}f(x)[\eta] = \lim_{t \to 0} \frac{f(x + t\eta) - f(x)}{t}. \tag{3.1}$$

When the domain of f is a manifold $\mathcal{M}$, the argument $x + t\eta$ in (3.1) does not make sense in general since $\mathcal{M}$ is not necessarily a vector space. This leads to the important concept of a tangent vector (Section 3.5). In order to define the notion of a steepest-descent direction, it will then remain to define the *length* of a tangent vector, a task carried out in Section 3.6 where the concept of a Riemannian manifold is introduced. This leads to a definition of the gradient of a function, the generalization of steepest-descent direction on a Riemannian manifold.

3.1 MANIFOLDS

We define the notion of a manifold in its full generality; then we consider the simple but important case of linear manifolds, a linear vector space interpreted as a manifold with Euclidean geometric structure. The manifold of $n \times p$ real matrices, from which all concrete examples in this book originate, is a linear manifold.

A d-dimensional manifold can be informally defined as a set $\mathcal{M}$ covered with a "suitable" collection of coordinate patches, or charts, that identify certain subsets of $\mathcal{M}$ with open subsets of $\mathbb{R}^d$. Such a collection of coordinate charts can be thought of as the basic structure required to do differential calculus on $\mathcal{M}$.

It is often cumbersome or impractical to use coordinate charts to (locally) turn computational problems on $\mathcal{M}$ into computational problems on $\mathbb{R}^d$. The numerical algorithms developed later in this book rely on exploiting the natural matrix structure of the manifolds associated with the examples of interest, rather than imposing a local $\mathbb{R}^d$ structure. Nevertheless, coordinate charts are an essential tool for addressing fundamental notions such as the differentiability of a function on a manifold.

3.1.1 Definitions: charts, atlases, manifolds

The abstract definition of a manifold relies on the concepts of charts and atlases.

Let $\mathcal{M}$ be a set. A bijection (one-to-one correspondence) φ of a subset $\mathcal{U}$ of $\mathcal{M}$ onto an open subset of $\mathbb{R}^d$ is called a d-dimensional *chart of the set $\mathcal{M}$*, denoted by $(\mathcal{U}, \varphi)$. When there is no risk of confusion, we will simply write φ for $(\mathcal{U}, \varphi)$. Given a chart $(\mathcal{U}, \varphi)$ and $x \in \mathcal{U}$, the elements of $\varphi(x) \in \mathbb{R}^d$ are called the *coordinates* of x in the chart $(\mathcal{U}, \varphi)$.

The interest of the notion of chart $(\mathcal{U}, \varphi)$ is that it makes it possible to study objects associated with $\mathcal{U}$ by bringing them to the subset $\varphi(\mathcal{U})$ of $\mathbb{R}^d$. For example, if f is a real-valued function on $\mathcal{U}$, then $f \circ \varphi^{-1}$ is a function from $\mathbb{R}^d$ to $\mathbb{R}$, with domain $\varphi(\mathcal{U})$, to which methods of real analysis apply. To take advantage of this idea, we must require that each point of the set $\mathcal{M}$ be at least in one chart domain; moreover, if a point x belongs to the domains of two charts $(\mathcal{U}_1, \varphi_1)$ and $(\mathcal{U}_2, \varphi_2)$, then the two charts must give compatible information: for example, if a real-valued function f is defined

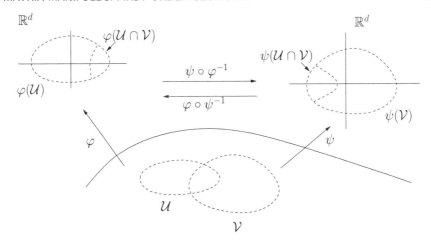

Figure 3.1 Charts.

on $\mathcal{U}_1 \cap \mathcal{U}_2$, then $f \circ \varphi_1^{-1}$ and $f \circ \varphi_2^{-1}$ should have the same differentiability properties on $\mathcal{U}_1 \cap \mathcal{U}_2$.

The following concept takes these requirements into account. A (C^∞) *atlas of* $\mathcal{M}$ *into* $\mathbb{R}^d$ is a collection of charts $(\mathcal{U}_\alpha, \varphi_\alpha)$ of the set $\mathcal{M}$ such that

1. $\bigcup_\alpha \mathcal{U}_\alpha = \mathcal{M}$,
2. for any pair α, β with $\mathcal{U}_\alpha \cap \mathcal{U}_\beta \neq \emptyset$, the sets $\varphi_\alpha(\mathcal{U}_\alpha \cap \mathcal{U}_\beta)$ and $\varphi_\beta(\mathcal{U}_\alpha \cap \mathcal{U}_\beta)$ are open sets in $\mathbb{R}^d$ and the change of coordinates

$$\varphi_\beta \circ \varphi_\alpha^{-1} : \mathbb{R}^d \to \mathbb{R}^d$$

(see Appendix A.3 for our conventions on functions) is *smooth* (class C^∞, i.e., differentiable for all degrees of differentiation) on its domain $\varphi_\alpha(\mathcal{U}_\alpha \cap \mathcal{U}_\beta)$; see illustration in Figure 3.1. We say that the elements of an atlas *overlap smoothly*.

Two atlases $\mathcal{A}_1$ and $\mathcal{A}_2$ are *equivalent* if $\mathcal{A}_1 \cup \mathcal{A}_2$ is an atlas; in other words, for every chart $(\mathcal{U}, \varphi)$ in $\mathcal{A}_2$, the set of charts $\mathcal{A}_1 \cup \{(\mathcal{U}, \varphi)\}$ is still an atlas. Given an atlas $\mathcal{A}$, let $\mathcal{A}^+$ be the set of all charts $(\mathcal{U}, \varphi)$ such that $\mathcal{A} \cup \{(\mathcal{U}, \varphi)\}$ is also an atlas. It is easy to see that $\mathcal{A}^+$ is also an atlas, called the *maximal atlas* (or *complete atlas*) generated by the atlas $\mathcal{A}$. Two atlases are equivalent if and only if they generate the same maximal atlas. A maximal atlas of a set $\mathcal{M}$ is also called a *differentiable structure* on $\mathcal{M}$.

In the literature, a manifold is sometimes simply defined as a set endowed with a differentiable structure. However, this definition does not exclude certain unconventional topologies. For example, it does not guarantee that convergent sequences have a single limit point (an example is given in Section 4.3.2). To avoid such counterintuitive situations, we adopt the following classical definition. A *(d-dimensional) manifold* is a couple $(\mathcal{M}, \mathcal{A}^+)$, where $\mathcal{M}$ is a set and $\mathcal{A}^+$ is a maximal atlas of $\mathcal{M}$ into $\mathbb{R}^d$, such that the topology induced by $\mathcal{A}^+$ is Hausdorff and second-countable. (These topological issues are discussed in Section 3.1.2.)

A maximal atlas of a set $\mathcal{M}$ that induces a second-countable Hausdorff topology is called a *manifold structure* on $\mathcal{M}$. Often, when $(\mathcal{M}, \mathcal{A}^+)$ is a manifold, we simply say "the manifold $\mathcal{M}$" when the differentiable structure is clear from the context, and we say "the set $\mathcal{M}$" to refer to $\mathcal{M}$ as a plain set without a particular differentiable structure. Note that it is not necessary to specify the whole maximal atlas to define a manifold structure: it is enough to provide an atlas that generates the manifold structure.

Given a manifold $(\mathcal{M}, \mathcal{A}^+)$, an atlas of the set $\mathcal{M}$ whose maximal atlas is $\mathcal{A}^+$ is called an *atlas of the manifold* $(\mathcal{M}, \mathcal{A}^+)$; a chart of the set $\mathcal{M}$ that belongs to $\mathcal{A}^+$ is called a *chart of the manifold* $(\mathcal{M}, \mathcal{A}^+)$, and its domain is a *coordinate domain* of the manifold. By a chart around a point $x \in \mathcal{M}$, we mean a chart of $(\mathcal{M}, \mathcal{A}^+)$ whose domain $\mathcal{U}$ contains x. The set $\mathcal{U}$ is then a *coordinate neighborhood* of x.

Given a chart φ on $\mathcal{M}$, the inverse mapping φ^{-1} is called a *local parameterization* of $\mathcal{M}$. A family of local parameterizations is equivalent to a family of charts, and the definition of a manifold may be given in terms of either.

3.1.2 The topology of a manifold*

Recall that the star in the section title indicates material that can be readily skipped at a first reading.

It can be shown that the collection of coordinate domains specified by a maximal atlas $\mathcal{A}^+$ of a set $\mathcal{M}$ forms a basis for a topology of the set $\mathcal{M}$. (We refer the reader to Section A.2 for a short introduction to topology.) We call this topology the *atlas topology* of $\mathcal{M}$ induced by $\mathcal{A}$. In the atlas topology, a subset $\mathcal{V}$ of $\mathcal{M}$ is open if and only if, for any chart $(\mathcal{U}, \varphi)$ in $\mathcal{A}^+$, $\varphi(\mathcal{V} \cap \mathcal{U})$ is an open subset of $\mathbb{R}^d$. Equivalently, a subset $\mathcal{V}$ of $\mathcal{M}$ is open if and only if, for each $x \in \mathcal{V}$, there is a chart $(\mathcal{U}, \varphi)$ in $\mathcal{A}^+$ such that $x \in \mathcal{U} \subset \mathcal{V}$. An atlas $\mathcal{A}$ of a set $\mathcal{M}$ is said to be *compatible* with a topology $\mathcal{T}$ on the set $\mathcal{M}$ if the atlas topology is equal to $\mathcal{T}$.

An atlas topology always satisfies separation axiom T_1, i.e., given any two distinct points x and y, there is an open set $\mathcal{U}$ that contains x and not y. (Equivalently, every singleton is a closed set.) But not all atlas topologies are *Hausdorff* (i.e., T_2): two distinct points do not necessarily have disjoint neighborhoods. Non-Hausdorff spaces can display unusual and counterintuitive behavior. From the perspective of numerical iterative algorithms the most worrying possibility is that a convergent sequence on a non-Hausdorff topological space may have several distinct limit points. Our definition of manifold rules out non-Hausdorff topologies.

A topological space is *second-countable* if there is a countable collection $\mathcal{B}$ of open sets such that every open set is the union of some subcollection of $\mathcal{B}$. Second-countability is related to *partitions of unity*, a crucial tool in resolving certain fundamental questions such as the existence of a Riemannian metric (Section 3.6) and the existence of an affine connection (Section 5.2).

The existence of partitions of unity subordinate to arbitrary open coverings is equivalent to the property of *paracompactness*. A set endowed with a Hausdorff atlas topology is paracompact (and has countably many components) if (and only if) it is second-countable. Since manifolds are assumed to be Hausdorff and second-countable, they admit partitions of unity.

For a manifold $(\mathcal{M}, \mathcal{A}^+)$, we refer to the atlas topology of $\mathcal{M}$ induced by $\mathcal{A}$ as the *manifold topology of* $\mathcal{M}$. Note that several statements in this book also hold without the Hausdorff and second-countable assumptions. These cases, however, are of marginal importance and will not be discussed.

Given a manifold $(\mathcal{M}, \mathcal{A}^+)$ and an open subset $\mathcal{X}$ of $\mathcal{M}$ (open is to be understood in terms of the manifold topology of $\mathcal{M}$), the collection of the charts of $(\mathcal{M}, \mathcal{A}^+)$ whose domain lies in $\mathcal{X}$ forms an atlas of $\mathcal{X}$. This defines a differentiable structure on $\mathcal{X}$ of the same dimension as $\mathcal{M}$. With this structure, $\mathcal{X}$ is called an *open submanifold* of $\mathcal{M}$.

A manifold is *connected* if it cannot be expressed as the disjoint union of two nonempty open sets. Equivalently (for a manifold), any two points can be joined by a piecewise smooth curve segment. The connected components of a manifold are open, thus they admit a natural differentiable structure as open submanifolds. The optimization algorithms considered in this book are iterative and oblivious to the existence of connected components other than the one to which the current iterate belongs. Therefore we have no interest in considering manifolds that are not connected.

3.1.3 How to recognize a manifold

Assume that a computational problem involves a search space $\mathcal{X}$. How can we check that $\mathcal{X}$ is a manifold? It should be clear from Section 3.1.1 that this question is not well posed: by definition, a manifold is not simply a set $\mathcal{X}$ but rather a couple $(\mathcal{X}, \mathcal{A}^+)$ where $\mathcal{X}$ is a set and $\mathcal{A}^+$ is a maximal atlas of $\mathcal{X}$ inducing a second-countable Hausdorff topology.

A well-posed question is to ask whether a given set $\mathcal{X}$ admits an atlas. There are sets that do not admit an atlas and thus cannot be turned into a manifold. A simple example is the set of rational numbers: this set does not even admit charts; otherwise, it would not be denumerable. Nevertheless, sets abound that admit an atlas. Even sets that do not "look" differentiable may admit an atlas. For example, consider the curve $\gamma : \mathbb{R} \to \mathbb{R}^2 : \gamma(t) = (t, |t|)$ and let $\mathcal{X}$ be the range of γ; see Figure 3.2. Consider the chart $\varphi : \mathcal{X} \to \mathbb{R} : (t, |t|) \mapsto t$. It turns out that $\mathcal{A} := \{(\mathcal{X}, \varphi)\}$ is an atlas of the set $\mathcal{X}$; therefore, $(\mathcal{X}, \mathcal{A}^+)$ is a manifold. The incorrect intuition that $\mathcal{X}$ cannot be a manifold because of its "corner" corresponds to the fact that $\mathcal{X}$ is not a *submanifold* of $\mathbb{R}^2$; see Section 3.3.

A set $\mathcal{X}$ may admit more than one maximal atlas. As an example, take the set $\mathbb{R}$ and consider the charts $\varphi_1 : x \mapsto x$ and $\varphi_2 : x \mapsto x^3$. Note that φ_1 and φ_2 are not compatible since the mapping $\varphi_1 \circ \varphi_2^{-1}$ is not differentiable at the origin. However, each chart individually forms an atlas of the set $\mathbb{R}$. These two atlases are not equivalent; they do not generate the same maximal

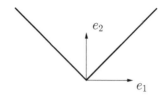

Figure 3.2 Image of the curve $\gamma : t \mapsto (t, |t|)$.

atlas. Nevertheless, the chart $x \mapsto x$ is clearly more natural than the chart $x \mapsto x^3$. Most manifolds of interest admit a differentiable structure that is the most "natural"; see in particular the notions of embedded and quotient matrix manifold in Sections 3.3 and 3.4.

3.1.4 Vector spaces as manifolds

Let $\mathcal{E}$ be a d-dimensional vector space. Then, given a basis $(e_i)_{i=1,\ldots,d}$ of $\mathcal{E}$, the function

$$\psi : \mathcal{E} \to \mathbb{R}^d : x \mapsto \begin{bmatrix} x^1 \\ \vdots \\ x^d \end{bmatrix}$$

such that $x = \sum_{i=1}^d x^i e_i$ is a chart of the set $\mathcal{E}$. All charts built in this way are compatible; thus they form an atlas of the set $\mathcal{E}$, which endows $\mathcal{E}$ with a manifold structure. Hence, every vector space is a *linear manifold* in a natural way.

Needless to say, the challenging case is the one where the manifold structure is *nonlinear*, i.e., manifolds that are not endowed with a vector space structure. The numerical algorithms considered in this book apply equally to linear and nonlinear manifolds and reduce to classical optimization algorithms when the manifold is linear.

3.1.5 The manifolds $\mathbb{R}^{n \times p}$ and $\mathbb{R}_*^{n \times p}$

Algorithms formulated on abstract manifolds are not strictly speaking numerical algorithms in the sense that they involve manipulation of differential-geometric objects instead of numerical calculations. Turning these abstract algorithms into numerical algorithms for specific optimization problems relies crucially on producing adequate numerical representations of the geometric objects that arise in the abstract algorithms. A significant part of this book is dedicated to building a toolbox of results that make it possible to perform this "geometric-to-numerical" conversion on matrix manifolds (i.e., manifolds obtained by taking embedded submanifolds and quotient manifolds of $\mathbb{R}^{n \times p}$). The process derives from the manifold structure of the set $\mathbb{R}^{n \times p}$ of $n \times p$ real matrices, discussed next.

The set $\mathbb{R}^{n \times p}$ is a vector space with the usual sum and multiplication by a scalar. Consequently, it has a natural linear manifold structure. A chart of this manifold is given by $\varphi : \mathbb{R}^{n \times p} \to \mathbb{R}^{np} : X \mapsto \text{vec}(X)$, where $\text{vec}(X)$ denotes the vector obtained by stacking the columns of X below one another. We will refer to the set $\mathbb{R}^{n \times p}$ with its linear manifold structure as the *manifold* $\mathbb{R}^{n \times p}$. Its dimension is np.

The manifold $\mathbb{R}^{n \times p}$ can be further turned into a Euclidean space with the inner product

$$\langle Z_1, Z_2 \rangle := \text{vec}(Z_1)^T \text{vec}(Z_2) = \text{tr}(Z_1^T Z_2). \tag{3.2}$$

The norm induced by the inner product is the *Frobenius norm* defined by

$$\|Z\|_F^2 = \text{tr}(Z^T Z),$$

i.e., $\|Z\|_F^2$ is the sum of the squares of the elements of Z. Observe that the manifold topology of $\mathbb{R}^{n \times p}$ is equivalent to its canonical topology as a Euclidean space (see Appendix A.2).

Let $\mathbb{R}_*^{n \times p}$ $(p \le n)$ denote the set of all $n \times p$ matrices whose columns are linearly independent. This set is an open subset of $\mathbb{R}^{n \times p}$ since its complement $\{X \in \mathbb{R}^{n \times p} : \det(X^T X) = 0\}$ is closed. Consequently, it admits a structure of an open submanifold of $\mathbb{R}^{n \times p}$. Its differentiable structure is generated by the chart $\varphi : \mathbb{R}_*^{n \times p} \to \mathbb{R}^{np} : X \mapsto \text{vec}(X)$. This manifold will be referred to as the *manifold* $\mathbb{R}_*^{n \times p}$, or the *noncompact Stiefel manifold* of full-rank $n \times p$ matrices.

In the particular case $p = 1$, the noncompact Stiefel manifold reduces to the Euclidean space $\mathbb{R}^n$ with the origin removed. When $p = n$, the noncompact Stiefel manifold becomes the general linear group GL_n, i.e., the set of all invertible $n \times n$ matrices.

Notice that the chart vec : $\mathbb{R}^{n \times p} \to \mathbb{R}^{np}$ is unwieldy, as it destroys the matrix structure of its argument; in particular, $\text{vec}(AB)$ cannot be written as a simple expression of $\text{vec}(A)$ and $\text{vec}(B)$. In this book, the emphasis is on preserving the matrix structure.

3.1.6 Product manifolds

Let $\mathcal{M}_1$ and $\mathcal{M}_2$ be manifolds of dimension d_1 and d_2, respectively. The set $\mathcal{M}_1 \times \mathcal{M}_2$ is defined as the set of pairs (x_1, x_2), where x_1 is in $\mathcal{M}_1$ and x_2 is in $\mathcal{M}_2$. If $(\mathcal{U}_1, \varphi_1)$ and $(\mathcal{U}_2, \varphi_2)$ are charts of the manifolds $\mathcal{M}_1$ and $\mathcal{M}_2$, respectively, then the mapping $\varphi_1 \times \varphi_2 : \mathcal{U}_1 \times \mathcal{U}_2 \to \mathbb{R}^{d_1} \times \mathbb{R}^{d_2} : (x_1, x_2) \mapsto (\varphi_1(x_1), \varphi_2(x_2))$ is a chart for the set $\mathcal{M}_1 \times \mathcal{M}_2$. All the charts thus obtained form an atlas for the set $\mathcal{M}_1 \times \mathcal{M}_2$. With the differentiable structure defined by this atlas, $\mathcal{M}_1 \times \mathcal{M}_2$ is called the *product* of the manifolds $\mathcal{M}_1$ and $\mathcal{M}_2$. Its manifold topology is equivalent to the product topology. Product manifolds will be useful in some later developments.

3.2 DIFFERENTIABLE FUNCTIONS

Mappings between manifolds appear in many places in optimization algorithms on manifolds. First of all, any optimization problem on a manifold $\mathcal{M}$ involves a cost function, which can be viewed as a mapping from the manifold $\mathcal{M}$ into the manifold $\mathbb{R}$. Other instances of mappings between manifolds are inclusions (in the theory of submanifolds; see Section 3.3), natural projections onto quotients (in the theory of quotient manifolds, see Section 3.4), and retractions (a fundamental tool in numerical algorithms on manifolds; see Section 4.1). This section introduces the notion of differentiability for functions between manifolds. The coordinate-free definition of a differential will come later, as it requires the concept of a tangent vector.

Let F be a function from a manifold $\mathcal{M}_1$ of dimension d_1 into another manifold $\mathcal{M}_2$ of dimension d_2. Let x be a point of $\mathcal{M}_1$. Choosing charts φ_1 and φ_2 around x and $F(x)$, respectively, the function F around x can be "read through the charts", yielding the function

$$\hat{F} = \varphi_2 \circ F \circ \varphi_1^{-1} : \mathbb{R}^{d_1} \to \mathbb{R}^{d_2}, \tag{3.3}$$

called a *coordinate representation* of F. (Note that the domain of $\hat{F}$ is in general a subset of $\mathbb{R}^{d_1}$; see Appendix A.3 for the conventions.)

We say that F is *differentiable* or *smooth* at x if $\hat{F}$ is of class C^∞ at $\varphi_1(x)$. It is easily verified that this definition does not depend on the choice of the charts chosen at x and $F(x)$. A function $F : \mathcal{M}_1 \to \mathcal{M}_2$ is said to be *smooth* if it is smooth at every point of its domain.

A (smooth) *diffeomorphism* $F : \mathcal{M}_1 \to \mathcal{M}_2$ is a bijection such that F and its inverse F^{-1} are both smooth. Two manifolds $\mathcal{M}_1$ and $\mathcal{M}_2$ are said to be *diffeomorphic* if there exists a diffeomorphism on $\mathcal{M}_1$ onto $\mathcal{M}_2$.

In this book, all functions are assumed to be smooth unless otherwise stated.

3.2.1 Immersions and submersions

The concepts of immersion and submersion will make it possible to define submanifolds and quotient manifolds in a concise way. Let $F : \mathcal{M}_1 \to \mathcal{M}_2$ be a differentiable function from a manifold $\mathcal{M}_1$ of dimension d_1 into a manifold $\mathcal{M}_2$ of dimension d_2. Given a point x of $\mathcal{M}_1$, the *rank* of F at x is the dimension of the range of $\mathrm{D}\hat{F}(\varphi_1(x))[\cdot] : \mathbb{R}^{d_1} \to \mathbb{R}^{d_2}$, where $\hat{F}$ is a coordinate representation (3.3) of F around x, and $\mathrm{D}\hat{F}(\varphi_1(x))$ denotes the *differential* of $\hat{F}$ at $\varphi_1(x)$ (see Section A.5). (Notice that this definition does not depend on the charts used to obtain the coordinate representation $\hat{F}$ of F.) The function F is called an *immersion* if its rank is equal to d_1 at each point of its domain (hence $d_1 \leq d_2$). If its rank is equal to d_2 at each point of its domain (hence $d_1 \geq d_2$), then it is called a *submersion*.

The function F is an immersion if and only if, around each point of its domain, it admits a coordinate representation that is the *canonical immersion* $(u^1, \ldots, u^{d_1}) \mapsto (u^1, \ldots, u^{d_1}, 0, \ldots, 0)$. The function F is a submersion if and only if, around each point of its domain, it admits the *canonical submersion*

$(u^1, \ldots, u^{d_1}) \mapsto (u^1, \ldots, u^{d_2})$ as a coordinate representation. A point $y \in \mathcal{M}_2$ is called a *regular value* of F if the rank of F is d_2 at every $x \in F^{-1}(y)$.

3.3 EMBEDDED SUBMANIFOLDS

A set $\mathcal{X}$ may admit several manifold structures. However, if the set $\mathcal{X}$ is a subset of a manifold $(\mathcal{M}, \mathcal{A}^+)$, then it admits at most one submanifold structure. This is the topic of this section.

3.3.1 General theory

Let $(\mathcal{M}, \mathcal{A}^+)$ and $(\mathcal{N}, \mathcal{B}^+)$ be manifolds such that $\mathcal{N} \subset \mathcal{M}$. The manifold $(\mathcal{N}, \mathcal{B}^+)$ is called an *immersed submanifold* of $(\mathcal{M}, \mathcal{A}^+)$ if the inclusion map $i : \mathcal{N} \to \mathcal{M} : x \mapsto x$ is an immersion.

Let $(\mathcal{N}, \mathcal{B}^+)$ be a submanifold of $(\mathcal{M}, \mathcal{A}^+)$. Since $\mathcal{M}$ and $\mathcal{N}$ are manifolds, they are also topological spaces with their manifold topology. If the manifold topology of $\mathcal{N}$ coincides with its subspace topology induced from the topological space $\mathcal{M}$, then $\mathcal{N}$ is called an *embedded submanifold*, a *regular submanifold*, or simply a *submanifold* of the manifold $\mathcal{M}$. Asking that a subset $\mathcal{N}$ of a manifold $\mathcal{M}$ be an embedded submanifold of $\mathcal{M}$ removes all freedom for the choice of a differentiable structure on $\mathcal{N}$:

Proposition 3.3.1 *Let $\mathcal{N}$ be a subset of a manifold $\mathcal{M}$. Then $\mathcal{N}$ admits at most one differentiable structure that makes it an embedded submanifold of $\mathcal{M}$.*

As a consequence of Proposition 3.3.1, when we say in this book that a subset of a manifold "is" a submanifold, we mean that it admits one (unique) differentiable structure that makes it an embedded submanifold. The manifold $\mathcal{M}$ in Proposition 3.3.1 is called the *embedding space*. When the embedding space is $\mathbb{R}^{n \times p}$ or an open subset of $\mathbb{R}^{n \times p}$, we say that $\mathcal{N}$ is a *matrix submanifold*.

To check whether a subset $\mathcal{N}$ of a manifold $\mathcal{M}$ is an embedded submanifold of $\mathcal{M}$ and to construct an atlas of that differentiable structure, one can use the next proposition, which states that every embedded submanifold is locally a coordinate slice. Given a chart $(\mathcal{U}, \varphi)$ of a manifold $\mathcal{M}$, a φ-*coordinate slice* of $\mathcal{U}$ is a set of the form $\varphi^{-1}(\mathbb{R}^m \times \{0\})$ that corresponds to all the points of $\mathcal{U}$ whose last $n - m$ coordinates in the chart φ are equal to zero.

Proposition 3.3.2 (submanifold property) *A subset $\mathcal{N}$ of a manifold $\mathcal{M}$ is a d-dimensional embedded submanifold of $\mathcal{M}$ if and only if, around each point $x \in \mathcal{N}$, there exists a chart $(\mathcal{U}, \varphi)$ of $\mathcal{M}$ such that $\mathcal{N} \cap \mathcal{U}$ is a φ-coordinate slice of $\mathcal{U}$, i.e.,*

$$\mathcal{N} \cap \mathcal{U} = \{x \in \mathcal{U} : \varphi(x) \in \mathbb{R}^d \times \{0\}\}.$$

In this case, the chart $(\mathcal{N} \cap \mathcal{U}, \varphi)$, where φ is seen as a mapping into $\mathbb{R}^d$, is a chart of the embedded submanifold $\mathcal{N}$.

The next propositions provide sufficient conditions for subsets of manifolds to be embedded submanifolds.

Proposition 3.3.3 (submersion theorem) *Let $F : \mathcal{M}_1 \to \mathcal{M}_2$ be a smooth mapping between two manifolds of dimension d_1 and d_2, $d_1 > d_2$, and let y be a point of $\mathcal{M}_2$. If y is a regular value of F (i.e., the rank of F is equal to d_2 at every point of $F^{-1}(y)$), then $F^{-1}(y)$ is a closed embedded submanifold of $\mathcal{M}_1$, and $\dim(F^{-1}(y)) = d_1 - d_2$.*

Proposition 3.3.4 (subimmersion theorem) *Let $F : \mathcal{M}_1 \to \mathcal{M}_2$ be a smooth mapping between two manifolds of dimension d_1 and d_2 and let y be a point of $F(\mathcal{M}_1)$. If F has constant rank $k < d_1$ in a neighborhood of $F^{-1}(y)$, then $F^{-1}(y)$ is a closed embedded submanifold of $\mathcal{M}_1$ of dimension $d_1 - k$.*

Functions on embedded submanifolds pose no particular difficulty. Let $\mathcal{N}$ be an embedded submanifold of a manifold $\mathcal{M}$. If f is a smooth function on $\mathcal{M}$, then $f|_{\mathcal{N}}$, the *restriction* of f to $\mathcal{N}$, is a smooth function on $\mathcal{N}$. Conversely, any smooth function on $\mathcal{N}$ can be written locally as a restriction of a smooth function defined on an open subset $\mathcal{U} \subset \mathcal{M}$.

3.3.2 The Stiefel manifold

The (orthogonal) Stiefel manifold is an embedded submanifold of $\mathbb{R}^{n \times p}$ that will appear frequently in our practical examples.

Let $\mathrm{St}(p, n)$ $(p \leq n)$ denote the set of all $n \times p$ orthonormal matrices; i.e.,

$$\mathrm{St}(p, n) := \{X \in \mathbb{R}^{n \times p} : X^T X = I_p\}, \tag{3.4}$$

where I_p denotes the $p \times p$ identity matrix. The set $\mathrm{St}(p, n)$ (endowed with its submanifold structure as discussed below) is called an *(orthogonal or compact) Stiefel manifold*. Note that the Stiefel manifold $\mathrm{St}(p, n)$ is distinct from the noncompact Stiefel manifold $\mathbb{R}^{n \times p}_*$ defined in Section 3.1.5.

Clearly, $\mathrm{St}(p, n)$ is a subset of the set $\mathbb{R}^{n \times p}$. Recall that the set $\mathbb{R}^{n \times p}$ admits a linear manifold structure as described in Section 3.1.5. To show that $\mathrm{St}(p, n)$ is an embedded submanifold of the manifold $\mathbb{R}^{n \times p}$, consider the function $F : \mathbb{R}^{n \times p} \to \mathcal{S}_{\mathrm{sym}}(p) : X \mapsto X^T X - I_p$, where $\mathcal{S}_{\mathrm{sym}}(p)$ denotes the set of all symmetric $p \times p$ matrices. Note that $\mathcal{S}_{\mathrm{sym}}(p)$ is a vector space. Clearly, $\mathrm{St}(p, n) = F^{-1}(0_p)$. It remains to show that F is a submersion at each point X of $\mathrm{St}(p, n)$. The fact that the domain of F is a vector space exempts us from having to read F through a chart: we simply need to show that for all $\widehat{Z}$ in $\mathcal{S}_{\mathrm{sym}}(p)$, there exists Z in $\mathbb{R}^{n \times p}$ such that $\mathrm{D}F(X)[Z] = \widehat{Z}$. We have (see Appendix A.5 for details on matrix differentiation)

$$\mathrm{D}F(X)[Z] = X^T Z + Z^T X.$$

It is easy to see that $\mathrm{D}F(X)\left[\frac{1}{2}X\widehat{Z}\right] = \widehat{Z}$ since $X^T X = I_p$ and $\widehat{Z}^T = \widehat{Z}$. This shows that F is full rank. It follows from Proposition 3.3.3 that the set $\mathrm{St}(p, n)$ defined in (3.4) is an embedded submanifold of $\mathbb{R}^{n \times p}$.

To obtain the dimension of $\mathrm{St}(p, n)$, observe that the vector space $\mathcal{S}_{\mathrm{sym}}(p)$ has dimension $\frac{1}{2}p(p+1)$ since a symmetric matrix is completely determined by its upper triangular part (including the diagonal). From Proposition 3.3.3, we obtain

$$\dim(\mathrm{St}(p, n)) = np - \tfrac{1}{2}p(p+1).$$

Since $\mathrm{St}(p, n)$ is an embedded submanifold of $\mathbb{R}^{n \times p}$, its topology is the subset topology induced by $\mathbb{R}^{n \times p}$. The manifold $\mathrm{St}(p, n)$ is closed: it is the inverse image of the closed set $\{0_p\}$ under the continuous function $F : \mathbb{R}^{n \times p} \mapsto \mathcal{S}_{\mathrm{sym}}(p)$. It is bounded: each column of $X \in \mathrm{St}(p, n)$ has norm 1, so the Frobenius norm of X is equal to $\sqrt{p}$. It then follows from the Heine-Borel theorem (see Section A.2) that the manifold $\mathrm{St}(p, n)$ is *compact*.

For $p = 1$, the Stiefel manifold $\mathrm{St}(p, n)$ reduces to the *unit sphere S^{n-1}* in $\mathbb{R}^n$. Notice that the superscript $n-1$ indicates the dimension of the manifold.

For $p = n$, the Stiefel manifold $\mathrm{St}(p, n)$ becomes the *orthogonal group O_n*. Its dimension is $\frac{1}{2}n(n-1)$.

3.4 QUOTIENT MANIFOLDS

Whereas the topic of submanifolds is covered in any introductory textbook on manifolds, the subject of quotient manifolds is less classical. We develop the theory in some detail because it has several applications in matrix computations, most notably in algorithms that involve subspaces of $\mathbb{R}^n$. Computations involving subspaces are usually carried out using matrices to represent the corresponding subspace generated by the span of its columns. The difficulty is that for one given subspace, there are infinitely many matrices that represent the subspace. It is then desirable to partition the set of matrices into classes of "equivalent" elements that represent the same object. This leads to the concept of quotient spaces and quotient manifolds. In this section, we first present the general theory of quotient manifolds, then we return to the special case of subspaces and their representations.

3.4.1 Theory of quotient manifolds

Let $\mathcal{M}$ be a manifold equipped with an *equivalence relation* $\sim$, i.e., a relation that is

1. reflexive: $x \sim x$ for all $x \in \mathcal{M}$,
2. symmetric: $x \sim y$ if and only if $y \sim x$ for all $x, y \in \mathcal{M}$,
3. transitive: if $x \sim y$ and $y \sim z$ then $x \sim z$ for all $x, y, z \in \mathcal{M}$.

The set

$$[x] := \{y \in \mathcal{M} : y \sim x\}$$

of all elements that are equivalent to a point x is called the *equivalence class* containing x. The set

$$\mathcal{M}/\!\sim\, := \{[x] : x \in \mathcal{M}\}$$

of all equivalence classes of $\sim$ in $\mathcal{M}$ is called the *quotient* of $\mathcal{M}$ by $\sim$. Notice that the points of $\mathcal{M}/\sim$ are subsets of $\mathcal{M}$. The mapping $\pi : \mathcal{M} \to \mathcal{M}/\sim$ defined by $x \mapsto [x]$ is called the *natural projection* or *canonical projection*. Clearly, $\pi(x) = \pi(y)$ if and only if $x \sim y$, so we have $[x] = \pi^{-1}(\pi(x))$. We will use $\pi(x)$ to denote $[x]$ viewed as a point of $\mathcal{M}/\sim$, and $\pi^{-1}(\pi(x))$ for $[x]$ viewed as a subset of $\mathcal{M}$. The set $\mathcal{M}$ is called the *total space* of the quotient $\mathcal{M}/\sim$.

Let $(\mathcal{M}, \mathcal{A}^+)$ be a manifold with an equivalence relation $\sim$ and let $\mathcal{B}^+$ be a manifold structure on the set $\mathcal{M}/\sim$. The manifold $(\mathcal{M}/\sim, \mathcal{B}^+)$ is called a *quotient manifold* of $(\mathcal{M}, \mathcal{A}^+)$ if the natural projection π is a submersion.

Proposition 3.4.1 *Let $\mathcal{M}$ be a manifold and let $\mathcal{M}/\sim$ be a quotient of $\mathcal{M}$. Then $\mathcal{M}/\sim$ admits at most one manifold structure that makes it a quotient manifold of $\mathcal{M}$.*

Given a quotient $\mathcal{M}/\sim$ of a manifold $\mathcal{M}$, we say that the set $\mathcal{M}/\sim$ *is a quotient manifold* if it admits a (unique) quotient manifold structure. In this case, we say that the equivalence relation $\sim$ is *regular*, and we refer to the set $\mathcal{M}/\sim$ endowed with this manifold structure as the manifold $\mathcal{M}/\sim$.

The following result gives a characterization of regular equivalence relations. Note that the *graph* of a relation $\sim$ is the set

$$\mathrm{graph}(\sim) := \{(x, y) \in \mathcal{M} \times \mathcal{M} : x \sim y\}.$$

Proposition 3.4.2 *An equivalence relation $\sim$ on a manifold $\mathcal{M}$ is regular (and thus $\mathcal{M}/\sim$ is a quotient manifold) if and only if the following conditions hold together:*

    (i) The graph of $\sim$ is an embedded submanifold of the product manifold $\mathcal{M} \times \mathcal{M}$.

    (ii) The projection $\pi_1 : \mathrm{graph}(\sim) \to \mathcal{M}$, $\pi_1(x, y) = x$ is a submersion.

    (iii) The graph of $\sim$ is a closed subset of $\mathcal{M} \times \mathcal{M}$ (where $\mathcal{M}$ is endowed with its manifold topology).

The dimension of $\mathcal{M}/\sim$ is given by

$$\dim(\mathcal{M}/\sim) = 2\dim(\mathcal{M}) - \dim(\mathrm{graph}(\sim)). \qquad (3.5)$$

The next proposition distinguishes the role of the three conditions in Proposition 3.4.2.

Proposition 3.4.3 *Conditions (i) and (ii) in Proposition 3.4.2 are necessary and sufficient for $\mathcal{M}/\sim$ to admit an atlas that makes π a submersion. Such an atlas is unique, and the atlas topology of $\mathcal{M}/\sim$ is identical to its quotient topology. Condition (iii) in Proposition 3.4.2 is necessary and sufficient for the quotient topology to be Hausdorff.*

The following result follows from Proposition 3.3.3 by using the fact that the natural projection to a quotient manifold is by definition a submersion.

Proposition 3.4.4 *Let $\mathcal{M}/\sim$ be a quotient manifold of a manifold $\mathcal{M}$ and let π denote the canonical projection. If $\dim(\mathcal{M}/\sim) < \dim(\mathcal{M})$, then each equivalence class $\pi^{-1}(\pi(x))$, $x \in \mathcal{M}$, is an embedded submanifold of $\mathcal{M}$ of dimension $\dim(\mathcal{M}) - \dim(\mathcal{M}/\sim)$.*

If $\dim(\mathcal{M}/\sim) = \dim(\mathcal{M})$, then each equivalence class $\pi^{-1}(\pi(x))$, $x \in \mathcal{M}$, is a discrete set of points. From now on we consider only the case $\dim(\mathcal{M}/\sim) < \dim(\mathcal{M})$.

When $\mathcal{M}$ is $\mathbb{R}^{n \times p}$ or a submanifold of $\mathbb{R}^{n \times p}$, we call $\mathcal{M}/\sim$ a *matrix quotient manifold*. For ease of reference, we will use the generic name *structure space* both for embedding spaces (associated with embedded submanifolds) and for total spaces (associated with quotient manifolds). We call a *matrix manifold* any manifold that is constructed from $\mathbb{R}^{n \times p}$ by the operations of taking embedded submanifolds and quotient manifolds. The major matrix manifolds that appear in this book are the noncompact Stiefel manifold (defined in Section 3.1.5), the orthogonal Stiefel manifold (Section 3.3.2), and the Grassmann manifold (Section 3.4.4). Other important matrix manifolds are the *oblique manifold*

$$\{X \in \mathbb{R}^{n \times p} : \mathrm{diag}(X^T X) = I_p\},$$

where $\mathrm{diag}(M)$ denotes the matrix M with all its off-diagonal elements assigned to zero; the generalized Stiefel manifold

$$\{X \in \mathbb{R}^{n \times p} : X^T B X = I\}$$

where B is a symmetric positive-definite matrix; the *flag manifolds*, which are quotients of $\mathbb{R}_*^{n \times p}$ where two matrices are equivalent when they are related by a right multiplication by a block upper triangular matrix with prescribed block size; and the manifold of *symplectic matrices*

$$\{X \in \mathbb{R}^{2n \times 2n} : X^T J X = J\},$$

where $J = \left[\begin{smallmatrix} 0_n & I_n \\ -I_n & 0_n \end{smallmatrix}\right]$.

3.4.2 Functions on quotient manifolds

A function f on $\mathcal{M}$ is termed *invariant under* $\sim$ if $f(x) = f(y)$ whenever $x \sim y$, in which case the function f induces a unique function $\tilde{f}$ on $\mathcal{M}/\sim$, called the *projection* of f, such that $f = \tilde{f} \circ \pi$.

The smoothness of $\tilde{f}$ can be checked using the following result.

Proposition 3.4.5 *Let $\mathcal{M}/\sim$ be a quotient manifold and let $\tilde{f}$ be a function on $\mathcal{M}/\sim$. Then $\tilde{f}$ is smooth if and only if $f := \tilde{f} \circ \pi$ is a smooth function on $\mathcal{M}$.*

3.4.3 The real projective space $\mathbb{RP}^{n-1}$

The real projective space $\mathbb{RP}^{n-1}$ is the set of all directions in $\mathbb{R}^n$, i.e., the set of all straight lines passing through the origin of $\mathbb{R}^n$. Let $\mathbb{R}^n_* := \mathbb{R}^n - \{0\}$ denote the Euclidean space $\mathbb{R}^n$ with the origin removed. Note that $\mathbb{R}^n_*$ is the $p = 1$ particularization of the noncompact Stiefel manifold $\mathbb{R}^{n \times p}_*$ (Section 3.1.5); hence $\mathbb{R}^n_*$ is an open submanifold of $\mathbb{R}^n$. The real projective space $\mathbb{RP}^{n-1}$ is naturally identified with the quotient $\mathbb{R}^n_*/\sim$, where the equivalence relation is defined by

$$x \sim y \qquad \Leftrightarrow \qquad \exists t \in \mathbb{R}_* : y = xt,$$

and we write

$$\mathbb{RP}^{n-1} \simeq \mathbb{R}^n_*/\sim$$

to denote the identification of the two sets.

The proof that $\mathbb{R}^n_*/\sim$ is a quotient manifold follows as a special case of Proposition 3.4.6 (stating that the Grassmann manifold is a matrix quotient manifold). The letters $\mathbb{RP}$ stand for "real projective", while the superscript $(n-1)$ is the dimension of the manifold. There are also complex projective spaces and more generally projective spaces over more abstract vector spaces.

3.4.4 The Grassmann manifold $\mathrm{Grass}(p, n)$

Let n be a positive integer and let p be a positive integer not greater than n. Let $\mathrm{Grass}(p, n)$ denote the set of all p-dimensional subspaces of $\mathbb{R}^n$. In this section, we produce a one-to-one correspondence between $\mathrm{Grass}(p, n)$ and a quotient manifold of $\mathbb{R}^{n \times p}$, thereby endowing $\mathrm{Grass}(p, n)$ with a matrix manifold structure.

Recall that the noncompact Stiefel manifold $\mathbb{R}^{n \times p}_*$ is the set of all $n \times p$ matrices with full column rank. Let $\sim$ denote the equivalence relation on $\mathbb{R}^{n \times p}_*$ defined by

$$X \sim Y \qquad \Leftrightarrow \qquad \mathrm{span}(X) = \mathrm{span}(Y), \qquad (3.6)$$

where $\mathrm{span}(X)$ denotes the subspace $\{X\alpha : \alpha \in \mathbb{R}^p\}$ spanned by the columns of $X \in \mathbb{R}^{n \times p}_*$. Since the fibers of $\mathrm{span}(\cdot)$ are the equivalence classes of $\sim$ and since $\mathrm{span}(\cdot)$ is onto $\mathrm{Grass}(p, n)$, it follows that $\mathrm{span}(\cdot)$ induces a one-to-one correspondence between $\mathrm{Grass}(p, n)$ and $\mathbb{R}^{n \times p}_*/\sim$.

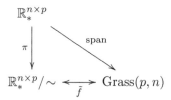

Before showing that the set $\mathbb{R}^{n \times p}_*/\sim$ is a quotient manifold, we introduce some notation and terminology. If a matrix X and a subspace $\mathcal{X}$ satisfy

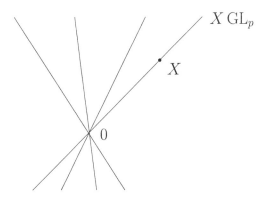

Figure 3.3 Schematic illustration of the representation of $\mathrm{Grass}(p, n)$ as the quotient space $\mathbb{R}_*^{n \times p}/\mathrm{GL}_p$. Each point is an n-by-p matrix. Each line is an equivalence class of the matrices that have the same span. Each line corresponds to an element of $\mathrm{Grass}(p, n)$. The figure corresponds to the case $n = 2$, $p = 1$.

$\mathcal{X} = \mathrm{span}(X)$, we say that $\mathcal{X}$ is the *span* of X, that X spans $\mathcal{X}$, or that X is a *matrix representation* of $\mathcal{X}$. The set of all matrix representations of $\mathrm{span}(X)$ is the equivalence class $\pi^{-1}(\pi(X))$. We have $\pi^{-1}(\pi(X)) = \{XM : M \in \mathrm{GL}_p\} =: X\mathrm{GL}_p$; indeed, the operations $X \mapsto XM$, $M \in \mathrm{GL}_p$, correspond to all possible changes of basis for $\mathrm{span}(X)$. We will thus use the notation $\mathbb{R}_*^{n \times p}/\mathrm{GL}_p$ for $\mathbb{R}_*^{n \times p}/\sim$. Therefore we have

$$\mathrm{Grass}(p, n) \simeq \mathbb{R}_*^{n \times p}/\mathrm{GL}_p.$$

A schematic illustration of the quotient $\mathbb{R}_*^{n \times p}/\mathrm{GL}_p$ is given in Figure 3.3.

The identification of $\mathbb{R}_*^{n \times p}/\mathrm{GL}_p$ with the set of p-dimensional subspaces (p-planes) in $\mathbb{R}^n$ makes this quotient particularly worth studying. Next, the quotient $\mathbb{R}_*^{n \times p}/\mathrm{GL}_p$ is shown to be a quotient manifold.

Proposition 3.4.6 (Grassmann manifold) *The quotient set $\mathbb{R}_*^{n \times p}/\mathrm{GL}_p$ (i.e., the quotient of $\mathbb{R}_*^{n \times p}$ by the equivalence relation defined in (3.6)) admits a (unique) structure of quotient manifold.*

Proof. We show that the conditions in Proposition 3.4.2 are satisfied. We first prove condition (ii). Let (X_0, Y_0) be in $\mathrm{graph}(\sim)$. Then there exists M such that $Y_0 = X_0 M$. Given any V in $\mathbb{R}^{n \times p}$, the curve $t \mapsto (X_0 + tV, (X_0 + tV)M)$ is into $\mathrm{graph}(\sim)$ and satisfies $\frac{\mathrm{d}}{\mathrm{d}t}(\pi_1(\gamma(t)))\big|_{t=0} = V$. This shows that π_1 is a submersion. For condition (iii), observe that the graph of $\sim$ is closed as it is the preimage of the closed set $\{0_{n \times p}\}$ under the continuous function $\mathbb{R}^{n \times p} \times \mathbb{R}_*^{n \times p} \to \mathbb{R}^{n \times p} : (X, Y) \mapsto (I - X(X^T X)^{-1} X^T)Y$. For condition (i), the idea is to produce submersions F_i with open domain $\Omega_i \subset (\mathbb{R}_*^{n \times p} \times \mathbb{R}_*^{n \times p})$ such that $\mathrm{graph}(\sim) \cap \Omega_i$ is the zero-level set of F_i and that the Ω_i's cover $\mathrm{graph}(\sim)$. It then follows from Proposition 3.3.3 that $\mathrm{graph}(\sim)$ is an embedded submanifold of $\mathbb{R}_*^{n\times p} \times \mathbb{R}_*^{n\times p}$. To this end, assume for a moment that we have a smooth function

$$\mathbb{R}_*^{n\times p} \to \mathrm{St}(n-p,n) : X \mapsto X_\perp \qquad (3.7)$$

such that $X^T X_\perp = 0$ for all X in an open domain $\tilde{\Omega}$ and consider

$$F : \tilde{\Omega} \times \mathbb{R}_*^{n\times p} \to \mathbb{R}^{(n-p)\times p} : (X,Y) \mapsto X_\perp^T Y.$$

Then $F^{-1}(0) = \mathrm{graph}(\sim) \cap \mathrm{dom}(F)$. Moreover, F is a submersion on its domain since for any $V \in \mathbb{R}^{(n-p)\times p}$,

$$DF(X,Y)[0, X_\perp V] = X_\perp^T(X_\perp V) = V.$$

It remains to define the smooth function (3.7). Depending on n and p, it may or may not be possible to define such a function on the whole $\mathbb{R}_*^{n\times p}$. However, there are always such functions, constructed as follows, whose domain $\tilde{\Omega}$ is open and dense in $\mathbb{R}_*^{n\times p}$. Let $E \in \mathbb{R}^{n\times(n-p)}$ be a constant matrix of the form

$$E = \left[e_{i_1} | \cdots | e_{i_{n-p}} \right],$$

where the e_i's are the canonical vectors in $\mathbb{R}^n$ (unit vectors with a 1 in the ith entry), and define $X_\perp$ as the orthonormal matrix obtained by taking the last $n-p$ columns of the Gram-Schmidt orthogonalization of the matrix $[X|E]$. This function is smooth on the domain $\tilde{\Omega} = \{X \in \mathbb{R}_*^{n\times p} : [X|E] \text{ full rank}\}$, which is an open dense subset of $\mathbb{R}_*^{n\times p}$. Consequently, $F(X,Y) = X_\perp^T Y$ is smooth (and submersive) on the domain $\Omega = \tilde{\Omega} \times \mathbb{R}_*^{n\times p}$. This shows that $\mathrm{graph}(\sim) \cap \Omega$ is an embedded submanifold of $(\mathbb{R}_*^{n\times p} \times \mathbb{R}_*^{n\times p})$. Taking other matrices E yields other domains Ω which together cover $(\mathbb{R}_*^{n\times p} \times \mathbb{R}_*^{n\times p})$, so $\mathrm{graph}(\sim)$ is an embedded submanifold of $(\mathbb{R}_*^{n\times p} \times \mathbb{R}_*^{n\times p})$, and the proof is complete. $\square$

Endowed with its quotient manifold structure, the set $\mathbb{R}_*^{n\times p}/\mathrm{GL}_p$ is called the *Grassmann manifold* of p-planes in $\mathbb{R}^n$ and denoted by $\mathrm{Grass}(p,n)$. The particular case $\mathrm{Grass}(1,n) = \mathbb{R}\mathrm{P}^n$ is the real projective space discussed in Section 3.4.3. From Proposition 3.3.3, we have that $\dim(\mathrm{graph}(\sim)) = 2np - (n-p)p$. It then follows from (3.5) that

$$\dim(\mathrm{Grass}(p,n)) = p(n-p).$$

3.5 TANGENT VECTORS AND DIFFERENTIAL MAPS

There are several possible approaches to generalizing the notion of a *directional derivative*

$$\mathrm{D}f(x)[\eta] = \lim_{t\to 0} \frac{f(x+t\eta) - f(x)}{t} \qquad (3.8)$$

to a real-valued function f defined on a manifold. A first possibility is to view η as a *derivation at x*, that is, an object that, when given a real-valued function f defined on a neighborhood of $x \in \mathcal{M}$, returns a real ηf, and that satisfies the properties of a derivation operation: linearity and the Leibniz rule (see Section 3.5.5). This "axiomatization" of the notion of a directional derivative is elegant and powerful, but it gives little intuition as to how a tangent vector could possibly be represented as a matrix array in a computer.

A second, perhaps more intuitive approach to generalizing the directional derivative (3.8) is to replace $t \mapsto (x+t\eta)$ by a smooth curve γ on $\mathcal{M}$ through x (i.e., $\gamma(0) = x$). This yields a well-defined directional derivative $\frac{\mathrm{d}(f(\gamma(t)))}{\mathrm{d}t}\Big|_{t=0}$. (Note that this is a classical derivative since the function $t \mapsto f(\gamma(t))$ is a smooth function from $\mathbb{R}$ to $\mathbb{R}$.) Hence we have an operation, denoted by $\dot{\gamma}(0)$, that takes a function f, defined locally in a neighbourhood of x, and returns the real number $\frac{\mathrm{d}(f(\gamma(t)))}{\mathrm{d}t}\Big|_{t=0}$.

These two approaches are reconciled by showing that every derivative along a curve defines a pointwise derivation and that every pointwise derivation can be realized as a derivative along a curve. The first claim is direct. The second claim can be proved using a local coordinate representation, a third approach used to generalize the notion of a directional derivative.

3.5.1 Tangent vectors

Let $\mathcal{M}$ be a manifold. A smooth mapping $\gamma : \mathbb{R} \to \mathcal{M}: t \mapsto \gamma(t)$ is termed a *curve in* $\mathcal{M}$. The idea of defining a derivative $\gamma'(t)$ as

$$\gamma'(t) := \lim_{\tau \to 0} \frac{\gamma(t+\tau) - \gamma(t)}{\tau} \tag{3.9}$$

requires a vector space structure to compute the difference $\gamma(t+\tau) - \gamma(t)$ and thus fails for an abstract nonlinear manifold. However, given a smooth real-valued function f on $\mathcal{M}$, the function $f \circ \gamma : t \mapsto f(\gamma(t))$ is a smooth function from $\mathbb{R}$ to $\mathbb{R}$ with a well-defined classical derivative. This is exploited in the following definition. Let x be a point on $\mathcal{M}$, let γ be a curve through x at $t = 0$, and let $\mathfrak{F}_x(\mathcal{M})$ denote the set of smooth real-valued functions defined on a neighborhood of x. The mapping $\dot{\gamma}(0)$ from $\mathfrak{F}_x(\mathcal{M})$ to $\mathbb{R}$ defined by

$$\dot{\gamma}(0)f := \frac{\mathrm{d}(f(\gamma(t)))}{\mathrm{d}t}\Big|_{t=0}, \quad f \in \mathfrak{F}_x(\mathcal{M}), \tag{3.10}$$

is called the *tangent vector to the curve* γ *at* $t = 0$.

We emphasize that $\dot{\gamma}(0)$ is defined as a mapping from $\mathfrak{F}_x(\mathcal{M})$ to $\mathbb{R}$ and not as the time derivative $\gamma'(0)$ as in (3.9), which in general is meaningless. However, when $\mathcal{M}$ is (a submanifold of) a vector space $\mathcal{E}$, the mapping $\dot{\gamma}(0)$ from $\mathfrak{F}_x(\mathcal{M})$ to $\mathbb{R}$ and the derivative $\gamma'(0) := \lim_{t \to 0} \frac{1}{t}(\gamma(t) - \gamma(0))$ are closely related: for all functions $\overline{f}$ defined in a neighborhood $\mathcal{U}$ of $\gamma(0)$ in $\mathcal{E}$, we have

$$\dot{\gamma}(0)f = \mathrm{D}\overline{f}(\gamma(0))[\gamma'(0)],$$

where f denotes the restriction of $\overline{f}$ to $\mathcal{U} \cap \mathcal{M}$; see Sections 3.5.2 and 3.5.7 for details. It is useful to keep this interpretation in mind because the derivative $\gamma'(0)$ is a more familiar mathematical object than $\dot{\gamma}(0)$.

We can now formally define the notion of a tangent vector.

Definition 3.5.1 (tangent vector) *A tangent vector ξ_x to a manifold $\mathcal{M}$ at a point x is a mapping from $\mathfrak{F}_x(\mathcal{M})$ to $\mathbb{R}$ such that there exists a curve γ on $\mathcal{M}$ with $\gamma(0) = x$, satisfying*

$$\xi_x f = \dot{\gamma}(0)f := \left.\frac{\mathrm{d}(f(\gamma(t)))}{\mathrm{d}t}\right|_{t=0}$$

for all $f \in \mathfrak{F}_x(\mathcal{M})$. Such a curve γ is said to realize *the tangent vector ξ_x.*

The point x is called the *foot* of the tangent vector ξ_x. We will often omit the subscript indicating the foot and simply write ξ for ξ_x.

Given a tangent vector ξ to $\mathcal{M}$ at x, there are infinitely many curves γ that realize ξ (i.e., $\dot{\gamma}(0) = \xi$). They can be characterized as follows in local coordinates.

Proposition 3.5.2 *Two curves γ_1 and γ_2 through a point x at $t = 0$ satisfy $\dot{\gamma}_1(0) = \dot{\gamma}_2(0)$ if and only if, given a chart $(\mathcal{U}, \varphi)$ with $x \in \mathcal{U}$, it holds that*

$$\left.\frac{\mathrm{d}(\varphi(\gamma_1(t)))}{\mathrm{d}t}\right|_{t=0} = \left.\frac{\mathrm{d}(\varphi(\gamma_2(t)))}{\mathrm{d}t}\right|_{t=0}.$$

Proof. The "only if" part is straightforward since each component of the vector-valued φ belongs to $\mathfrak{F}_x(\mathcal{M})$. For the "if" part, given any $f \in \mathfrak{F}_x(\mathcal{M})$, we have

$$\dot{\gamma}_1(0)f = \left.\frac{\mathrm{d}(f(\gamma_1(t)))}{\mathrm{d}t}\right|_{t=0} = \left.\frac{\mathrm{d}((f \circ \varphi^{-1})(\varphi(\gamma_1(t))))}{\mathrm{d}t}\right|_{t=0}$$

$$= \left.\frac{\mathrm{d}((f \circ \varphi^{-1})(\varphi(\gamma_2(t))))}{\mathrm{d}t}\right|_{t=0} = \dot{\gamma}_2(0)f.$$

$\square$

The *tangent space* to $\mathcal{M}$ at x, denoted by $T_x\mathcal{M}$, is the set of all tangent vectors to $\mathcal{M}$ at x. This set admits a structure of *vector space* as follows. Given $\dot{\gamma}_1(0)$ and $\dot{\gamma}_2(0)$ in $T_x\mathcal{M}$ and a, b in $\mathbb{R}$, define

$$(a\dot{\gamma}_1(0) + b\dot{\gamma}_2(0))\, f := a\,(\dot{\gamma}_1(0)f) + b\,(\dot{\gamma}_2(0)f).$$

To show that $(a\dot{\gamma}_1(0) + b\dot{\gamma}_2(0))$ is a well-defined tangent vector, we need to show that there exists a curve γ such that $\dot{\gamma}(0) = a\dot{\gamma}_1(0) + b\dot{\gamma}_2(0)$. Such a curve is obtained by considering a chart $(\mathcal{U}, \varphi)$ with $x \in \mathcal{U}$ and defining $\gamma(t) = \varphi^{-1}(a\varphi(\gamma_1(t) + b\varphi(\gamma_2(t))$. It is readily checked that this γ satisfies the required property.

The property that the tangent space $T_x\mathcal{M}$ is a vector space is very important. In the same way that the derivative of a real-valued function provides a local linear approximation of the function, the tangent space $T_x\mathcal{M}$ provides a local vector space approximation of the manifold. In particular, in Section 4.1, we define mappings, called *retractions*, between $\mathcal{M}$ and $T_x\mathcal{M}$, which can be used to locally transform an optimization problem on the manifold $\mathcal{M}$ into an optimization problem on the more friendly vector space $T_x\mathcal{M}$.

Using a coordinate chart, it is possible to show that the dimension of the vector space $T_x\mathcal{M}$ is equal to d, the dimension of the manifold $\mathcal{M}$: given a chart $(\mathcal{U}, \varphi)$ at x, a basis of $T_x\mathcal{M}$ is given by $(\dot\gamma_1(0), \ldots, \dot\gamma_d(0))$, where $\gamma_i(t) := \varphi^{-1}(\varphi(x) + te_i)$, with e_i denoting the ith canonical vector of $\mathbb{R}^d$. Notice that $\dot\gamma_i(0)f = \partial_i(f \circ \varphi^{-1})(\varphi(x))$, where ∂_i denotes the partial derivative with respect to the ith component:

$$\partial_i h(x) := \lim_{t \to 0} \frac{h(x + te_i) - h(x)}{t}.$$

One has, for any tangent vector $\dot\gamma(0)$, the decomposition

$$\dot\gamma(0) = \sum_i (\dot\gamma(0)\varphi_i)\dot\gamma_i(0),$$

where φ_i denotes the ith component of φ. This provides a way to define the coordinates of tangent vectors at x using the chart $(\mathcal{U}, \varphi)$, by defining the element of $\mathbb{R}^d$

$$\begin{pmatrix} \dot\gamma(0)\varphi_1 \\ \vdots \\ \dot\gamma(0)\varphi_d \end{pmatrix}$$

as the representation of the tangent vector $\dot\gamma(0)$ in the chart $(\mathcal{U}, \varphi)$.

3.5.2 Tangent vectors to a vector space

Let $\mathcal{E}$ be a vector space and let x be a point of $\mathcal{E}$. As pointed out in Section 3.1.4, $\mathcal{E}$ admits a linear manifold structure. Strictly speaking, a tangent vector ξ to $\mathcal{E}$ at x is a mapping

$$\xi : \mathfrak{F}_x(\mathcal{E}) \to \mathbb{R} : f \mapsto \xi f = \left. \frac{d(f(\gamma(t)))}{dt} \right|_{t=0},$$

where γ is a curve in $\mathcal{E}$ with $\gamma(0) = x$. Defining $\gamma'(0) \in \mathcal{E}$ as in (3.9), we have

$$\xi f = Df(x)[\gamma'(0)].$$

Moreover, $\gamma'(0)$ does not depend on the curve γ that realizes ξ. This defines a canonical linear one-to-one correspondence $\xi \mapsto \gamma'(0)$, which identifies $T_x\mathcal{E}$ with $\mathcal{E}$:

$$T_x\mathcal{E} \simeq \mathcal{E}. \qquad (3.11)$$

Since tangent vectors are local objects (a tangent vector at a point x acts on smooth real-valued functions defined in any neighborhood of x), it follows that if $\mathcal{E}_*$ is an open submanifold of $\mathcal{E}$, then

$$T_x\mathcal{E}_* \simeq \mathcal{E} \qquad (3.12)$$

for all $x \in \mathcal{E}_*$. A schematic illustration is given in Figure 3.4.

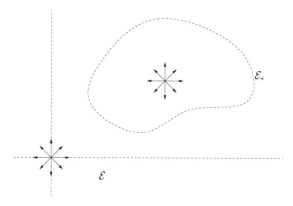

Figure 3.4 Tangent vectors to an open subset $\mathcal{E}_*$ of a vector space $\mathcal{E}$.

3.5.3 Tangent bundle

Given a manifold $\mathcal{M}$, let $T\mathcal{M}$ be the set of all tangent vectors to $\mathcal{M}$:

$$T\mathcal{M} := \bigcup_{x \in \mathcal{M}} T_x\mathcal{M}.$$

Since each $\xi \in T\mathcal{M}$ is in one and only one tangent space $T_x\mathcal{M}$, it follows that $\mathcal{M}$ is a quotient of $T\mathcal{M}$ with natural projection

$$\pi : T\mathcal{M} \to \mathcal{M} : \xi \in T_x\mathcal{M} \mapsto x,$$

i.e., $\pi(\xi)$ is the foot of ξ. The set $T\mathcal{M}$ admits a natural manifold structure as follows. Given a chart $(\mathcal{U}, \varphi)$ of $\mathcal{M}$, the mapping

$$\xi \in T_x\mathcal{M} \mapsto (\varphi_1(x), \ldots, \varphi_d(x), \xi\varphi_1, \ldots, \xi\varphi_d)^T$$

is a chart of the set $T\mathcal{M}$ with domain $\pi^{-1}(\mathcal{U})$. It can be shown that the collection of the charts thus constructed forms an atlas of the set $T\mathcal{M}$, turning it into a manifold called the *tangent bundle* of $\mathcal{M}$.

3.5.4 Vector fields

A *vector field* ξ on a manifold $\mathcal{M}$ is a smooth function from $\mathcal{M}$ to the tangent bundle $T\mathcal{M}$ that assigns to each point $x \in \mathcal{M}$ a tangent vector $\xi_x \in T_x\mathcal{M}$. On a submanifold of a vector space, a vector field can be pictured as a collection of arrows, one at each point of $\mathcal{M}$. Given a vector field ξ on $\mathcal{M}$ and a (smooth) real-valued function $f \in \mathfrak{F}(\mathcal{M})$, we let ξf denote the real-valued function on $\mathcal{M}$ defined by

$$(\xi f)(x) := \xi_x(f)$$

for all x in $\mathcal{M}$. The addition of two vector fields and the multiplication of a vector field by a function $f \in \mathfrak{F}(\mathcal{M})$ are defined as follows:

$$(f\xi)_x := f(x)\xi_x,$$
$$(\xi + \zeta)_x := \xi_x + \zeta_x \qquad \text{for all } x \in \mathcal{M}.$$

Smoothness is preserved by these operations. We let $\mathfrak{X}(\mathcal{M})$ denote the set of smooth vector fields endowed with these two operations.

Let $(\mathcal{U}, \varphi)$ be a chart of the manifold $\mathcal{M}$. The vector field E_i on $\mathcal{U}$ defined by

$$(E_i f)(x) := \partial_i (f \circ \varphi^{-1})(\varphi(x)) = \mathrm{D}(f \circ \varphi^{-1})(\varphi(x))[e_i]$$

is called the *ith coordinate vector field* of $(\mathcal{U}, \varphi)$. These coordinate vector fields are smooth, and every vector field ξ admits the decomposition

$$\xi = \sum_i (\xi \varphi_i) E_i$$

on $\mathcal{U}$. (A pointwise version of this result was given in Section 3.5.1.)

If the manifold is an n-dimensional vector space $\mathcal{E}$, then, given a basis $(e_i)_{i=1,\ldots,d}$ of $\mathcal{E}$, the vector fields E_i, $i = 1, \ldots, n$, defined by

$$(E_i f)(x) := \lim_{t \to 0} \frac{f(x + t e_i) - f(x)}{t} = \mathrm{D}f(x)[e_i]$$

form a basis of $\mathfrak{X}(\mathcal{E})$.

3.5.5 Tangent vectors as derivations*

Let x and η be elements of $\mathbb{R}^n$. The derivative mapping that, given a real-valued function f on $\mathbb{R}^n$, returns the real $\mathrm{D}f(x)[\eta]$ can be axiomatized as follows on manifolds. Let $\mathcal{M}$ be a manifold and recall that $\mathfrak{F}(\mathcal{M})$ denotes the set of all smooth real-valued functions on $\mathcal{M}$. Note that $\mathfrak{F}(\mathcal{M}) \subset \mathfrak{F}_x(\mathcal{M})$ for all $x \in \mathcal{M}$. A *derivation at* $x \in \mathcal{M}$ is a mapping ξ_x from $\mathfrak{F}(\mathcal{M})$ to $\mathbb{R}$ that is

1. $\mathbb{R}$-linear: $\xi_x(af + bg) = a\xi_x(f) + b\xi_x(g)$, and
2. Leibnizian: $\xi_x(fg) = \xi_x(f)g(x) + f(x)\xi_x(g)$, for all a, $b \in \mathbb{R}$ and f, $g \in \mathfrak{F}(\mathcal{M})$.

With the operations

$$(\xi_x + \zeta_x)f := \xi_x(f) + \zeta_x(f),$$
$$(a\xi_x)f := a\xi_x(f) \qquad \text{for all } f \in \mathfrak{F}(\mathcal{M}), \ a \in \mathbb{R},$$

the set of all derivations at x becomes a vector space. It can also be shown that a derivation ξ_x at x is a local notion: if two real-valued functions f and g are equal on a neighborhood of x, then $\xi_x(f) = \xi_x(g)$.

The concept of a tangent vector at x, as defined in Section 3.5.1, and the notion of a derivation at x are equivalent in the following sense: (i) Given a curve γ on $\mathcal{M}$ through x at $t = 0$, the mapping $\dot{\gamma}(0)$ from $\mathfrak{F}(\mathcal{M}) \subseteq \mathfrak{F}_x(\mathcal{M})$ to $\mathbb{R}$, defined in (3.10), is a derivation at x. (ii) Given a derivation ξ at x, there exists a curve γ on $\mathcal{M}$ through x at $t = 0$ such that $\dot{\gamma}(0) = \xi$. For example, the curve γ defined by $\gamma(t) = \varphi^{-1}(\varphi(0) + t \sum_i (\xi(\varphi_i)e_i))$ satisfies the property.

A (global) *derivation* on $\mathfrak{F}(\mathcal{M})$ is a mapping $\mathcal{D} : \mathfrak{F}(\mathcal{M}) \to \mathfrak{F}(\mathcal{M})$ that is

1. $\mathbb{R}$-linear: $\mathcal{D}(af + bg) = a\,\mathcal{D}(f) + b\,\mathcal{D}(g)$, $(a,\ b \in \mathbb{R})$, and
2. Leibnizian: $\mathcal{D}(fg) = \mathcal{D}(f)g + f\,\mathcal{D}(g)$.

Every vector field $\xi \in \mathfrak{X}(\mathcal{M})$ defines a derivation $f \mapsto \xi f$. Conversely, every derivation on $\mathfrak{F}(\mathcal{M})$ can be realized as a vector field. (Viewing vector fields as derivations comes in handy in understanding Lie brackets; see Section 5.3.1.)

3.5.6 Differential of a mapping

Let $F : \mathcal{M} \to \mathcal{N}$ be a smooth mapping between two manifolds $\mathcal{M}$ and $\mathcal{N}$. Let ξ_x be a tangent vector at a point x of $\mathcal{M}$. It can be shown that the mapping $DF(x)[\xi_x]$ from $\mathfrak{F}_{F(x)}(\mathcal{N})$ to $\mathbb{R}$ defined by

$$(DF(x)[\xi])\,f := \xi(f \circ F) \tag{3.13}$$

is a tangent vector to $\mathcal{N}$ at $F(x)$. The tangent vector $DF(x)[\xi_x]$ is realized by $F \circ \gamma$, where γ is any curve that realizes ξ_x. The mapping

$$DF(x) : T_x\mathcal{M} \to T_{F(x)}\mathcal{N} : \xi \mapsto DF(x)[\xi]$$

is a linear mapping called the *differential* (or *differential map*, *derivative*, or *tangent map*) of F at x (see Figure 3.5).

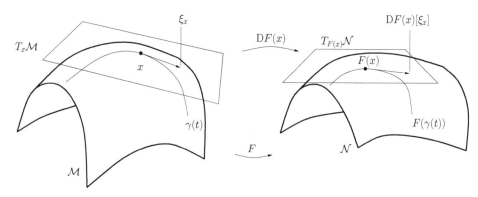

Figure 3.5 Differential map of F at x.

Note that F is an immersion (respectively, submersion) if and only if $DF(x) : T_x\mathcal{M} \to T_{F(x)}\mathcal{N}$ is an injection (respectively, surjection) for every $x \in \mathcal{M}$.

If $\mathcal{N}$ is a vector space $\mathcal{E}$, then the canonical identification $T_{F(x)}\mathcal{E} \simeq \mathcal{E}$ yields

$$DF(x)[\xi_x] = \sum_i (\xi_x F^i)e_i, \tag{3.14}$$

where $F(x) = \sum_i F^i(x)e_i$ is the decomposition of $F(x)$ in a basis $(e_i)_{i=1,\dots,n}$ of $\mathcal{E}$.

If $\mathcal{N} = \mathbb{R}$, then $F \in \mathfrak{F}_x(\mathcal{M})$, and we simply have

$$DF(x)[\xi_x] = \xi_x F \tag{3.15}$$

using the identification $T_x\mathbb{R} \simeq \mathbb{R}$. We will often use $\mathrm{D}F(x)[\xi_x]$ as an alternative notation for $\xi_x F$, as it better emphasizes the derivative aspect.

If $\mathcal{M}$ and $\mathcal{N}$ are linear manifolds, then, with the identification $T_x\mathcal{M} \simeq \mathcal{M}$ and $T_y\mathcal{N} \simeq \mathcal{N}$, $\mathrm{D}F(x)$ reduces to its classical definition

$$\mathrm{D}F(x)[\xi_x] = \lim_{t \to 0} \frac{F(x + t\xi_x) - F(x)}{t}. \tag{3.16}$$

Given a differentiable function $F : \mathcal{M} \mapsto \mathcal{N}$ and a vector field ξ on $\mathcal{M}$, we let $\mathrm{D}F[\xi]$ denote the mapping

$$\mathrm{D}F[\xi] : \mathcal{M} \to T\mathcal{N} : x \mapsto \mathrm{D}F(x)[\xi_x].$$

In particular, given a real-valued function f on $\mathcal{M}$ and a vector field ξ on $\mathcal{M}$,

$$\mathrm{D}f[\xi] = \xi f.$$

3.5.7 Tangent vectors to embedded submanifolds

We now investigate the case where $\mathcal{M}$ is an embedded submanifold of a vector space $\mathcal{E}$. Let γ be a curve in $\mathcal{M}$, with $\gamma(0) = x$. Define

$$\gamma'(0) := \lim_{t \to 0} \frac{\gamma(t) - \gamma(0)}{t},$$

where the subtraction is well defined since $\gamma(t)$ belongs to the vector space $\mathcal{E}$ for all t. (Strictly speaking, one should write $i(\gamma(t)) - i(\gamma(0))$, where i is the natural inclusion of $\mathcal{M}$ in $\mathcal{E}$; the inclusion is omitted to simplify the notation.) It follows that $\gamma'(0)$ thus defined is an element of $T_x\mathcal{E} \simeq \mathcal{E}$ (see Figure 3.6). Since γ is a curve in $\mathcal{M}$, it also induces a tangent vector $\dot{\gamma}(0) \in T_x\mathcal{M}$. Not

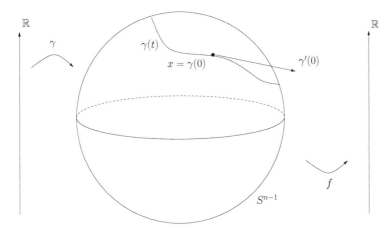

Figure 3.6 Curves and tangent vectors on the sphere. Since S^{n-1} is an embedded submanifold of $\mathbb{R}^n$, the tangent vector $\dot{\gamma}(0)$ can be pictured as the directional derivative $\gamma'(0)$.

surprisingly, $\gamma'(0)$ and $\dot{\gamma}(0)$ are closely related: If $\overline{f}$ is a real-valued function in a neighborhood $\mathcal{U}$ of x in $\mathcal{E}$ and f denotes the restriction of $\overline{f}$ to $\mathcal{U} \cap \mathcal{M}$ (which is a neighborhood of x in $\mathcal{M}$ since $\mathcal{M}$ is embedded), then we have

$$\dot{\gamma}(0)f = \left.\frac{\mathrm{d}}{\mathrm{d}t} f(\gamma(t))\right|_{t=0} = \left.\frac{\mathrm{d}}{\mathrm{d}t} \overline{f}(\gamma(t))\right|_{t=0} = \mathrm{D}\overline{f}(x)\left[\gamma'(0)\right]. \qquad (3.17)$$

This yields a natural identification of $T_x\mathcal{M}$ with the set

$$\{\gamma'(0) : \gamma \text{ curve in } \mathcal{M},\ \gamma(0) = x\}, \qquad (3.18)$$

which is a linear subspace of the vector space $T_x\mathcal{E} \simeq \mathcal{E}$. In particular, when $\mathcal{M}$ is a matrix submanifold (i.e., the embedding space is $\mathbb{R}^{n \times p}$), we have $T_x\mathcal{E} = \mathbb{R}^{n \times p}$, hence the tangent vectors to $\mathcal{M}$ are naturally represented by $n \times p$ matrix arrays.

Graphically, a tangent vector to a submanifold of a vector space can be thought of as an "arrow" tangent to the manifold. It is convenient to keep this intuition in mind when dealing with more abstract manifolds; however, one should bear in mind that the notion of a tangent arrow cannot always be visualized meaningfully in this manner, in which case one must return to the definition of tangent vectors as objects that, given a real-valued function, return a real number, as stated in Definition 3.5.1.

In view of the identification of $T_x\mathcal{M}$ with (3.18), we now write $\dot{\gamma}(t)$, $\gamma'(t)$, and $\frac{\mathrm{d}}{\mathrm{d}t}\gamma(t)$ interchangeably. We also use the equality sign, such as in (3.19) below, to denote the identification of $T_x\mathcal{M}$ with (3.18).

When $\mathcal{M}$ is (locally or globally) defined as a level set of a constant-rank function $F : \mathcal{E} \mapsto \mathbb{R}^n$, we have

$$T_x\mathcal{M} = \ker(\mathrm{D}F(x)). \qquad (3.19)$$

In other words, the tangent vectors to $\mathcal{M}$ at x correspond to those vectors ξ that satisfy $\mathrm{D}F(x)[\xi] = 0$. Indeed, if γ is a curve in $\mathcal{M}$ with $\gamma(0) = x$, we have $F(\gamma(t)) = 0$ for all t, hence

$$\mathrm{D}F(x)\left[\dot{\gamma}(0)\right] = \left.\frac{\mathrm{d}(F(\gamma(t)))}{\mathrm{d}t}\right|_{t=0} = 0,$$

which shows that $\dot{\gamma}(0) \in \ker(\mathrm{D}F(x))$. By counting dimensions using Proposition 3.3.4, it follows that $T_x\mathcal{M}$ and $\ker(\mathrm{D}F(x))$ are two vector spaces of the same dimension with one included in the other. This proves the equality (3.19).

Example 3.5.1 *Tangent space to a sphere*
Let $t \mapsto x(t)$ be a curve in the unit sphere S^{n-1} through x_0 at $t = 0$. Since $x(t) \in S^{n-1}$ for all t, we have

$$x^T(t)x(t) = 1$$

for all t. Differentiating this equation with respect to t yields

$$\dot{x}^T(t)x(t) + x^T(t)\dot{x}(t) = 0,$$

hence $\dot{x}(0)$ is an element of the set

$$\{z \in \mathbb{R}^n : x_0^T z = 0\}. \qquad (3.20)$$

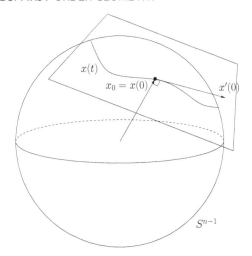

Figure 3.7 Tangent space on the sphere. Since S^{n-1} is an embedded submanifold of $\mathbb{R}^n$, the tangent space $T_x S^{n-1}$ can be pictured as the hyperplane tangent to the sphere at x, with origin at x.

This shows that $T_{x_0} S^{n-1}$ is a subset of (3.20). Conversely, let z belong to the set (3.20). Then the curve $t \mapsto x(t) := (x_0 + tz)/\|x_0 + tz\|$ is on S^{n-1} and satisfies $\dot{x}(0) = z$. Hence (3.20) is a subset of $T_{x_0} S^{n-1}$. In conclusion,

$$T_x S^{n-1} = \{z \in \mathbb{R}^n : x^T z = 0\}, \tag{3.21}$$

which is the set of all vectors orthogonal to x in $\mathbb{R}^n$; see Figure 3.7.

More directly, consider the function $F : \mathbb{R}^n \to \mathbb{R} : x \mapsto x^T x - 1$. Since $S^{n-1} = \{x \in \mathbb{R}^n : F(x) = 0\}$ and since F is full rank on S^{n-1}, it follows from (3.19) that

$$T_x S^{n-1} = \ker(DF(x)) = \{z \in \mathbb{R}^n : x^T z + z^T x = 0\} = \{z \in \mathbb{R}^n : x^T z = 0\},$$

as in (3.21).

Example 3.5.2 *Orthogonal Stiefel manifold*

We consider the orthogonal Stiefel manifold

$$\mathrm{St}(p, n) = \{X \in \mathbb{R}^{n \times p} : X^T X = I_p\}$$

as an embedded submanifold of the Euclidean space $\mathbb{R}^{n \times p}$ (see Section 3.3.2). Let X_0 be an element of $\mathrm{St}(p, n)$ and let $t \mapsto X(t)$ be a curve in $\mathrm{St}(p, n)$ through X_0 at $t = 0$; i.e., $X(t) \in \mathbb{R}^{n \times p}$, $X(0) = X_0$, and

$$X^T(t) X(t) = I_p \tag{3.22}$$

for all t. It follows by differentiating (3.22) that

$$\dot{X}^T(t) X(t) + X^T(t) \dot{X}(t) = 0. \tag{3.23}$$

We deduce that $\dot{X}(0)$ belongs to the set

$$\{Z \in \mathbb{R}^{n \times p} : X_0^T Z + Z^T X_0 = 0\}. \tag{3.24}$$

We have thus shown that $T_{X_0} \operatorname{St}(p, n)$ is a subset of (3.24). It is possible to conclude, as in the previous example, by showing that for all Z in (3.24) there is a curve in $\operatorname{St}(p, n)$ through X_0 at t such that $\dot{X}(0) = Z$. A simpler argument is to invoke (3.19) by pointing out that (3.24) is the kernel of $DF(X_0)$, where $F : X \mapsto X^T X$, so that I_p is a regular value of F and $F^{-1}(I_p) = \operatorname{St}(p, n)$. In conclusion, the set described in (3.24) is the tangent space to $\operatorname{St}(p, n)$ at X_0. That is,

$$T_X \operatorname{St}(p, n) = \{Z \in \mathbb{R}^{n \times p} : X^T Z + Z^T X = 0\}.$$

We now propose an alternative characterization of $T_X \operatorname{St}(p, n)$. Without loss of generality, since $\dot{X}(t)$ is an element of $\mathbb{R}^{n \times p}$ and $X(t)$ has full rank, we can set

$$\dot{X}(t) = X(t)\Omega(t) + X_{\perp}(t)K(t), \tag{3.25}$$

where $X_{\perp}(t)$ is any $n \times (n - p)$ matrix such that $\operatorname{span}(X_{\perp}(t))$ is the orthogonal complement of $\operatorname{span}(X(t))$. Replacing (3.25) in (3.23) yields

$$\Omega(t)^T + \Omega(t) = 0;$$

i.e., $\Omega(t)$ is a skew-symmetric matrix. Counting dimensions, we deduce that

$$T_X \operatorname{St}(p, n) = \{X\Omega + X_{\perp}K : \Omega^T = -\Omega, \ K \in \mathbb{R}^{(n-p) \times p}\}.$$

Observe that the two characterizations of $T_X \operatorname{St}(p, n)$ are facilitated by the embedding of $\operatorname{St}(p, n)$ in $\mathbb{R}^{n \times p}$: $T_X \operatorname{St}(p, n)$ is identified with a linear subspace of $\mathbb{R}^{n \times p}$.

Example 3.5.3 *Orthogonal group*

Since the orthogonal group O_n is $\operatorname{St}(p, n)$ with $p = n$, it follows from the previous section that

$$T_U O_n = \{Z = U\Omega : \Omega^T = -\Omega\} = U\mathcal{S}_{\text{skew}}(n), \tag{3.26}$$

where $\mathcal{S}_{\text{skew}}(n)$ denotes the set of all skew-symmetric $n \times n$ matrices.

3.5.8 Tangent vectors to quotient manifolds

We have seen that tangent vectors of a submanifold embedded in a vector space $\mathcal{E}$ can be viewed as tangent vectors to $\mathcal{E}$ and pictured as arrows in $\mathcal{E}$ tangent to the submanifold. The situation of a quotient $\mathcal{E}/\sim$ of a vector space $\mathcal{E}$ is more abstract. Nevertheless, the structure space $\mathcal{E}$ also offers convenient representations of tangent vectors to the quotient.

For generality, we consider an abstract manifold $\overline{\mathcal{M}}$ and a quotient manifold $\mathcal{M} = \overline{\mathcal{M}}/\sim$ with canonical projection π. Let ξ be an element of $T_x\mathcal{M}$ and let $\overline{x}$ be an element of the equivalence class $\pi^{-1}(x)$. Any element $\overline{\xi}$ of

$T_{\overline{x}}\overline{\mathcal{M}}$ that satisfies $\mathrm{D}\pi(\overline{x})[\overline{\xi}] = \xi$ can be considered a representation of ξ. Indeed, for any smooth function $f : \mathcal{M} \to \mathbb{R}$, the function $\overline{f} := f \circ \pi : \overline{\mathcal{M}} \to \mathbb{R}$ is smooth (Proposition 3.4.5), and one has

$$\mathrm{D}\overline{f}(\overline{x})[\overline{\xi}] = \mathrm{D}f(\pi(\overline{x}))[\mathrm{D}\pi(\overline{x})[\overline{\xi}]] = \mathrm{D}f(x)[\xi].$$

A difficulty with this approach is that there are infinitely many valid representations $\overline{\xi}$ of ξ at $\overline{x}$.

It is desirable to identify a unique "lifted" representation of tangent vectors of $T_x\mathcal{M}$ in $T_{\overline{x}}\overline{\mathcal{M}}$ in order that we can use the lifted tangent vector representation unambiguously in numerical computations. Recall from Proposition 3.4.4 that the equivalence class $\pi^{-1}(x)$ is an embedded submanifold of $\overline{\mathcal{M}}$. Hence $\pi^{-1}(x)$ admits a tangent space

$$\mathcal{V}_{\overline{x}} = T_{\overline{x}}(\pi^{-1}(x))$$

called the *vertical space* at $\overline{x}$. A mapping $\mathcal{H}$ that assigns to each element $\overline{x}$ of $\overline{\mathcal{M}}$ a subspace $\mathcal{H}_{\overline{x}}$ of $T_{\overline{x}}\overline{\mathcal{M}}$ complementary to $\mathcal{V}_{\overline{x}}$ (i.e., such that $\mathcal{H}_{\overline{x}} \oplus \mathcal{V}_{\overline{x}} = T_{\overline{x}}\overline{\mathcal{M}}$) is called a *horizontal distribution* on $\overline{\mathcal{M}}$. Given $\overline{x} \in \overline{\mathcal{M}}$, the subspace $\mathcal{H}_{\overline{x}}$ of $T_{\overline{x}}\overline{\mathcal{M}}$ is then called the *horizontal space* at $\overline{x}$; see Figure 3.8. Once $\overline{\mathcal{M}}$ is endowed with a horizontal distribution, there exists one and only one element $\overline{\xi}_{\overline{x}}$ that belongs to $\mathcal{H}_{\overline{x}}$ and satisfies $\mathrm{D}\pi(\overline{x})[\overline{\xi}_{\overline{x}}] = \xi$. This unique vector $\overline{\xi}_{\overline{x}}$ is called the *horizontal lift* of ξ at $\overline{x}$.

In particular, when the structure space is (a subset of) $\mathbb{R}^{n \times p}$, the horizontal lift $\overline{\xi}_{\overline{x}}$ is an $n \times p$ matrix, which lends itself to representation in a computer as a matrix array.

Example 3.5.4 *Real projective space*

Recall from Section 3.4.3 that the projective space $\mathbb{RP}^{n-1}$ is the quotient $\mathbb{R}^n_*/\sim$, where $x \sim y$ if and only if there is an $\alpha \in \mathbb{R}_*$ such that $y = x\alpha$. The equivalence class of a point x of $\mathbb{R}^n_*$ is

$$[x] = \pi^{-1}(\pi(x)) = x\mathbb{R}_* := \{x\alpha : \alpha \in \mathbb{R}_*\}.$$

The vertical space at a point $x \in \mathbb{R}^n_*$ is

$$\mathcal{V}_x = x\mathbb{R} := \{x\alpha : \alpha \in \mathbb{R}\}.$$

A suitable choice of horizontal distribution is

$$\mathcal{H}_x := (\mathcal{V}_x)^\perp := \{z \in \mathbb{R}^n : x^T z = 0\}. \tag{3.27}$$

(This horizontal distribution will play a particular role in Section 3.6.2 where the projective space is turned into a Riemannian quotient manifold.)

A tangent vector $\xi \in T_{\pi(x)}\mathbb{RP}^{n-1}$ is represented by its horizontal lift $\overline{\xi}_x \in \mathcal{H}_x$ at a point $x \in \mathbb{R}^n_*$. It would be equally valid to use another representation $\overline{\xi}_y \in \mathcal{H}_y$ of the same tangent vector at another point $y \in \mathbb{R}^n_*$ such that $x \sim y$. The two representations $\overline{\xi}_x$ and $\overline{\xi}_y$ are not equal as vectors in $\mathbb{R}^n$ but are related by a scaling factor, as we now show. First, note that $x \sim y$ if and only if there exists a nonzero scalar α such that $y = \alpha x$. Let $f : \mathbb{RP}^{n-1} \to \mathbb{R}$ be an arbitrary smooth function and define $\overline{f} := f \circ \pi : \mathbb{R}^n_* \to \mathbb{R}$. Consider the function $g : x \mapsto \alpha x$, where α is an arbitrary nonzero scalar. Since

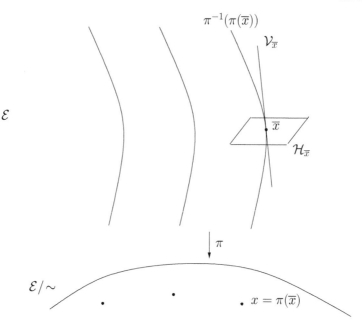

Figure 3.8 Schematic illustration of a quotient manifold. An equivalence class $\pi^{-1}(\pi(\overline{x}))$ is pictured as a subset of the total space $\mathcal{E}$ and corresponds to the single point $\pi(\overline{x})$ in the quotient manifold $\mathcal{E}/\sim$. At $\overline{x}$, the tangent space to the equivalence class is the vertical space $\mathcal{V}_{\overline{x}}$, and the horizontal space $\mathcal{H}_{\overline{x}}$ is chosen as a complement of the vertical space.

$\pi(g(x)) = \pi(x)$ for all x, we have $\overline{f}(g(x)) = \overline{f}(x)$ for all x, and it follows by taking the differential of both sides that

$$\mathrm{D}\overline{f}(g(x))[\mathrm{D}g(x)[\overline{\xi}_x]] = \mathrm{D}\overline{f}(x)[\overline{\xi}_x]. \tag{3.28}$$

By the definition of $\overline{\xi}_x$, we have $\mathrm{D}\overline{f}(x)[\overline{\xi}_x] = \mathrm{D}f(\pi(x))[\xi]$. Moreover, we have $\mathrm{D}g(x)[\overline{\xi}_x] = \alpha\overline{\xi}_x$. Thus (3.28) yields $\mathrm{D}\overline{f}(\alpha x)[\alpha\overline{\xi}_x]] = \mathrm{D}f(\pi(\alpha x))[\xi]$. This result, since it is valid for any smooth function f, implies that $\mathrm{D}\pi(\alpha x)[\alpha\overline{\xi}_x] = \xi$. This, along with the fact that $\alpha\overline{\xi}_x$ is an element of $\mathcal{H}_{\alpha x}$, implies that $\alpha\overline{\xi}_x$ is the horizontal lift of ξ at αx, i.e.,

$$\overline{\xi}_{\alpha x} = \alpha\overline{\xi}_x.$$

Example 3.5.5   *Grassmann manifolds*

   Tangent vectors to the Grassmann manifolds and their matrix representations are presented in Section 3.6.

3.6 RIEMANNIAN METRIC, DISTANCE, AND GRADIENTS

Tangent vectors on manifolds generalize the notion of a directional derivative. In order to characterize which direction of motion from x produces the steepest increase in f, we further need a notion of length that applies to tangent vectors. This is done by endowing every tangent space $T_x\mathcal{M}$ with an *inner product* $\langle \cdot, \cdot \rangle_x$, i.e., a bilinear, symmetric positive-definite form. The inner product $\langle \cdot, \cdot \rangle_x$ induces a norm,

$$\|\xi_x\|_x := \sqrt{\langle \xi_x, \xi_x \rangle_x},$$

on $T_x\mathcal{M}$. (The subscript x may be omitted if there is no risk of confusion.) The (normalized) direction of steepest ascent is then given by

$$\arg\max_{\xi \in T_x\mathcal{M}:\|\xi_x\|=1} \mathrm{D}f(x)[\xi_x].$$

A manifold whose tangent spaces are endowed with a smoothly varying inner product is called a *Riemannian manifold*. The smoothly varying inner product is called the *Riemannian metric*. We will use interchangeably the notation

$$g(\xi_x, \zeta_x) = g_x(\xi_x, \zeta_x) = \langle \xi_x, \zeta_x \rangle = \langle \xi_x, \zeta_x \rangle_x$$

to denote the inner product of two elements ξ_x and ζ_x of $T_x\mathcal{M}$. Strictly speaking, a Riemannian manifold is thus a couple $(\mathcal{M}, g)$, where $\mathcal{M}$ is a manifold and g is a Riemannian metric on $\mathcal{M}$. Nevertheless, when the Riemannian metric is unimportant or clear from the context, we simply talk about "the Riemannian manifold $\mathcal{M}$". A vector space endowed with an inner product is a particular Riemannian manifold called *Euclidean space*. Any (second-countable Hausdorff) manifold admits a Riemannian structure.

Let $(\mathcal{U}, \varphi)$ be a chart of a Riemannian manifold $(\mathcal{M}, g)$. The components of g in the chart are given by

$$g_{ij} := g(E_i, E_j),$$

where E_i denotes the ith coordinate vector field (see Section 3.5.4). Thus, for vector fields $\xi = \sum_i \xi^i E_i$ and $\zeta = \sum_i \zeta^i E_i$, we have

$$g(\xi, \zeta) = \langle \xi, \eta \rangle = \sum_{i,j} g_{ij} \xi^i \zeta^j.$$

Note that the g_{ij}'s are real-valued functions on $\mathcal{U} \subseteq \mathcal{M}$. One can also define the real-valued functions $g_{ij} \circ \varphi^{-1}$ on $\varphi(\mathcal{U}) \subseteq \mathbb{R}^d$; we use the same notation g_{ij} for both. We also use the notation $G : \hat{x} \mapsto G_{\hat{x}}$ for the matrix-valued function such that the (i, j) element of $G_{\hat{x}}$ is $g_{ij}|_{\hat{x}}$. If we let $\hat{\xi}_{\hat{x}} = \mathrm{D}\varphi \left(\varphi^{-1}(\hat{x}) \right)[\xi_x]$ and $\hat{\zeta}_{\hat{x}} = \mathrm{D}\varphi \left(\varphi^{-1}(\hat{x}) \right)[\zeta_x]$, with $\hat{x} = \varphi(x)$, denote the representations of ξ_x and ζ_x in the chart, then we have, in matrix notation,

$$g(\xi_x, \zeta_x) = \langle \xi_x, \zeta_x \rangle = \hat{\xi}_{\hat{x}}^T G_{\hat{x}} \hat{\zeta}_{\hat{x}}. \tag{3.29}$$

Note that G is a symmetric, positive definite matrix at every point.

The *length of a curve* $\gamma : [a, b] \to \mathcal{M}$ on a Riemannian manifold $(\mathcal{M}, g)$ is defined by

$$L(\gamma) = \int_a^b \sqrt{g(\dot{\gamma}(t), \dot{\gamma}(t))} \, \mathrm{d}t.$$

The *Riemannian distance* on a connected Riemannian manifold $(\mathcal{M}, g)$ is

$$\mathrm{dist} : \mathcal{M} \times \mathcal{M} \to \mathbb{R} : \mathrm{dist}(x, y) = \inf_\Gamma L(\gamma) \tag{3.30}$$

where Γ is the set of all curves in $\mathcal{M}$ joining points x and y. Assuming (as usual) that $\mathcal{M}$ is Hausdorff, it can be shown that the Riemannian distance defines a *metric*; i.e.,

1. $\mathrm{dist}(x, y) \geq 0$, with $\mathrm{dist}(x, y) = 0$ if and only if $x = y$ (positive-definiteness);
2. $\mathrm{dist}(x, y) = \mathrm{dist}(y, x)$ (symmetry);
3. $\mathrm{dist}(x, z) + \mathrm{dist}(z, y) \geq \mathrm{dist}(x, y)$ (triangle inequality).

Metrics and Riemannian metrics should not be confused. A metric is an abstraction of the notion of distance, whereas a Riemannian metric is an inner product on tangent spaces. There is, however, a link since any Riemannian metric induces a distance, the Riemannian distance.

Given a smooth scalar field f on a Riemannian manifold $\mathcal{M}$, the *gradient* of f at x, denoted by $\mathrm{grad}\, f(x)$, is defined as the unique element of $T_x\mathcal{M}$ that satisfies

$$\langle \mathrm{grad}\, f(x), \xi \rangle_x = \mathrm{D}f(x)[\xi], \quad \forall \xi \in T_x\mathcal{M}. \tag{3.31}$$

The coordinate expression of $\mathrm{grad}\, f$ is, in matrix notation,

$$\widehat{\mathrm{grad}\, f}(\hat{x}) = G_{\hat{x}}^{-1} \, \mathrm{Grad}\, \widehat{f}(\hat{x}), \tag{3.32}$$

where G is the matrix-valued function defined in (3.29) and Grad denotes the *Euclidean gradient* in $\mathbb{R}^d$,

$$\mathrm{Grad}\, \widehat{f}(\hat{x}) := \begin{pmatrix} \partial_1 \widehat{f}(\hat{x}) \\ \vdots \\ \partial_d \widehat{f}(\hat{x}) \end{pmatrix}.$$

(Indeed, from (3.29) and (3.32), we have $\langle \mathrm{grad}\, f, \xi \rangle = \hat{\xi}^T G(G^{-1} \, \mathrm{Grad}\, \widehat{f}) = \hat{\xi}^T \mathrm{Grad}\, \widehat{f} = \mathrm{D}\widehat{f}[\hat{\xi}] = \mathrm{D}f[\xi]$ for any vector field ξ.)

The gradient of a function has the following remarkable steepest-ascent properties (see Figure 3.9):

- The direction of $\mathrm{grad}\, f(x)$ is the steepest-ascent direction of f at x:

$$\frac{\mathrm{grad}\, f(x)}{\|\mathrm{grad}\, f(x)\|} = \underset{\xi \in T_x\mathcal{M}: \|\xi\|=1}{\arg\max} \, \mathrm{D}f(x)[\xi].$$

- The norm of $\mathrm{grad}\, f(x)$ gives the steepest slope of f at x:

$$\|\mathrm{grad}\, f(x)\| = \mathrm{D}f(x)\left[\frac{\mathrm{grad}\, f(x)}{\|\mathrm{grad}\, f(x)\|}\right].$$

These two properties are important in the scope of optimization methods.

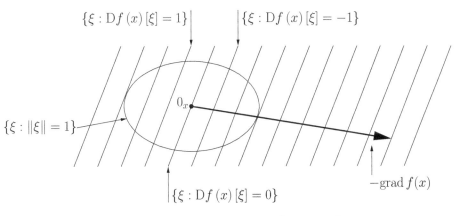

$\{\xi : \mathrm{D}f(x)[\xi] = 1\}$ $\{\xi : \mathrm{D}f(x)[\xi] = -1\}$

$\{\xi : \|\xi\| = 1\}$

0_x

$-\mathrm{grad}\, f(x)$

$\{\xi : \mathrm{D}f(x)[\xi] = 0\}$

Figure 3.9 Illustration of steepest descent.

3.6.1 Riemannian submanifolds

If a manifold $\overline{\mathcal{M}}$ is endowed with a Riemannian metric, one would expect that manifolds generated from $\overline{\mathcal{M}}$ (such as submanifolds and quotient manifolds) can inherit a Riemannian metric in a natural way. This section considers the case of embedded submanifolds; quotient manifolds are dealt with in the next section.

Let $\mathcal{M}$ be an embedded submanifold of a Riemannian manifold $\overline{\mathcal{M}}$. Since every tangent space $T_x\mathcal{M}$ can be regarded as a subspace of $T_x\overline{\mathcal{M}}$, the Riemannian metric $\overline{g}$ of $\overline{\mathcal{M}}$ induces a Riemannian metric g on $\mathcal{M}$ according to

$$g_x(\xi, \zeta) = \overline{g}_x(\xi, \zeta), \quad \xi,\ \zeta \in T_x\mathcal{M},$$

where ξ and ζ on the right-hand side are viewed as elements of $T_x\overline{\mathcal{M}}$. This turns $\mathcal{M}$ into a Riemannian manifold. Endowed with this Riemannian metric, $\mathcal{M}$ is called a *Riemannian submanifold* of $\overline{\mathcal{M}}$. The orthogonal complement of $T_x\mathcal{M}$ in $T_x\overline{\mathcal{M}}$ is called the *normal space to $\mathcal{M}$ at x* and is denoted by $(T_x\mathcal{M})^{\perp}$:

$$(T_x\mathcal{M})^{\perp} = \{\xi \in T_x\overline{\mathcal{M}} : \overline{g}_x(\xi, \zeta) = 0 \text{ for all } \zeta \in T_x\mathcal{M}\}.$$

Any element $\xi \in T_x\overline{\mathcal{M}}$ can be uniquely decomposed into the sum of an element of $T_x\mathcal{M}$ and an element of $(T_x\mathcal{M})^{\perp}$:

$$\xi = \mathrm{P}_x\xi + \mathrm{P}_x^{\perp}\xi,$$

where P_x denotes the orthogonal projection onto $T_x\mathcal{M}$ and $\mathrm{P}_x^{\perp}$ denotes the orthogonal projection onto $(T_x\mathcal{M})^{\perp}$.

Example 3.6.1 *Sphere*
  On the unit sphere S^{n-1} considered a Riemannian submanifold of $\mathbb{R}^n$, the inner product inherited from the standard inner product on $\mathbb{R}^n$ is given by

$$\langle \xi, \eta \rangle_x := \xi^T \eta. \tag{3.33}$$

The normal space is

$$(T_x S^{n-1})^{\perp} = \{x\alpha : \alpha \in \mathbb{R}\},$$

and the projections are given by

$$P_x \xi = (I - xx^T)\xi, \qquad P_x^{\perp} \xi = xx^T \xi$$

for $x \in S^{n-1}$.

Example 3.6.2   *Orthogonal Stiefel manifold*

Recall that the tangent space to the orthogonal Stiefel manifold $\mathrm{St}(p,n)$ is

$$T_X \mathrm{St}(p,n) = \{X\Omega + X_{\perp}K : \Omega^T = -\Omega, \ K \in \mathbb{R}^{(n-p) \times p}\}.$$

The Riemannian metric inherited from the embedding space $\mathbb{R}^{n \times p}$ is

$$\langle \xi, \eta \rangle_X := \mathrm{tr}(\xi^T \eta). \tag{3.34}$$

If $\xi = X\Omega_\xi + X_{\perp}K_\xi$ and $\eta = X\Omega_\eta + X_{\perp}K_\eta$, then $\langle \xi, \eta \rangle_X = \mathrm{tr}(\Omega_\xi^T \Omega_\eta + K_\xi^T K_\eta)$. In view of the identity $\mathrm{tr}(S^T \Omega) = 0$ for all $S \in \mathcal{S}_{\mathrm{sym}}(p)$, $\Omega \in \mathcal{S}_{\mathrm{skew}}(p)$, the normal space is

$$(T_X \mathrm{St}(p,n))^{\perp} = \{XS : S \in \mathcal{S}_{\mathrm{sym}}(p)\}.$$

The projections are given by

$$P_X \xi = (I - XX^T)\xi + X \mathrm{skew}(X^T \xi), \tag{3.35}$$

$$P_X^{\perp} \xi = X \mathrm{sym}(X^T \xi), \tag{3.36}$$

where $\mathrm{sym}(A) := \frac{1}{2}(A + A^T)$ and $\mathrm{skew}(A) := \frac{1}{2}(A - A^T)$ denote the components of the decomposition of A into the sum of a symmetric term and a skew-symmetric term.

Let $\overline{f}$ be a cost function defined on a Riemannian manifold $\overline{\mathcal{M}}$ and let f denote the restriction of $\overline{f}$ to a Riemannian submanifold $\mathcal{M}$. The gradient of f is equal to the projection of the gradient of $\overline{f}$ onto $T_x\mathcal{M}$:

$$\mathrm{grad}\, f(x) = P_x \, \mathrm{grad}\, \overline{f}(x). \tag{3.37}$$

Indeed, $P_x \, \mathrm{grad}\, \overline{f}(x)$ belongs to $T_x\mathcal{M}$ and (3.31) is satisfied since, for all $\zeta \in T_x\mathcal{M}$, we have $\langle P_x \, \mathrm{grad}\, \overline{f}(x), \zeta \rangle = \langle \mathrm{grad}\, \overline{f}(x) - P_x^{\perp} \mathrm{grad}\, \overline{f}(x), \zeta \rangle = \langle \mathrm{grad}\, \overline{f}(x), \zeta \rangle = \mathrm{D}\overline{f}(x)[\zeta] = \mathrm{D}f(x)[\zeta]$.

3.6.2 Riemannian quotient manifolds

We now consider the case of a quotient manifold $\mathcal{M} = \overline{\mathcal{M}}/\sim$, where the structure space $\overline{\mathcal{M}}$ is endowed with a Riemannian metric $\overline{g}$. The horizontal space $\mathcal{H}_{\overline{x}}$ at $\overline{x} \in \overline{\mathcal{M}}$ is canonically chosen as the orthogonal complement in $T_{\overline{x}}\overline{\mathcal{M}}$ of the vertical space $\mathcal{V}_{\overline{x}} = T_{\overline{x}}\pi^{-1}(x)$, namely,

$$\mathcal{H}_{\overline{x}} := (T_{\overline{x}}\mathcal{V}_{\overline{x}})^{\perp} = \{\eta_{\overline{x}} \in T_{\overline{x}}\overline{\mathcal{M}} : \overline{g}(\chi_{\overline{x}}, \eta_{\overline{x}}) = 0 \text{ for all } \chi_{\overline{x}} \in \mathcal{V}_{\overline{x}}\}.$$

Recall that the horizontal lift at $\overline{x} \in \pi^{-1}(x)$ of a tangent vector $\xi_x \in T_x\mathcal{M}$ is the unique tangent vector $\overline{\xi}_{\overline{x}} \in \mathcal{H}_{\overline{x}}$ that satisfies $\mathrm{D}\pi(\overline{x})[\overline{\xi}_{\overline{x}}]$. If, for every

$x \in \mathcal{M}$ and every ξ_x, $\zeta_x \in T_x\mathcal{M}$, the expression $\overline{g}_{\overline{x}}(\overline{\xi}_{\overline{x}}, \overline{\zeta}_{\overline{x}})$ does *not* depend on $\overline{x} \in \pi^{-1}(x)$, then

$$g_x(\xi_x, \zeta_x) := \overline{g}_{\overline{x}}(\overline{\xi}_{\overline{x}}, \overline{\zeta}_{\overline{x}}) \tag{3.38}$$

defines a Riemannian metric on $\mathcal{M}$. Endowed with this Riemannian metric, $\mathcal{M}$ is called a *Riemannian quotient manifold* of $\overline{\mathcal{M}}$, and the natural projection $\pi : \overline{\mathcal{M}} \to \mathcal{M}$ is a *Riemannian submersion*. (In other words, a Riemannian submersion is a submersion of Riemannian manifolds such that $\mathrm{D}\pi$ preserves inner products of vectors normal to fibers.)

Riemannian quotient manifolds are interesting because several differential objects on the quotient manifold can be represented by corresponding objects in the structure space in a natural manner (see in particular Section 5.3.4). Notably, if $\overline{f}$ is a function on $\overline{\mathcal{M}}$ that induces a function f on $\mathcal{M}$, then one has

$$\overline{\mathrm{grad}\, f}_{\overline{x}} = \mathrm{grad}\, \overline{f}(\overline{x}). \tag{3.39}$$

Note that $\mathrm{grad}\, \overline{f}(\overline{x})$ belongs to the horizontal space: since $\overline{f}$ is constant on each equivalence class, it follows that $\overline{g}_{\overline{x}}(\mathrm{grad}\, \overline{f}(\overline{x}), \xi) \equiv \mathrm{D}\overline{f}(\overline{x})\,[\xi] = 0$ for all vertical vectors ξ, hence $\mathrm{grad}\, \overline{f}(\overline{x})$ is orthogonal to the vertical space.

We use the notation $\mathrm{P}^h_{\overline{x}}\xi_{\overline{x}}$ and $\mathrm{P}^v_{\overline{x}}\xi_{\overline{x}}$ for the projection of $\xi_{\overline{x}} \in T_{\overline{x}}\overline{\mathcal{M}}$ onto $\mathcal{H}_{\overline{x}}$ and $\mathcal{V}_{\overline{x}}$.

Example 3.6.3 *Projective space*
On the projective space $\mathbb{RP}^{n-1}$, the definition

$$\langle \xi, \eta \rangle_{x\mathbb{R}} := \frac{1}{x^T x} \xi_x^T \overline{\eta}_x$$

turns the canonical projection $\pi : \mathbb{R}^n_* \to \mathbb{RP}^{n-1}$ into a Riemannian submersion.

Example 3.6.4 *Grassmann manifolds*
We show that the Grassmann manifold $\mathrm{Grass}(p,n) = \mathbb{R}^{n \times p}_* / \mathrm{GL}_p$ admits a structure of a Riemannian quotient manifold when $\mathbb{R}^{n \times p}_*$ is endowed with the Riemannian metric

$$\overline{g}_Y(Z_1, Z_2) = \mathrm{tr}\left((Y^T Y)^{-1} Z_1^T Z_2\right).$$

The vertical space at Y is by definition the tangent space to the equivalence class $\pi^{-1}(\pi(Y)) = \{YM : M \in \mathbb{R}^{p \times p}_*\}$, which yields

$$\mathcal{V}_Y = \{YM : M \in \mathbb{R}^{p \times p}\}.$$

The horizontal space at Y is then defined as the orthogonal complement of the vertical space with respect to the metric $\overline{g}$. This yields

$$\mathcal{H}_Y = \{Z \in \mathbb{R}^{n \times p} : Y^T Z = 0\}, \tag{3.40}$$

and the orthogonal projection onto the horizontal space is given by

$$\mathrm{P}^h_Y Z = (I - Y(Y^T Y)^{-1} Y^T)Z. \tag{3.41}$$

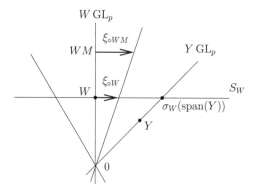

Figure 3.10 Grass(p, n) is shown as the quotient $\mathbb{R}_*^{n \times p}/\mathrm{GL}_p$ for the case $p = 1$, $n = 2$. Each point, the origin excepted, is an element of $\mathbb{R}_*^{n \times p} = \mathbb{R}^2 - \{0\}$. Each line is an equivalence class of elements of $\mathbb{R}_*^{n \times p}$ that have the same span. So each line through the origin corresponds to an element of Grass(p, n). The affine subspace S_W is an affine cross section as defined in (3.43). The relation (3.42) satisfied by the horizontal lift $\bar{\xi}$ of a tangent vector $\xi \in T_W$ Grass(p, n) is also illustrated. This figure can help to provide insight into the general case, however, one nonetheless has to be careful when drawing conclusions from it. For example, in general there does not exist a submanifold of $\mathbb{R}^{n \times p}$ that is orthogonal to the fibers Y GL$_p$ at each point, although it is obviously the case for $p = 1$ (any centered sphere in $\mathbb{R}^n$ will do).

Given $\xi \in T_{\mathrm{span}(Y)}$ Grass(p, n), there exists a unique horizontal lift $\bar{\xi}_Y \in T_Y \mathbb{R}_*^{n \times p}$ satisfying

$$\mathrm{D}\pi(Y)[\bar{\xi}_Y] = \xi.$$

In order to show that Grass(p, n) admits a structure of a Riemannian quotient manifold of $(\mathbb{R}_*^{n \times p}, \bar{g})$, we have to show that

$$\bar{g}(\bar{\xi}_{YM}, \bar{\zeta}_{YM}) = \bar{g}(\bar{\xi}_Y, \bar{\zeta}_Y)$$

for all $M \in \mathbb{R}_*^{p \times p}$. This relies on the following result.

Proposition 3.6.1 *Given $Y \in \mathbb{R}_*^{n \times p}$ and $\xi \in T_{\mathrm{span}(Y)}$ Grass(p, n), we have*

$$\bar{\xi}_{YM} = \bar{\xi}_Y \cdot M \tag{3.42}$$

for all $M \in \mathbb{R}_*^{p \times p}$, where the center dot (usually omitted) denotes matrix multiplication.

Proof. Let $W \in \mathbb{R}_*^{n \times p}$. Let $\mathcal{U}_W = \{\mathrm{span}(Y) : W^T Y \text{ invertible}\}$. Notice that $\mathcal{U}_W$ is the set of all the p-dimensional subspaces $\mathcal{Y}$ of $\mathbb{R}^n$ that do not contain any direction orthogonal to $\mathrm{span}(W)$. Consider the mapping

$$\sigma_W : \mathcal{U}_W \to \mathbb{R}_*^{n \times p} : \mathrm{span}(Y) \mapsto Y(W^T Y)^{-1} W^T W;$$

see Figure 3.10. One has $\pi(\sigma_W(\mathcal{Y})) = \mathrm{span}(\sigma_W(\mathcal{Y})) = \mathcal{Y}$ for all $\mathcal{Y} \in \mathcal{U}_W$; i.e., σ_W is a right inverse of π. Consequently, $\mathrm{D}\pi(\sigma_W(\mathcal{Y})) \circ \mathrm{D}\sigma_W(\mathcal{Y}) = \mathrm{id}$. Moreover, the range of σ_W is

$$\mathcal{S}_W := \{Y \in \mathbb{R}_*^{n \times p} : W^T(Y - W) = 0\}, \tag{3.43}$$

from which it follows that the range of $\mathrm{D}\sigma_W(\mathcal{Y}) = \{Z \in \mathbb{R}^{n \times p} : W^T Z = 0\} = \mathcal{H}_W$. In conclusion,

$$\mathrm{D}\sigma_W(\mathcal{W})[\xi] = \overline{\xi}_W.$$

Now, $\sigma_{WM}(\mathcal{Y}) = \sigma_W(\mathcal{Y})M$ for all $M \in \mathbb{R}_*^{p \times p}$ and all $\mathcal{Y} \in \mathcal{U}_W$. It follows that

$$\overline{\xi}_{WM} = \mathrm{D}\sigma_{WM}(\mathcal{W})[\xi] = \mathrm{D}(\sigma_W \cdot M)(\mathcal{W})[\xi] = \mathrm{D}\sigma_W(\mathcal{W})[\xi] \cdot M = \overline{\xi}_W \cdot M,$$

where the center dot denotes the matrix multiplication. $\qquad\square$

Using this result, we have

$$\begin{aligned}
\overline{g}_{YM}(\overline{\xi}_{YM}, \overline{\zeta}_{YM}) &= \overline{g}_{YM}(\overline{\xi}_Y M, \overline{\zeta}_Y M) \\
&= \mathrm{tr}\left(((YM)^T YM)^{-1}(\overline{\xi}_Y M)^T(\overline{\zeta}_Y M)\right) \\
&= \mathrm{tr}\left(M^{-1}(Y^T Y)^{-1}M^{-T}M^T \overline{\xi}_Y^{\,T} \overline{\zeta}_Y M\right) \\
&= \mathrm{tr}\left((Y^T Y)^{-1}\overline{\xi}_Y^{\,T} \overline{\zeta}_Y\right) \\
&= \overline{g}_Y(\overline{\xi}_Y, \overline{\zeta}_Y).
\end{aligned}$$

This shows that $\mathrm{Grass}(p, n)$, endowed with the Riemannian metric

$$g_{\mathrm{span}(Y)}(\xi, \zeta) := \overline{g}_Y(\overline{\xi}_Y, \overline{\zeta}_Y), \tag{3.44}$$

is a Riemannian quotient manifold of $(\mathbb{R}_*^{n \times p}, \overline{g})$. In other words, the canonical projection $\pi : \mathbb{R}_*^{n \times p} \to \mathrm{Grass}(p, n)$ is a Riemannian submersion from $(\mathbb{R}_*^{n \times p}, \overline{g})$ to $(\mathrm{Grass}(p, n), g)$.

3.7 NOTES AND REFERENCES

Differential geometry textbooks that we have referred to when writing this book include Abraham et al. [AMR88], Boothby [Boo75], Brickell and Clark [BC70], do Carmo [dC92], Kobayashi and Nomizu [KN63], O'Neill [O'N83], Sakai [Sak96], and Warner [War83]. Some material was also borrowed from the course notes of M. De Wilde at the University of Liège [DW92]. Do Carmo [dC92] is well suited for engineers, as it does not assume any background in abstract topology; the prequel [dC76] on the differential geometry of curves and surfaces makes the introduction even smoother. Abraham et al. [AMR88] and Brickell and Clark [BC70] cover global analysis questions (submanifolds, quotient manifolds) at an introductory level. Brickell and Clark [BC70] has a detailed treatment of the topology of manifolds. O'Neill [O'N83] is an excellent reference for Riemannian connections of submanifolds and quotient manifolds (Riemannian submersions). Boothby [Boo75] provides an excellent introduction to differential geometry with a perspective on Lie theory, and Warner [War83] covers more advanced material in this direction. Other references on differential geometry include the classic works of Kobayashi and Nomizu [KN63], Helgason [Hel78], and Spivak [Spi70]. We also mention Darling [Dar94], which introduces abstract manifold theory only after covering Euclidean spaces and their submanifolds.

Several equivalent ways of defining a manifold can be found in the literature. The definition in do Carmo [dC92] is based on local parameterizations. O'Neill [O'N83, p. 22] points out that for a Hausdorff manifold (with countably many components), being second-countable is equivalent to being paracompact. (In abstract topology, a space X is *paracompact* if every open covering of X has a locally finite open refinement that covers X.) A differentiable manifold $\mathcal{M}$ admits a partition of unity if and only if it is paracompact [BC70, Th. 3.4.4]. The material on the existence and uniqueness of atlases has come chiefly from Brickell and Clark [BC70]. A function with constant rank on its domain is called a *subimmersion* in most textbooks. The terms "canonical immersion" and "canonical submersion" have been borrowed from Guillemin and Pollack [GP74, p. 14]. The manifold topology of an immersed submanifold is always finer than its topology as a subspace [BC70], but they need not be the same topology. (When they are, the submanifold is called *embedded.*) Examples of subsets of a manifold that do not admit a submanifold structure, and examples of immersed submanifolds that are not embedded, can be found in most textbooks on differential geometry, such as do Carmo [dC92]. Proposition 3.3.1, on the uniqueness of embedded submanifold structures, is proven in Brickell and Clark [BC70] and O'Neill [O'N83]. Proposition 3.3.3 can be found in several textbooks without the condition $d_1 > d_2$. In the case where $d_1 = d_2$, $F^{-1}(y)$ is a discrete set of points [BC70, Prop. 6.2.1]. In several references, embedded submanifolds are called *regular submanifolds* or simply *submanifolds*. Proposition 3.3.2, on coordinate slices, is sometimes used to define the notion of an embedded submanifold, such as in Abraham *et al.* [AMR88]. Our definition of a regular equivalence relation follows that of Abraham *et al.* [AMR88]. The characterization of quotient manifolds in Proposition 3.4.2 can be found in Abraham *et al.* [AMR88, p. 208]. A shorter proof of Proposition 3.4.6 (showing that $\mathbb{R}_*^{n \times p}/\mathrm{GL}_p$ admits a structure of quotient manifold, the Grassmann manifold) can be given using the theory of homogeneous spaces, see Boothby [Boo75] or Warner [War83].

Most textbooks define tangent vectors as derivations. Do Carmo [dC92] introduces tangent vectors to curves, as in Section 3.5.1. O'Neill [O'N83] proposes both definitions. A tangent vector at a point x of a manifold can also be defined as an equivalence class of all curves that realize the same derivation: $\gamma_1 \sim \gamma_2$ if and only if, in a chart (U, φ) around $x = \gamma_1(0) = \gamma_2(0)$, we have $(\varphi \circ \gamma_1)'(0) = (\varphi \circ \gamma_2)'(0)$. This notion does not depend on the chart since, if $(\mathcal{V}, \psi)$ is another chart around x, then

$$(\psi \circ \gamma)'(0) = (\psi \circ \varphi^{-1})'(\varphi(m)) \cdot (\varphi \circ \gamma)'(0).$$

This is the approach taken, for example, by Gallot *et al.* [GHL90].

The notation $DF(x)[\xi]$ is not standard. Most textbooks use $dF_x \xi$ or $F_{*x} \xi$. Our notation is slightly less compact but makes it easier to distinguish the three elements F, x, and ξ of the expression and has proved more flexible when undertaking explicit computations involving matrix manifolds.

An alternative way to define smoothness of a vector field is to require that the function ξf be smooth for every $f \in \mathfrak{F}(\mathcal{M})$; see O'Neill [O'N83]. In the parlance of abstract algebra, the set $\mathfrak{F}(\mathcal{M})$ of all smooth real-valued functions on $\mathcal{M}$, endowed with the usual operations of addition and multiplication, is a *commutative ring*, and the set $\mathfrak{X}(\mathcal{M})$ of vector fields is a *module* over $\mathfrak{F}(\mathcal{M})$ [O'N83]. Formula (3.26) for the tangent space to the orthogonal group can also be obtained by treating O_n as a Lie group: the operation of left multiplication by U, $L_U : X \mapsto UX$, sends the neutral element I to U, and the differential of L_U at I sends $T_I O_n = \mathfrak{o}(n) = \mathcal{S}_{\text{skew}}(n)$ to $U \mathcal{S}_{\text{skew}}(n)$; see, e.g., Boothby [Boo75] or Warner [War83]. For a proof that the Riemannian distance satisfies the three axioms of a metric, see O'Neill [O'N83, Prop. 5.18]. The axiom that fails to hold in general for non-Hausdorff manifolds is that $\text{dist}(x, y) = 0$ if and only if $x = y$. An example can be constructed from the material in Section 4.3.2. Riemannian submersions are covered in some detail in Cheeger and Ebin [CE75], do Carmo [dC92], Klingenberg [Kli82], O'Neill [O'N83], and Sakai [Sak96]. The term "Riemannian quotient manifold" is new.

The Riemannian metric given in (3.44) is the essentially unique rotation-invariant Riemannian metric on the Grassmann manifold [Lei61, AMS04]. More information on Grassmann manifolds can be found in Ferrer *et al.* [FGP94], Edelman *et al.* [EAS98], Absil *et al.* [AMS04], and references therein.

In order to define the steepest-descent direction of a real-valued function f on a manifold $\mathcal{M}$, it is enough to endow the tangent spaces to $\mathcal{M}$ with a norm. Under smoothness assumptions, this turns $\mathcal{M}$ into a *Finsler manifold*. Finsler manifolds have received little attention in the literature in comparison with the more restrictive notion of Riemannian manifolds. For recent work on Finsler manifolds, see Bao *et al.* [BCS00].

Chapter Four

Line-Search Algorithms on Manifolds

Line-search methods in $\mathbb{R}^n$ are based on the update formula

$$x_{k+1} = x_k + t_k \eta_k, \qquad (4.1)$$

where $\eta_k \in \mathbb{R}^n$ is the *search direction* and $t_k \in \mathbb{R}$ is the *step size*. The goal of this chapter is to develop an analogous theory for optimization problems posed on nonlinear manifolds.

The proposed generalization of (4.1) to a manifold $\mathcal{M}$ consists of selecting η_k as a tangent vector to $\mathcal{M}$ at x_k and performing a search along a curve in $\mathcal{M}$ whose tangent vector at $t = 0$ is η_k. The selection of the curve relies on the concept of retraction, introduced in Section 4.1. The choice of a computationally efficient retraction is an important decision in the design of high-performance numerical algorithms on nonlinear manifolds. Several practical examples are given for the matrix manifolds associated with the main examples of interest considered in this book.

This chapter also provides the convergence theory of line-search algorithms defined on Riemannian manifolds. Several example applications related to the eigenvalue problem are presented.

4.1 RETRACTIONS

Conceptually, the simplest approach to optimizing a differentiable function is to continuously translate a test point $x(t)$ in the direction of steepest descent, $-\operatorname{grad} f(x)$, on the constraint set until one reaches a point where the gradient vanishes. Points x where $\operatorname{grad} f(x) = 0$ are called *stationary points* or *critical points* of f. A numerical implementation of the continuous gradient descent approach requires the construction of a curve γ such that $\dot{\gamma}(t) = -\operatorname{grad} f(\gamma(t))$ for all t. Except in very special circumstances, the construction of such a curve using numerical methods is impractical. The closest numerical analogy is the class of optimization methods that use *line-search* procedures, namely, iterative algorithms that, given a point x, compute a descent direction $\eta := -\operatorname{grad} f(x)$ (or some approximation of the gradient) and move in the direction of η until a "reasonable" decrease in f is found. In $\mathbb{R}^n$, the concept of moving in the direction of a vector is straightforward. On a manifold, the notion of moving in the direction of a tangent vector, while staying on the manifold, is generalized by the notion of a retraction mapping.

Conceptually, a retraction R at x, denoted by R_x, is a mapping from $T_x\mathcal{M}$ to $\mathcal{M}$ with a local rigidity condition that preserves gradients at x; see Figure 4.1.

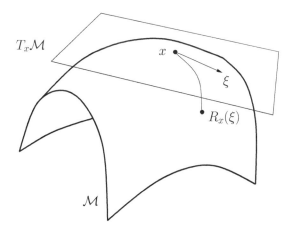

Figure 4.1 Retraction.

Definition 4.1.1 (retraction) *A retraction on a manifold $\mathcal{M}$ is a smooth mapping R from the tangent bundle $T\mathcal{M}$ onto $\mathcal{M}$ with the following properties. Let R_x denote the restriction of R to $T_x\mathcal{M}$.*

(i) $R_x(0_x) = x$, *where 0_x denotes the zero element of $T_x\mathcal{M}$.*
(ii) *With the canonical identification $T_{0_x}T_x\mathcal{M} \simeq T_x\mathcal{M}$, R_x satisfies*

$$\mathrm{D}R_x(0_x) = \mathrm{id}_{T_x\mathcal{M}}, \tag{4.2}$$

where $\mathrm{id}_{T_x\mathcal{M}}$ denotes the identity mapping on $T_x\mathcal{M}$.

We generally assume that the domain of R is the whole tangent bundle $T\mathcal{M}$. This property holds for all practical retractions considered in this book.

Concerning condition (4.2), notice that, since R_x is a mapping from $T_x\mathcal{M}$ to $\mathcal{M}$ sending 0_x to x, it follows that $\mathrm{D}R_x(0_x)$ is a mapping from $T_{0_x}(T_x\mathcal{M})$ to $T_x\mathcal{M}$ (see Section 3.5.6). Since $T_x\mathcal{M}$ is a vector space, there is a natural identification $T_{0_x}(T_x\mathcal{M}) \simeq T_x\mathcal{M}$ (see Section 3.5.2). We refer to the condition $\mathrm{D}R_x(0_x) = \mathrm{id}_{T_x\mathcal{M}}$ as the *local rigidity* condition. Equivalently, for every tangent vector ξ in $T_x\mathcal{M}$, the curve $\gamma_\xi : t \mapsto R_x(t\xi)$ satisfies $\dot{\gamma}_\xi(0) = \xi$. Moving along this curve γ_ξ is thought of as moving in the direction ξ while constrained to the manifold $\mathcal{M}$.

Besides turning elements of $T_x\mathcal{M}$ into points of $\mathcal{M}$, a second important purpose of a retraction R_x is to transform cost functions defined in a neighborhood of $x \in \mathcal{M}$ into cost functions defined on the vector space $T_x\mathcal{M}$. Specifically, given a real-valued function f on a manifold $\mathcal{M}$ equipped with a retraction R, we let $\widehat{f} = f \circ R$ denote the *pullback* of f through R. For $x \in \mathcal{M}$, we let

$$\widehat{f}_x = f \circ R_x \tag{4.3}$$

denote the restriction of $\widehat{f}$ to $T_x\mathcal{M}$. Note that $\widehat{f}_x$ is a real-valued function on a vector space. Observe that because of the local rigidity condition (4.2), we have (with the canonical identification (3.11)) $\mathrm{D}\widehat{f}_x(0_x) = \mathrm{D}f(x)$. If $\mathcal{M}$ is endowed with a Riemannian metric (and thus $T_x\mathcal{M}$ with an inner product), we have

$$\mathrm{grad}\,\widehat{f}_x(0_x) = \mathrm{grad}\,f(x). \tag{4.4}$$

All the main examples that are considered in this book (and most matrix manifolds of interest) admit a Riemannian metric. Every manifold that admits a Riemannian metric also admits a retraction defined by the *Riemannian exponential mapping* (see Section 5.4 for details). The domain of the exponential mapping is not necessarily the whole $T\mathcal{M}$. When it is, the Riemannian manifold is called *complete*. The Stiefel and Grassmann manifolds, endowed with the Riemannian metrics defined in Section 3.6, are complete.

The Riemannian exponential mapping is, in the geometric sense, the most natural retraction to use on a Riemannian manifold and featured heavily in the early literature on the development of numerical algorithms on Riemannian manifolds. Unfortunately, the Riemannian exponential mapping is itself defined as the solution of a nonlinear ordinary differential equation that, in general, poses significant numerical challenges to compute cheaply. In most cases of interest in this book, the solution of the Riemannian exponential can be expressed in terms of classical analytic functions with matrix arguments. However, the evaluation of matrix analytic functions is also a challenging problem and usually computationally intensive to solve. Indeed, computing the exponential may turn out to be more difficult than the original Riemannian optimization problem under consideration (see Section 7.5.2 for an example). These drawbacks are an invitation to consider alternatives in the form of approximations to the exponential that are computationally cheap without jeopardizing the convergence properties of the optimization schemes. Retractions provide a framework for analyzing such alternatives. All the algorithms in this book make use of retractions in one form or another, and the convergence analysis is carried out for general retractions.

In the remainder of this Section 4.1, we show how several structures (embedded submanifold, quotient manifold) and mathematical objects (local coordinates, projections, factorizations) can be exploited to define retractions.

4.1.1 Retractions on embedded submanifolds

Let $\mathcal{M}$ be an embedded submanifold of a vector space $\mathcal{E}$. Recall that $T_x\mathcal{M}$ can be viewed as a linear subspace of $T_x\mathcal{E}$ (Section 3.5.7) which itself can be identified with $\mathcal{E}$ (Section 3.5.2). This allows us, slightly abusing notation, to consider the sum $x + \xi$ of a point x of $\mathcal{M}$, viewed as an element of $\mathcal{E}$, and a tangent vector $\xi \in T_x\mathcal{M}$, viewed as an element of $T_x\mathcal{E} \simeq \mathcal{E}$. In this setting, it is tempting to define a retraction along the following lines. Given x in $\mathcal{M}$ and $\xi \in T_x\mathcal{M}$, compute $R_x(\xi)$ by

1. moving along ξ to get the point $x + \xi$ in the linear embedding space;
2. "projecting" the point $x + \xi$ back to the manifold $\mathcal{M}$.

The issue is to define a projection that (i) turns the procedure into a well-defined retraction and (ii) is computationally efficient. In the embedded submanifolds of interest in this book, as well as in several other cases, the second step can be based on matrix decompositions. Examples of such decompositions include QR factorization and polar decomposition. The purpose of the present section is to develop a general theory of decomposition-based retractions. With this theory at hand, it will be straightforward to show that several mappings constructed along the above lines are well-defined retractions.

Let $\mathcal{M}$ be an embedded manifold of a vector space $\mathcal{E}$ and let $\mathcal{N}$ be an abstract manifold such that $\dim(\mathcal{M}) + \dim(\mathcal{N}) = \dim(\mathcal{E})$. Assume that there is a diffeomorphism

$$\phi : \mathcal{M} \times \mathcal{N} \to \mathcal{E}_* : (F, G) \mapsto \phi(F, G),$$

where $\mathcal{E}_*$ is an open subset of $\mathcal{E}$ (thus $\mathcal{E}_*$ is an open submanifold of $\mathcal{E}$), with a neutral element $I \in \mathcal{N}$ satisfying

$$\phi(F, I) = F, \quad \forall F \in \mathcal{M}.$$

(The letter I is chosen in anticipation that the neutral element will be the identity matrix of a matrix manifold $\mathcal{N}$ in cases of interest.)

Proposition 4.1.2 *Under the above assumptions on ϕ, the mapping*

$$R_X(\xi) := \pi_1(\phi^{-1}(X + \xi)),$$

where $\pi_1 : \mathcal{M} \times \mathcal{N} \to \mathcal{M} : (F, G) \mapsto F$ is the projection onto the first component, defines a retraction on $\mathcal{M}$.

Proof. Since $\mathcal{E}_*$ is open, it follows that $X + \xi$ belongs to $\mathcal{E}_*$ for all ξ in some neighborhood of 0_X. Since ϕ^{-1} is defined on the whole $\mathcal{E}_*$, it follows that $R_X(\xi)$ is defined for all ξ in a neighborhood of the origin of $T_X\mathcal{M}$. Smoothness of R and the property $R_X(0_X) = X$ are direct. For the local rigidity property, first note that for all $\xi \in T_X\mathcal{M}$, we have

$$D_1\phi(X, I)[\xi] = D\phi(X, I)[(\xi, 0)] = \xi.$$

Since $\pi_1 \circ \phi^{-1}(\phi(F, I)) = F$, it follows that, for all $\xi \in T_X\mathcal{M}$,

$$\xi = D(\pi_1 \circ \phi^{-1})(\phi(X, I))\, [D_1\phi(X, I)[\xi]] = D(\pi_1 \circ \phi^{-1})(X)[\xi] = DR_X(0_X)[\xi],$$

which proves the claim that R_X is a retraction. $\qquad\qquad \square$

Example 4.1.1 **Retraction on the sphere S^{n-1}**
Let $\mathcal{M} = S^{n-1}$, let $\mathcal{N} = \{x \in \mathbb{R} : x > 0\}$, and consider the mapping

$$\phi : \mathcal{M} \times \mathcal{N} \to \mathbb{R}^n_* : (x, r) \mapsto xr.$$

It is straightforward to verify that ϕ is a diffeomorphism. Proposition 4.1.2 yields the retraction

$$R_x(\xi) = \frac{x + \xi}{\|x + \xi\|},$$

defined for all $\xi \in T_x S^{n-1}$. Note that $R_x(\xi)$ is the point of S^{n-1} that minimizes the distance to $x + \xi$.

Example 4.1.2 *Retraction on the orthogonal group*

Let $\mathcal{M} = O_n$ be the orthogonal group. The QR decomposition *of a matrix $A \in \mathbb{R}_*^{n \times n}$ is the decomposition of A as $A = QR$, where Q belongs to O_n and R belongs to $\mathcal{S}_{\mathrm{upp+}}(n)$, the set of all upper triangular matrices with strictly positive diagonal elements. The inverse of QR decomposition is the mapping*

$$\phi : O_n \times \mathcal{S}_{\mathrm{upp+}}(n) \to \mathbb{R}_*^{n \times n} : (Q, R) \mapsto QR. \tag{4.5}$$

We let qf $:= \pi_1 \circ \phi^{-1}$ *denote the mapping that sends a matrix to the Q factor of its QR decomposition. The mapping* qf *can be computed using the Gram-Schmidt orthonormalization.*

We have to check that ϕ satisfies all the assumptions of Proposition 4.1.2. The identity matrix is the neutral element: $\phi(Q, I) = Q$ for all $Q \in O_n$. It follows from the existence and uniqueness properties of the QR decomposition that ϕ is bijective. The mapping ϕ is smooth since it is the restriction of a smooth map (matrix product) to a submanifold. Concerning ϕ^{-1}, notice that its first matrix component Q is obtained by a Gram-Schmidt process, which is C^∞ on the set of full-rank matrices. Since the second component R is obtained as $Q^{-1}M$, it follows that ϕ^{-1} is C^∞. In conclusion, the assumptions of Proposition 4.1.2 hold for (4.5), and consequently,

$$R_X(X\Omega) := \mathrm{qf}(X + X\Omega) = \mathrm{qf}(X(I + \Omega)) = X\mathrm{qf}(I + \Omega)$$

is a retraction on the orthogonal group O_n.

A second possibility is to consider the polar decomposition *of a matrix $A = QP$, where $Q \in O_n$ and $P \in \mathcal{S}_{\mathrm{sym+}}(n)$, the set of all symmetric positive-definite matrices of size n. The inverse of polar decomposition is a mapping*

$$\phi : O_n \times \mathcal{S}_{\mathrm{sym+}}(n) \to \mathbb{R}_*^{n \times n} : (Q, P) \mapsto QP.$$

We have $\phi^{-1}(A) = (A(A^T A)^{-1/2}, (A^T A)^{1/2})$. This shows that ϕ is a diffeomorphism, and thus

$$R_X(X\Omega) := X(I + \Omega)((X(I + \Omega))^T X(I + \Omega))^{-1/2}$$
$$= X(I + \Omega)(I - \Omega^2)^{-1/2} \tag{4.6}$$

is a retraction on O_n. Computing this retraction requires an eigenvalue decomposition of the $n \times n$ symmetric matrix $(I - \Omega^2)$. Note that it does not make sense to use this retraction in the context of an eigenvalue algorithm on O_n since the computational cost of computing a single retraction is comparable to that for solving the original optimization problem.

A third possibility is to use Givens rotations. *For an $n \times n$ skew-symmetric matrix Ω, let $\mathrm{Giv}(\Omega) = \prod_{1 \le i < j \le n} G(i, j, \Omega_{ij})$, where the order of multiplication is any fixed order and where $G(i, j, \theta)$ is the Givens rotation of angle θ in the (i, j) plane, namely, $G(i, j, \theta)$ is the identity matrix with the substitutions $e_i^T G(i, j, \theta) e_i = e_j^T G(i, j, \theta) e_j = \cos(\theta)$ and $e_i^T G(i, j, \theta) e_j = -e_j^T G(i, j, \theta) e_i = \sin(\theta)$. Then the mapping $R : TO_n \to O_n$ defined by*

$$R_X(X\Omega) = X \, \mathrm{Giv}(\Omega)$$

is a retraction on O_n.

Another retraction on O_n, based on the Cayley transform, *is given by*

$$R_X(X\Omega) = X(I - \tfrac{1}{2}\Omega)^{-1}(I + \tfrac{1}{2}\Omega).$$

Anticipating the material in Chapter 5, we point out that the Riemannian exponential mapping on O_n (viewed as a Riemannian submanifold of $\mathbb{R}^{n \times n}$) is given by

$$\mathrm{Exp}_X(X\Omega) = X \exp(\Omega),$$

where $\exp$ *denotes the matrix exponential defined by* $\exp(\Omega) := \sum_{i=0}^{\infty} \frac{1}{i!}\Omega^i$. *Note that Riemannian exponential mappings are always retractions (Proposition 5.4.1). Algorithms for accurately evaluating the exponential have a numerical cost at best similar to those for evaluating (4.6). However, there are several computationally efficient Lie group-based algorithms for approximating the exponential that fit the definition of a retraction (see pointers in Notes and References).*

Example 4.1.3 *Retraction on the Stiefel manifold*

Consider the Stiefel manifold $\mathrm{St}(p, n) = \{X \in \mathbb{R}^{n \times p} : X^T X = I_p\}$. *The retraction based on the polar decomposition is*

$$R_X(\xi) = (X + \xi)(I_p + \xi^T \xi)^{-1/2}, \tag{4.7}$$

where we have used the fact that ξ, as an element of $T_X \mathrm{St}(p, n)$, satisfies $X^T\xi + \xi^T X = 0$. When p is small, the numerical cost of evaluating (4.7) is reasonable since it involves the eigenvalue decomposition of the small $p \times p$ matrix $(I_p + \xi^T\xi)^{-1/2}$ along with matrix linear algebra operations that require only $O(np^2)$ additions and multiplications.

Much as in the case of the orthogonal group, an alternative to choice (4.7) is

$$R_X(\xi) := \mathrm{qf}(X + \xi), \tag{4.8}$$

where $\mathrm{qf}(A)$ denotes the Q factor of the decomposition of $A \in \mathbb{R}_*^{n \times p}$ as $A = QR$, where Q belongs to $\mathrm{St}(p, n)$ and R is an upper triangular $n \times p$ matrix with strictly positive diagonal elements. Computing $R_X(\xi)$ can be done in a finite number of basic arithmetic operations (addition, subtraction, multiplication, division) and square roots using, e.g., the modified Gram-Schmidt algorithm.

4.1.2 Retractions on quotient manifolds

We now consider the case of a quotient manifold $\mathcal{M} = \overline{\mathcal{M}}/\sim$. Recall the notation π for the canonical projection and $\overline{\xi}_{\overline{x}}$ for the horizontal lift at $\overline{x}$ of a tangent vector $\xi \in T_{\pi(\overline{x})}\mathcal{M}$.

Proposition 4.1.3 *Let $\mathcal{M} = \overline{\mathcal{M}}/\sim$ be a quotient manifold with a prescribed horizontal distribution. Let $\overline{R}$ be a retraction on $\overline{\mathcal{M}}$ such that for all $x \in \mathcal{M}$ and $\xi \in T_x\mathcal{M}$,*

$$\pi(\overline{R}_{\overline{x}_a}(\overline{\xi}_{\overline{x}_a})) = \pi(\overline{R}_{\overline{x}_b}(\overline{\xi}_{\overline{x}_b})) \tag{4.9}$$

for all $\overline{x}_a, \overline{x}_b \in \pi^{-1}(x)$. Then

$$R_x(\xi) := \pi(\overline{R}_{\overline{x}}(\overline{\xi}_{\overline{x}})) \tag{4.10}$$

defines a retraction on $\mathcal{M}$.

Proof. Equation (4.9) guarantees that R is well defined as a mapping from $T\mathcal{M}$ to $\mathcal{M}$. Since $\overline{R}$ is a retraction, it also follows that the property $R_x(0_x) = x$ is satisfied. Finally, the local rigidity condition holds since, given $\overline{x} \in \pi^{-1}(x)$,

$$\mathrm{D}R_x(0_x)[\eta] = \mathrm{D}\pi(\overline{x}) \circ \mathrm{D}\overline{R}_{\overline{x}}(0_{\overline{x}})[\overline{\eta}_{\overline{x}}] = \mathrm{D}\pi(\overline{x})[\overline{\eta}_{\overline{x}}] = \eta$$

for all $\eta \in T_x\mathcal{M}$, by definition of the horizontal lift. $\square$

From now on we consider the case where the structure space $\overline{\mathcal{M}}$ is an open, dense (not necessarily proper) subset of a vector space $\overline{\mathcal{E}}$. Assume that a horizontal distribution $\mathcal{H}$ has been selected that endows every tangent vector to $\mathcal{M}$ with a horizontal lift. The natural choice for $\overline{R}$ is then

$$\overline{R}_y(\zeta) = y + \overline{\zeta}_y.$$

However, this choice does not necessarily satisfy (4.9). In other words, if x and y satisfy $\pi(x) = \pi(y)$, the property $\pi(x + \overline{\xi}_x) = \pi(y + \overline{\xi}_y)$ may fail to hold.

As an example, take the quotient of $\mathbb{R}^2$ for which the graphs of the curves $x_1 = a + a^3 x_2^2$ are equivalence classes, where $a \in \mathbb{R}$ parameterizes the set of all equivalence classes. Define the horizontal distribution as the constant subspace $e_1\mathbb{R}$. Given a tangent vector ξ to the quotient at the equivalence class $e_2\mathbb{R}$ (corresponding to $a = 0$), we obtain that the horizontal lift $\overline{\xi}_{(0,x_2)}$ is a constant $(C, 0)$ independent of x_2. It is clear that the equivalence class of $(0, x_2) + \overline{\xi}_{(0,x_2)} = (C, x_2)$ depends on x_2.

If we further require the equivalence classes to be the orbits of a Lie group acting linearly on $\overline{\mathcal{M}}$, with a horizontal distribution that is invariant by the Lie group action, then condition (4.9) holds. In particular, this is the case for the main examples considered in this book.

Example 4.1.4 *Retraction on the projective space*

Consider the real projective space $\mathbb{RP}^{n-1} = \mathbb{R}^n_* / \mathbb{R}_*$ with the horizontal distribution defined in (3.27). A retraction can be defined as

$$R_{\pi(y)}\xi = \pi(y + \overline{\xi}_y),$$

where $\overline{\xi}_y \in \mathbb{R}^n$ is the horizontal lift of $\xi \in T_{\pi(y)}\mathbb{RP}^{n-1}$ at y.

Example 4.1.5 *Retraction on the Grassmann manifold*

Consider the Grassmann manifold $\mathrm{Grass}(p, n) = \mathbb{R}^{n \times p}_* / \mathrm{GL}_p$ with the horizontal distribution defined in (3.40). It can be checked using the homogeneity property of horizontal lifts (Proposition 3.6.1) that

$$R_{\mathrm{span}(Y)}(\xi) = \mathrm{span}(Y + \overline{\xi}_Y) \tag{4.11}$$

is well-defined. Hence (4.11) defines a retraction on $\mathrm{Grass}(p, n)$.

Note that the matrix $Y + \bar{\xi}_Y$ is in general not orthonormal. In particular, if Y is orthonormal, then $Y + \bar{\xi}_Y$ is not orthonormal unless $\xi = 0$. In the scope of a numerical algorithm, in order to avoid ill-conditioning, it may be advisable to use $\mathrm{qf}\left(Y + \bar{\xi}_Y\right)$ instead of $Y + \bar{\xi}_Y$ as a basis for the subspace $R_{\mathrm{span}(Y)}(\xi)$.

4.1.3 Retractions and local coordinates*

In this section it is shown that every smooth manifold can be equipped with "local" retractions derived from its coordinate charts and that every retraction generates an atlas of the manifold. These operations, however, may pose computational challenges.

For every point x of a smooth manifold $\mathcal{M}$, there exists a smooth map $\mu_x : \mathbb{R}^d \mapsto \mathcal{M}$, $\mu_x(0) = x$, that is a local diffeomorphism around $0 \in \mathbb{R}^d$; the map μ_x is called a *local parameterization around* x and can be thought of as the inverse of a coordinate chart around $x \in \mathcal{M}$. If $\mathcal{U}$ is a neighborhood of a point x_* of $\mathcal{M}$, and $\mu : \mathcal{U} \times \mathbb{R}^d \to \mathcal{M}$ is a smooth map such that $\mu(x, z) = \mu_x(z)$ for all $x \in \mathcal{U}$ and $z \in \mathbb{R}^d$, then $\{\mu_x\}_{x \in \mathcal{M}}$ is called a *locally smooth family of parameterizations around* x_*. Note that a locally smooth parameterization μ around x_* can be constructed from a single chart around x_* by defining $\mu_x(z) = \varphi^{-1}(z + \varphi(x))$.

If $\{\mu_x\}_{x \in \mathcal{M}}$ is a locally smooth family of parameterizations around x_*, then the mappings

$$R_x : T_x\mathcal{M} \to \mathcal{M} : \xi \mapsto \mu_x(\mathrm{D}\mu_x^{-1}(x)[\xi])$$

define a retraction R whose domain is in general not the whole $T\mathcal{M}$. (It is readily checked that R_x satisfies the requirements in Definition 4.1.1.) Conversely, to define a smooth family of parameterizations around x_* from a retraction R, we can select smooth vector fields $\xi_1, \ldots, \xi_d$ on $\mathcal{M}$ such that, for all x in a neighborhood of x_*, $(\xi_1(x), \ldots, \xi_d(x))$ forms a basis of $T_x\mathcal{M}$, and then define

$$\mu_x(u_1, \ldots, u_d) = R_x(u_1\xi_1(x) + \cdots + u_d\xi_d(x)).$$

Note, however, that such a basis $\xi_1, \ldots, \xi_d$ of vector fields can in general be defined only locally. Moreover, producing the ξ's in practical cases may be tedious. For example, on the unit sphere S^{n-1}, the set $T_x S^{n-1}$ is a vector space of dimension $(n - 1)$ identified with $x_\perp := \{y \in \mathbb{R}^n : x^T y = 0\}$; however, when n is large, generating and storing a basis of $x_\perp$ is impractical, as this requires $(n-1)$ vectors of n components. In other words, even though the $(n - 1)$-dimensional vector space $T_x S^{n-1}$ is known to be isomorphic to $\mathbb{R}^{n-1}$, creating an explicit isomorphism is computationally difficult. In comparison, it is computationally inexpensive to generate an element of $x_\perp$ (using the projection onto $x_\perp$) and to perform in $x_\perp$ the usual operations of addition and multiplication by a scalar.

In view of the discussion above, one could anticipate difficulty in dealing with pullback cost functions $\hat{f}_x := f \circ R_x$ because they are defined on vector spaces $T_x\mathcal{M}$ that we may not want to explicitly represent as $\mathbb{R}^d$. Fortunately, many classical optimization techniques can be defined on abstract vector spaces, especially when the vector space has a structure of Euclidean space, which is the case for $T_x\mathcal{M}$ when $\mathcal{M}$ is Riemannian. We refer the reader to Appendix A for elements of calculus on abstract Euclidean spaces.

4.2 LINE-SEARCH METHODS

Line-search methods on manifolds are based on the update formula

$$x_{k+1} = R_{x_k}(t_k\eta_k),$$

where η_k is in $T_{x_k}\mathcal{M}$ and t_k is a scalar. Once the retraction R is chosen, the two remaining issues are to select the search direction η_k and then the step length t_k. To obtain global convergence results, some restrictions must be imposed on η_k and t_k.

Definition 4.2.1 (gradient-related sequence) *Given a cost function f on a Riemannian manifold $\mathcal{M}$, a sequence $\{\eta_k\}$, $\eta_k \in T_{x_k}\mathcal{M}$, is gradient-related if, for any subsequence $\{x_k\}_{k\in\mathcal{K}}$ of $\{x_k\}$ that converges to a non-critical point of f, the corresponding subsequence $\{\eta_k\}_{k\in\mathcal{K}}$ is bounded and satisfies*

$$\limsup_{k\to\infty,\ k\in\mathcal{K}}\ \langle\operatorname{grad} f(x_k), \eta_k\rangle < 0.$$

The next definition, related to the choice of t_k, relies on Armijo's backtracking procedure.

Definition 4.2.2 (Armijo point) *Given a cost function f on a Riemannian manifold $\mathcal{M}$ with retraction R, a point $x \in \mathcal{M}$, a tangent vector $\eta \in T_x\mathcal{M}$, and scalars $\bar{\alpha} > 0$, $\beta, \sigma \in (0,1)$, the Armijo point is $\eta^A = t^A\eta = \beta^m\bar{\alpha}\eta$, where m is the smallest nonnegative integer such that*

$$f(x) - f(R_x(\beta^m\bar{\alpha}\eta)) \geq -\sigma\,\langle\operatorname{grad} f(x), \beta^m\bar{\alpha}\eta\rangle_x.$$

The real t^A is the Armijo step size.

We propose the accelerated Riemannian line-search framework described in Algorithm 1.

The motivation behind Algorithm 1 is to set a framework that is sufficiently general to encompass many methods of interest while being sufficiently restrictive to satisfy certain fundamental convergence properties (proven in the next sections). In particular, it is clear that the choice $x_{k+1} = R_{x_k}(t_k^A\eta_k)$ in Step 3 of Algorithm 1 satisfies (4.12), but this choice is not mandatory. The loose condition (4.12) leaves a lot of leeway for exploiting problem-related information that may lead to a more efficient algorithm. In particular, the choice $x_{k+1} = R_{x_k}(t_k^*\eta_k)$, where $t_k^* = \arg\min_t f(R_{x_k}(t\eta_k))$, satisfies (4.12) and is a reasonable choice if this exact line search can be carried out efficiently.

Algorithm 1 Accelerated Line Search (ALS)

Require: Riemannian manifold $\mathcal{M}$; continuously differentiable scalar field
    f on $\mathcal{M}$; retraction R from $T\mathcal{M}$ to $\mathcal{M}$; scalars $\overline{\alpha} > 0$, $c, \beta, \sigma \in (0, 1)$.
Input: Initial iterate $x_0 \in \mathcal{M}$.
Output: Sequence of iterates $\{x_k\}$.
  1: **for** $k = 0, 1, 2, \ldots$ **do**
  2:    Pick η_k in $T_{x_k}\mathcal{M}$ such that the sequence $\{\eta_i\}_{i=0,1,\ldots}$ is gradient-related
      (Definition 4.2.1).
  3:    Select x_{k+1} such that

$$f(x_k) - f(x_{k+1}) \geq c\left(f(x_k) - f(R_{x_k}(t_k^A \eta_k))\right), \qquad (4.12)$$

      where t_k^A is the Armijo step size (Definition 4.2.2) for the given
      $\overline{\alpha}, \beta, \sigma, \eta_k$.
  4: **end for**

If there exists a computationally efficient procedure to minimize $f \circ R_{x_k}$ on a two-dimensional subspace of $T_{x_k}\mathcal{M}$, then a possible choice for x_{k+1} in Step 3 is $R_{x_k}(\xi_k)$, with ξ_k defined by

$$\xi_k := \arg\min_{\xi \in \mathcal{S}_k} f(R_{x_k}(\xi)), \quad \mathcal{S}_k := \mathrm{span}\left\{\eta_k, R_{x_k}^{-1}(x_{k-1})\right\}, \qquad (4.13)$$

where $\mathrm{span}\{u, v\} = \{au + bv : a, b \in \mathbb{R}\}$. This is a minimization over a two-dimensional subspace $\mathcal{S}_k$ of $T_{x_k}\mathcal{M}$. It is clear that $\mathcal{S}_k$ contains the Armijo point associated with η_k, since η_k is in $\mathcal{S}_k$. It follows that the bound (4.12) on x_{k+1} holds with $c = 1$. This "two-dimensional subspace acceleration" is well defined on a Riemannian manifold as long as x_k is sufficiently close to x_{k-1} that $R_{x_k}^{-1}(x_{k-1})$ is well defined. The approach is very efficient in the context of eigensolvers (see Section 4.6).

4.3 CONVERGENCE ANALYSIS

In this section, we define and discuss the notions of convergence and limit points on manifolds, then we give a global convergence result for Algorithm 1.

4.3.1 Convergence on manifolds

The notion of convergence on manifolds is a straightforward generalization of the $\mathbb{R}^n$ case. An infinite sequence $\{x_k\}_{k=0,1,\ldots}$ of points of a manifold $\mathcal{M}$ is said to be *convergent* if there exists a chart $(\mathcal{U}, \psi)$ of $\mathcal{M}$, a point $x_* \in \mathcal{U}$, and a $K > 0$ such that x_k is in $\mathcal{U}$ for all $k \geq K$ and such that the sequence $\{\psi(x_k)\}_{k=K,K+1,\ldots}$ converges to $\psi(x_*)$. The point $\psi^{-1}(\lim_{k\to\infty} \psi(x_k))$ is called the *limit* of the convergent sequence $\{x_k\}_{k=0,1,\ldots}$. Every convergent sequence of a (Hausdorff) manifold has one and only one limit point. (The Hausdorff assumption is crucial here. Multiple distinct limit points are possible for non-Hausdorff topologies; see Section 4.3.2.)

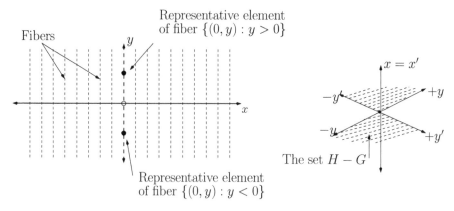

Figure 4.2 Left: A few equivalence classes of the quotient defined in Section 4.3.2. Right: The graph $\mathcal{G}$ consists of all the points in $\mathcal{H} \equiv \mathbb{R}^3$ that do not lie on the dashed planes indicated.

An equivalent and more concise definition is that a sequence on a manifold is convergent if it is convergent in the manifold topology, i.e., there is a point x_* such that every neighborhood of x_* contains all but finitely many points of the sequence.

Given a sequence $\{x_k\}_{k=0,1,\dots}$, we say that x is an *accumulation point* or a *limit point* of the sequence if there exists a subsequence $\{x_{j_k}\}_{k=0,1,\dots}$ that converges to x. The set of accumulation points of a sequence is called the *limit set* of the sequence.

4.3.2 A topological curiosity*

We present a non-Hausdorff quotient and a convergent sequence with two limit points.

Consider the set $\overline{\mathcal{M}} = \mathbb{R}^2_*$, i.e., the real plane with the origin excerpted. Consider the equivalence relation $\sim$ on $\overline{\mathcal{M}}$, where $(x, y) \sim (x', y')$ if and only if $x = x'$ and the straight line between (x, y) and (x', y') lies wholly in $\mathbb{R}^2_*$. In other words, the equivalence classes of $\sim$ are the two vertical half-lines $\{(0, y) : y > 0\}$ and $\{(0, y) : y < 0\}$ and all the vertical lines $\{(x, y) : y \in \mathbb{R}\}$, $x \neq 0$; see Figure 4.2.

Using Proposition 3.4.3, we show that $\overline{\mathcal{M}}/\sim$ admits a (unique) differentiable structure that makes the natural projection π a submersion, and we show that the topology induced by this differentiable structure is not Hausdorff. Consider the graph $\mathcal{G} = \{((x, y), (x', y')) : (x, y) \sim (x', y')\} \subset \overline{\mathcal{M}} \times \overline{\mathcal{M}}$. Set $\mathcal{H} = \{((x, y), (x', y')) : x = x'\}$ and observe that $\mathcal{G} \subseteq \mathcal{H}$ and $\mathcal{H}$ is an embedded submanifold of $\overline{\mathcal{M}} \times \overline{\mathcal{M}}$. The set $\mathcal{H} - \mathcal{G} = \{((x, y), (x', y')) : x = x' = 0, \operatorname{sign}(y) \neq \operatorname{sign}(y')\}$ is a closed subset of $\mathcal{H}$. It follows that $\mathcal{G}$ is an open submanifold of $\mathcal{H}$ and consequently an embedded submanifold of $\overline{\mathcal{M}} \times \overline{\mathcal{M}}$. It is straightforward to verify that $\pi_1 : \mathcal{G} \to \overline{\mathcal{M}}$ is a submersion. However, $\mathcal{G}$ is open in $\mathcal{H}$, hence $\mathcal{G}$ is not closed in $\overline{\mathcal{M}} \times \overline{\mathcal{M}}$. The conclusion follows from

Proposition 3.4.3.

To help the intuition, we produce a diffeomorphism between $\overline{\mathcal{M}}/\sim$ and a subset of $\overline{\mathcal{M}}$. Let $\mathcal{X}_0 = \{(x,0) : x \neq 0\}$ denote the horizontal axis of the real plane with the origin excluded. The quotient set $\overline{\mathcal{M}}/\sim$ is in one-to-one correspondence with $\mathcal{N} := \mathcal{X}_0 \cup \{(0,1),(0,-1)\}$ through the mapping Φ that sends each equivalence class to its element contained in $\mathcal{N}$. Let $\mathcal{U}_+ := \mathcal{X}_0 \cup \{(0,1)\}$ and $\mathcal{U}_- := \mathcal{X}_0 \cup \{(0,-1)\}$. Define charts ψ_+ and ψ_- of the set $\mathcal{N}$ into $\mathbb{R}$ with domains $\mathcal{U}_+$ and $\mathcal{U}_-$ by $\psi_\pm((x,0)) = x$ for all $x \neq 0$ and $\psi_+((0,1)) = 0$, $\psi_-((0,-1)) = 0$. These charts form an atlas of the set $\mathcal{N}$ and thus define a differentiable structure on $\mathcal{N}$. It is easy to check that the mapping $\Phi \circ \pi : \overline{\mathcal{M}} \to \mathcal{N}$, where $\pi : \overline{\mathcal{M}} \to \overline{\mathcal{M}}/\sim$ is the natural projection, is a submersion. In view of Proposition 3.4.3, this implies that the sets $\overline{\mathcal{M}}/\sim$ and $\mathcal{N}$, endowed with their differentiable structures, are diffeomorphic.

It is easy to produce a convergent sequence on $\mathcal{N}$ with two limit points. The sequence $\{(1/k,0)\}_{k=1,2,\ldots}$ converges to $(0,1)$ since $\{\psi_+(1/k,0)\}$ converges to $\psi_+(0,1)$. It also converges to $(0,-1)$ since $\{\psi_-(1/k,0)\}$ converges to $\psi_-(0,-1)$.

4.3.3 Convergence of line-search methods

We give a convergence result for the line-search method defined in Algorithm 1. The statement and the proof are inspired by the classical theory in $\mathbb{R}^n$. However, even when applied to $\mathbb{R}^n$, our statement is more general than the standard results. First, the line search is not necessarily done along a straight line. Second, points other than the Armijo point can be selected; for example, using a minimization over a subspace containing the Armijo point.

Theorem 4.3.1 *Let $\{x_k\}$ be an infinite sequence of iterates generated by Algorithm 1. Then every accumulation point of $\{x_k\}$ is a critical point of the cost function f.*

Proof. By contradiction. Suppose that there is a subsequence $\{x_k\}_{k \in \mathcal{K}}$ converging to some x_* with $\operatorname{grad} f(x_*) \neq 0$. Since $\{f(x_k)\}$ is nonincreasing, it follows that the whole sequence $\{f(x_k)\}$ converges to $f(x_*)$. Hence $f(x_k) - f(x_{k+1})$ goes to zero. By construction of the algorithm,

$$f(x_k) - f(x_{k+1}) \geq -c\sigma\alpha_k \langle \operatorname{grad} f(x_k), \eta_k \rangle_{x_k}.$$

Since $\{\eta_k\}$ is gradient-related, we must have $\{\alpha_k\}_{k \in \mathcal{K}} \to 0$. The α_k's are determined from the Armijo rule, and it follows that for all k greater than some $\overline{k}$, $\alpha_k = \beta^{m_k}\overline{\alpha}$, where m_k is an integer greater than zero. This means that the update $\frac{\alpha_k}{\beta}\eta_k$ did not satisfy the Armijo condition. Hence

$$f(x_k) - f\left(R_{x_k}\left(\frac{\alpha_k}{\beta}\eta_k\right)\right) < -\sigma\frac{\alpha_k}{\beta}\langle \operatorname{grad} f(x_k), \eta_k \rangle_{x_k}, \quad \forall k \in \mathcal{K}, \; k \geq \overline{k}.$$

Denoting

$$\tilde{\eta}_k = \frac{\eta_k}{\|\eta_k\|} \quad \text{and} \quad \tilde{\alpha}_k = \frac{\alpha_k\|\eta_k\|}{\beta}, \tag{4.14}$$

the inequality above reads

$$\frac{\widehat{f}_{x_k}(0) - \widehat{f}_{x_k}(\tilde{\alpha}_k \tilde{\eta}_k)}{\tilde{\alpha}_k} < -\sigma \langle \operatorname{grad} f(x_k), \tilde{\eta}_k \rangle_{x_k}, \quad \forall k \in \mathcal{K}, \ k \geq \overline{k},$$

where $\widehat{f}$ is defined as in (4.3). The mean value theorem ensures that there exists $t \in [0, \tilde{\alpha}_k]$ such that

$$- \operatorname{D}\widehat{f}_{x_k}(t \tilde{\eta}_k)[\tilde{\eta}_k] < -\sigma \langle \operatorname{grad} f(x_k), \tilde{\eta}_k \rangle_{x_k}, \quad \forall k \in \mathcal{K}, \ k \geq \overline{k}, \qquad (4.15)$$

where the differential is taken with respect to the Euclidean structure on $T_{x_k}\mathcal{M}$. Since $\{\alpha_k\}_{k \in \mathcal{K}} \to 0$ and since η_k is gradient-related, hence bounded, it follows that $\{\tilde{\alpha}_k\}_{k \in \mathcal{K}} \to 0$. Moreover, since $\tilde{\eta}_k$ has unit norm, it thus belongs to a compact set, and therefore there exists an index set $\tilde{\mathcal{K}} \subseteq \mathcal{K}$ such that $\{\tilde{\eta}_k\}_{k \in \tilde{\mathcal{K}}} \to \tilde{\eta}_*$ for some $\tilde{\eta}_*$ with $\|\tilde{\eta}_*\| = 1$. We now take the limit in (4.15) over $\tilde{\mathcal{K}}$. Since the Riemannian metric is continuous (by definition), and $f \in C^1$, and $D\widehat{f}_{x_k}(0)[\tilde{\eta}_k] = \langle \operatorname{grad} f(x_k), \tilde{\eta}_k \rangle_{x_k}$—see (3.31) and (4.4)—we obtain

$$-\langle \operatorname{grad} f(x_*), \tilde{\eta}_* \rangle_{x_*} \leq -\sigma \langle \operatorname{grad} f(x_*), \tilde{\eta}_* \rangle_{x_*}.$$

Since $\sigma < 1$, it follows that $\langle \operatorname{grad} f(x_*), \tilde{\eta}_* \rangle_{x_*} \geq 0$. On the other hand, from the fact that $\{\eta_k\}$ is gradient-related, one obtains that $\langle \operatorname{grad} f(x_*), \tilde{\eta}_* \rangle_{x_*} < 0$, a contradiction. $\qquad \square$

More can be said under compactness assumptions using a standard argument.

Corollary 4.3.2 *Let $\{x_k\}$ be an infinite sequence of iterates generated by Algorithm 1. Assume that the level set $\mathcal{L} = \{x \in \mathcal{M} : f(x) \leq f(x_0)\}$ is compact (which holds in particular when $\mathcal{M}$ itself is compact). Then $\lim_{k \to \infty} \|\operatorname{grad} f(x_k)\| = 0$.*

Proof. By contradiction, assume the contrary. Then there is a subsequence $\{x_k\}_{k \in \mathcal{K}}$ and $\epsilon > 0$ such that $\|\operatorname{grad} f(x_k)\| > \epsilon$ for all $k \in \mathcal{K}$. Because f is nonincreasing on $\{x_k\}$, it follows that $x_k \in \mathcal{L}$ for all k. Since $\mathcal{L}$ is compact, $\{x_k\}_{k \in \mathcal{K}}$ has an accumulation point x_* in $\mathcal{L}$. By the continuity of $\operatorname{grad} f$, one has $\|\operatorname{grad} f(x_*)\| \geq \epsilon$; i.e., x_* is not critical, a contradiction to Theorem 4.3.1. $\square$

4.4 STABILITY OF FIXED POINTS

Theorem 4.3.1 states that only critical points of the cost function f can be accumulation points of sequences $\{x_k\}$ generated by Algorithm 1. This result gives useful information on the behavior of Algorithm 1. Still, it falls short of what one would expect of an optimization method. Indeed, Theorem 4.3.1 does not specify whether the accumulation points are local minimizers, local maximizers, or *saddle points* (critical points that are neither local minimizers nor local maximizers).

Unfortunately, avoiding saddle points and local maximizers is too much to ask of a method that makes use of only first-order information on the cost function. We illustrate this with a very simple example. Let x_* be any critical point of a cost function f and consider the sequence $\{(x_k, \eta_k)\}$, $x_k = x_*$, $\eta_k = 0$. This sequence satisfies the requirements of Algorithm 1, and $\{x_k\}$ trivially converges to x_* even if x_* is a saddle point or a local minimizer.

Nevertheless, it is observed in practice that unless the initial point x_0 is carefully crafted, methods within the framework of Algorithm 1 do produce sequences whose accumulation points are local minima of the cost function. This observation is supported by the following stability analysis of critical points.

Let F be a mapping from $\mathcal{M}$ to $\mathcal{M}$. A point $x_* \in \mathcal{M}$ is a *fixed point* of F if $F(x_*) = x_*$. Let $F^{(n)}$ denote the result of n applications of F to x, i.e.,

$$F^{(1)}(x) = F(x), \quad F^{(i+1)}(x) = F(F^{(i)}(x)), \ i = 1, 2, \ldots.$$

A fixed point x_* of F is a *stable point* of F if, for every neighborhood $\mathcal{U}$ of x_*, there exists a neighborhood $\mathcal{V}$ of x_* such that, for all $x \in \mathcal{V}$ and all positive integer n, it holds that $F^{(n)}(x) \in \mathcal{U}$. The fixed point x_* is *asymptotically stable* if it is stable and, moreover, $\lim_{n \to \infty} F^{(n)}(x) = x_*$ for all x sufficiently close to x_*. The fixed point x_* is *unstable* if it is not stable; in other words, there exists a neighborhood $\mathcal{U}$ of x_* such that, for all neighborhood $\mathcal{V}$ of x_*, there is a point $x \in \mathcal{V}$ such that $F^{(n)}(x) \notin \mathcal{U}$ for some n. We say that F is a *descent mapping* for a cost function f if

$$f(F(x)) \leq f(x) \quad \text{for all } x \in \mathcal{M}.$$

Theorem 4.4.1 (unstable fixed points) *Let $F : \mathcal{M} \to \mathcal{M}$ be a descent mapping for a smooth cost function f and assume that for every $x \in \mathcal{M}$, all the accumulation points of $\{F^{(k)}(x)\}_{k=1,2,\ldots}$ are critical points of f. Let x_* be a fixed point of F (thus x_* is a critical point of f). Assume that x_* is not a local minimum of f. Further assume that there is a compact neighborhood $\mathcal{U}$ of x_* where, for every critical point y of f in $\mathcal{U}$, $f(y) = f(x_*)$. Then x_* is an unstable point for F.*

Proof. Since x_* is not a local minimum of f, it follows that every neighborhood $\mathcal{V}$ of x_* contains a point y with $f(y) < f(x_*)$. Consider the sequence $y_k := F^{(k)}(y)$. Suppose for the purpose of establishing a contradiction that $y_k \in \mathcal{U}$ for all k. Then, by compactness, $\{y_k\}$ has an accumulation point z in $\mathcal{U}$. By assumption, z is a critical point of f, hence $f(z) = f(x_*)$. On the other hand, since F is a descent mapping, it follows that $f(z) \leq f(y) < f(x_*)$, a contradiction. $\square$

The assumptions made about f and F in Theorem 4.4.1 may seem complicated, but they are satisfied in many circumstances. The conditions on F are satisfied by any method in the class of Algorithm 1. As for the condition on the critical points of f, it holds for example when f is real analytic. (This property can be recovered from Łojasiewicz's gradient inequality: if f is real analytic around x_*, then there are constants $c > 0$ and $\mu \in [0, 1)$ such that

$$\|\text{grad} f(x)\| \geq c|f(x) - f(x_*)|^\mu$$

for all x in some neighborhood of x_*.)

We now give a stability result.

Theorem 4.4.2 (capture theorem) *Let* $F : \mathcal{M} \to \mathcal{M}$ *be a descent mapping for a smooth cost function* f *and assume that, for every* $x \in \mathcal{M}$*, all the accumulation points of* $\{F^{(k)}(x)\}_{k=1,2,\ldots}$ *are critical points of* f*. Let* x_* *be a local minimizer and an isolated critical point of* f*. Assume further that* $\mathrm{dist}(F(x), x)$ *goes to zero as* x *goes to* x_**. Then* x_* *is an asymptotically stable point of* F*.*

Proof. Let $\mathcal{U}$ be a neighborhood of x_*. Since x_* is an isolated local minimizer of f, it follows that there exists a closed ball

$$\overline{B}_\epsilon(x_*) := \{x \in \mathcal{M} : \mathrm{dist}(x, x_*) \leq \epsilon\}$$

such that $\overline{B}_\epsilon(x_*) \subset \mathcal{U}$ and $f(x) > f(x_*)$ for all $x \in \overline{B}_\epsilon(x_*) - \{x_*\}$. In view of the condition on $\mathrm{dist}(F(x), x)$, there exists $\delta > 0$ such that, for all $x \in B_\delta(x_*)$, $F(x) \in \overline{B}_\epsilon(x_*)$. Let α be the minimum of f on the compact set $\overline{B}_\epsilon(x_*) - B_\delta(x_*)$. Let

$$\mathcal{V} = \{x \in \overline{B}_\epsilon(x_*) : f(x) < \alpha\}.$$

This set is included in $B_\delta(x_*)$. Hence, for every x in $\mathcal{V}$, it holds that $F(x) \in \overline{B}_\epsilon(x_*)$, and it also holds that $f(F(x)) \leq f(x) < \alpha$ since F is a descent mapping. It follows that $F(x) \in \mathcal{V}$ for all $x \in \mathcal{V}$, hence $F^{(n)}(x) \in \mathcal{V} \subset \mathcal{U}$ for all $x \in \mathcal{V}$ and all n. This is stability. Moreover, since by assumption x_* is the only critical point of f in $\mathcal{V}$, it follows that $\lim_{n \to \infty} F^{(n)}(x) = x_*$ for all $x \in \mathcal{V}$, which shows asymptotic stability. $\qquad\square$

The additional condition on $\mathrm{dist}(F(x), x)$ in Theorem 4.4.2 is not satisfied by every instance of Algorithm 1 because our accelerated line-search framework does not put any restriction on the step length. The distance condition is satisfied, for example, when η_k is selected such that $\|\eta_k\| \leq c\|\mathrm{grad}\, f(x_k)\|$ for some constant c and x_{k+1} is selected as the Armijo point.

In this section, we have assumed for simplicity that the next iterate depends only on the current iterate: $x_{k+1} = F(x_k)$. It is possible to generalize the above result to the case where x_{k+1} depends on x_k and on some "memory variables": $(x_{k+1}, y_{k+1}) = F(x_k, y_k)$.

4.5 SPEED OF CONVERGENCE

We have seen that, under reasonable assumptions, if the first iterate of Algorithm 1 is sufficiently close to an isolated local minimizer x_* of f, then the generated sequence $\{x_k\}$ converges to x_*. In this section, we address the issue of how fast the sequence converges to x_*.

4.5.1 Order of convergence

A sequence $\{x_k\}_{k=0,1,\ldots}$ of points of $\mathbb{R}^n$ is said to converge linearly to a point x_* if there exists a constant $c \in (0, 1)$ and an integer $K \geq 0$ such that, for all $k \geq K$, it holds that $\|x_{k+1} - x_*\| \leq c\|x_k - x_*\|$. In order to generalize this notion to manifolds, it is tempting to fall back to the $\mathbb{R}^n$ definition using charts and state that a sequence $\{x_k\}_{k=0,1,\ldots}$ of points of a manifold $\mathcal{M}$ converges linearly to a point $x_* \in \mathcal{M}$ if, given a chart $(\mathcal{U}, \psi)$ with $x \in \mathcal{U}$, the sequence $\{\psi(x_k)\}_{k=0,1,\ldots}$ converges linearly to $\psi(x_*)$. Unfortunately, the notion is not independent of the chart used. For example, let $\mathcal{M}$ be the set $\mathbb{R}^n$ with its canonical manifold structure and consider the sequence $\{x_k\}_{k=0,1,\ldots}$ defined by $x_k = 2^{-k}e_1$ if k is even and by $x_k = 2^{-k+2}e_2$ if k is odd. In the identity chart, this sequence is not linearly convergent because of the requirement that the constant c be smaller than 1. However, in the chart defined by $\psi(xe_1 + ye_2) = xe_1 + (y/4)e_2$, the sequence converges linearly with constant $c = \frac{1}{2}$.

If $\mathcal{M}$ is a Riemannian manifold, however, then the induced Riemannian distance makes it possible to define linear convergence as follows.

Definition 4.5.1 (linear convergence) *Let $\mathcal{M}$ be a Riemannian manifold and let* dist *denote the Riemannian distance on $\mathcal{M}$. We say that a sequence $\{x_k\}_{k=0,1,\ldots}$ converges linearly to a point $x_* \in \mathcal{M}$ if there exists a constant $c \in (0,1)$ and an integer $K \geq 0$ such that, for all $k \geq K$, it holds that*

$$\mathrm{dist}(x_{k+1}, x_*) \leq c \, \mathrm{dist}(x_k, x_*). \tag{4.16}$$

The limit

$$\limsup_{k \to \infty} \frac{\mathrm{dist}(x_{k+1}, x_*)}{\mathrm{dist}(x_k, x_*)}$$

is called the linear convergence factor *of the sequence. An iterative algorithm on $\mathcal{M}$ is said to* converge locally linearly *to a point x_* if there exists a neighborhood $\mathcal{U}$ of x_* and a constant $c \in (0,1)$ such that, for every initial point $x_0 \in \mathcal{U}$, the sequence $\{x_k\}$ generated by the algorithm satisfies (4.16).*

A convergent sequence $\{x_k\}$ on a Riemannian manifold $\mathcal{M}$ converges linearly to x_* with constant c if and only if

$$\|R_{x_*}^{-1}(x_{k+1}) - R_{x_*}^{-1}(x_*)\| \leq c \|R_{x_*}^{-1}(x_k) - R_{x_*}^{-1}(x_*)\|$$

for all k sufficiently large, where R is any retraction on $\mathcal{M}$ and $\|\cdot\|$ denotes the norm on $T_{x_*}\mathcal{M}$ defined by the Riemannian metric. (To see this, let Exp_{x_*} denote the exponential mapping introduced in Section 5.4, restricted to a neighborhood $\hat{\mathcal{U}}$ of 0_{x_*} in $T_{x_*}\mathcal{M}$ such that $\mathcal{U} := \mathrm{Exp}_{x_*}(\hat{\mathcal{U}})$ is a normal neighborhood of x_*. We have $\mathrm{dist}(x, x_*) = \|\mathrm{Exp}_{x_*}^{-1}(x) - \mathrm{Exp}_{x_*}^{-1}(x_*)\| = \|\mathrm{Exp}_{x_*}^{-1}(x)\|$ for all $x \in \mathcal{U}$. Moreover, since Exp is a retraction, we have $\mathrm{D}(R_{x_*}^{-1} \circ \mathrm{Exp}_{x_*})(0_{x_*}) = \mathrm{id}$. Hence $\|R_{x_*}^{-1}(x) - R_{x_*}^{-1}(x_*)\| = \|\mathrm{Exp}_{x_*}^{-1}(x) - \mathrm{Exp}_{x_*}^{-1}(x_*)\| + o(\|\mathrm{Exp}_{x_*}^{-1}(x) - \mathrm{Exp}_{x_*}^{-1}(x_*)\|) = \mathrm{dist}(x, x_*) + o(\mathrm{dist}(x, x_*)).$)

In contrast to linear convergence, the notions of superlinear convergence and order of convergence can be defined on a manifold independently of any other structure.

Definition 4.5.2 *Let $\mathcal{M}$ be a manifold and let $\{x_k\}_{k=0,1,\ldots}$ be a sequence on $\mathcal{M}$ converging to x_*. Let $(\mathcal{U}, \psi)$ be a chart of $\mathcal{M}$ with $x \in \mathcal{U}$. If*

$$\lim_{k \to \infty} \frac{\|\psi(x_{k+1}) - \psi(x_*)\|}{\|\psi(x_k) - \psi(x_*)\|} = 0,$$

then $\{x_k\}$ is said to converge superlinearly *to x_*. If there exist constants $p > 0$, $c \geq 0$, and $K \geq 0$ such that, for all $k \geq K$, there holds*

$$\|\psi(x_{k+1}) - \psi(x_*)\| \leq c\|\psi(x_k) - \psi(x_*)\|^p, \tag{4.17}$$

then $\{x_k\}$ is said to converge to x_* with order at least p. An iterative algorithm on a manifold $\mathcal{M}$ is said to converge locally to a point x_* with order at least p if there exists a chart $(\mathcal{U}, \psi)$ at x_* and a constant $c > 0$ such that, for every initial point $x_0 \in \mathcal{U}$, the sequence $\{x_k\}$ generated by the algorithm satisfies (4.17). If $p = 2$, the convergence is said to be quadratic, *and* cubic *if $p = 3$.*

Since by definition charts overlap smoothly, it can be shown that the definitions above do not depend on the choice of the chart $(\mathcal{U}, \psi)$. (The multiplicative constant c depends on the chart, but for any chart, there exists such a constant.)

Theorem 4.5.3 below gives calculus-based local convergence results for iterative methods defined by $x_{k+1} = F(x_k)$, where the iteration mapping $F : \mathcal{M} \to \mathcal{M}$ has certain smoothness properties.

Theorem 4.5.3 *Let $F : \mathcal{M} \to \mathcal{M}$ be a C^1 mapping whose domain and range include a neighborhood of a fixed point x_* of F.*

- *(i) If $\mathrm{D}F(x_*) = 0$, then the iterative algorithm with iteration mapping F converges locally superlinearly to x_*.*
- *(ii) If $\mathrm{D}F(x_*) = 0$ and F is C^2, then the iterative algorithm with mapping F converges locally quadratically to x_*.*

Although Theorem 4.5.3 is very powerful for smooth iteration mappings, it is rarely useful for practical line-search and trust-region methods because of the nondifferentiability of the step selection process.

4.5.2 Rate of convergence of line-search methods*

In this section we give an asymptotic convergence bound for Algorithm 1 when η_k is chosen as $-\mathrm{grad}\, f(x_k)$, without any further assumption on how x_{k+1} is selected.

The result invokes the smallest and largest eigenvalues of the Hessian of f at a critical point x_*. We have not yet given a definition for the Hessian of a cost function on a Riemannian manifold. (This is done in Section 5.5.) Nevertheless, regardless of this definition, it makes sense to talk about the eigenvalues of the Hessian at a critical point because of the following results.

Lemma 4.5.4 *Let $f : \mathbb{R}^n \to \mathbb{R}$ and $x_* \in \mathbb{R}^n$ such that $\mathrm{D}f(x_*) = 0$. Let $F : \mathbb{R}^n \to \mathbb{R}^n$ and $y_* \in \mathbb{R}^n$ such that $F(y_*) = x_*$ and that the Jacobian matrix of F at y_*,*

$$\mathrm{J}_F(y_*) := \begin{bmatrix} \partial_1 F^1(y_*) & \cdots & \partial_n F^1(y_*) \\ \vdots & \ddots & \vdots \\ \partial_1 F^n(y_*) & \cdots & \partial_n F^n(y_*) \end{bmatrix},$$

is orthogonal (i.e., $\mathrm{J}_F^T(y_*)\mathrm{J}_F(y_*) = I$). Let H be the Hessian matrix of f at x_*; i.e., $H_{ij} = \partial_i \partial_j f(x_*)$. Let $\hat{H}$ be the Hessian matrix of $f \circ F$ at y_*. Then $\lambda(H) = \lambda(\hat{H})$; i.e., the spectrum of H and the spectrum of $\hat{H}$ are the same.

Proof. Since $\partial_j(f \circ F)(y) = \sum_k \partial_k f(F(y)) \, \partial_j F^k(y)$, we have

$$\hat{H}_{ij} = \partial_i \partial_j (f \circ F)(y_*)$$

$$= \sum_{k,\ell} \partial_\ell \partial_k f(F(y_*)) \, \partial_i F^\ell(y_*) \, \partial_j F^k(y_*) + \sum_k \partial_k f(F(y_*)) \, \partial_i \partial_j F^k(y_*).$$

Since x_* is a critical point of f, it follows that $\partial_k f(F(y_*)) = 0$. Hence we have, in matrix notation,

$$\hat{H} = \mathrm{J}_F^T(y_*)H\mathrm{J}_F(y_*) = \mathrm{J}_F^{-1}(y_*)H\mathrm{J}_F(y_*).$$

This shows that H and $\hat{H}$ have the same spectrum because they are related by a similarity transformation. $\square$

Corollary 4.5.5 *Let f be a cost function on a Riemannian manifold $(\mathcal{M}, g)$ and let $x_* \in \mathcal{M}$ be a critical point of f, i.e., $\mathrm{grad}\, f(x_*) = 0$. Let $(\mathcal{U}, \psi)$ be any chart such that $x_* \in \mathcal{U}$ and that the representation of g_{x_*} in the chart is the identity, i.e., $g_{ij} = \delta_{ij}$ at x_*. Then the spectrum of the Hessian matrix of $f \circ \psi^{-1}$ at $\psi(x_*)$ does not depend on the choice of ψ.*

We can now state the main result of this section. When reading the theorem below, it is useful to note that $0 < r_* < 1$ since $\beta, \sigma \in (0, 1)$. Also, in common instances of Algorithm 1, the constant c in the descent condition (4.12) is equal to 1, hence (4.18) reduces to $f(x_{k+1}) - f(x_*) \leq r\,(f(x_k) - f(x_*))$.

Theorem 4.5.6 *Let $\{x_k\}$ be an infinite sequence of iterates generated by Algorithm 1 with $\eta_k := -\mathrm{grad}\, f(x_k)$, converging to a point x_*. (By Theorem 4.3.1, x_* is a critical point of f.) Let $\lambda_{H,\min}$ and $\lambda_{H,\max}$ be the smallest and largest eigenvalues of the Hessian of f at x_*. Assume that $\lambda_{H,\min} > 0$ (hence x_* is a local minimizer of f). Then, given r in the interval $(r_*, 1)$ with $r_* = 1 - \min\left(2\sigma\bar{\alpha}\lambda_{H,\min}, 4\sigma(1-\sigma)\beta\frac{\lambda_{H,\min}}{\lambda_{H,\max}}\right)$, there exists an integer $K \geq 0$ such that*

$$f(x_{k+1}) - f(x_*) \leq (r + (1-r)(1-c))\,(f(x_k) - f(x_*)) \qquad (4.18)$$

for all $k \geq K$, where c is the parameter in Algorithm 1.

Proof. Let $(\mathcal{U}, \psi)$ be a chart of the manifold $\mathcal{M}$ with $x_* \in \mathcal{U}$. We use the notation $\zeta_x := -\operatorname{grad} f(x)$. Coordinate expressions are denoted with a hat, e.g., $\hat{x} := \psi(x)$, $\hat{\mathcal{U}} = \psi(\mathcal{U})$, $\hat{f}(\hat{x}) := f(x)$, $\hat{\zeta}_{\hat{x}} := D\psi(x)[\zeta_x]$, $\hat{R}_{\hat{x}}(\hat{\zeta}) := \psi(R_x(\zeta))$. We also let $y_{\hat{x}}$ denote the Euclidean gradient of $\hat{f}$ at $\hat{x}$, i.e., $y_{\hat{x}} := (\partial_1 \hat{f}(\hat{x}), \ldots, \partial_d \hat{f}(\hat{x}))^T$. We let $G_{\hat{x}}$ denote the matrix representation of the Riemannian metric in the coordinates, and we let $H_{\hat{x}_*}$ denote the Hessian matrix of $\hat{f}$ at $\hat{x}_*$. Without loss of generality, we assume that $\hat{x}_* = 0$ and that $G_{\hat{x}_*} = I$, the identity matrix.

The major work is to obtain, at a current iterate x, a suitable upper bound on $f(R_x(t^A \zeta_x))$, where t^A is the Armijo step (so $t^A \zeta_x$ is the Armijo point). The Armijo condition is

$$f(R_x(t^A \zeta_x)) \le f(x) - \sigma \langle \zeta_x, t^A \zeta_x \rangle$$
$$\le f(x) - \sigma t^A \langle \zeta_x, \zeta_x \rangle. \qquad (4.19)$$

We first give a lower bound on $\langle \zeta_x, \zeta_x \rangle$ in terms of $f(x)$. Recall from (3.32) that $\hat{\zeta}_{\hat{x}} = G_{\hat{x}}^{-1} y_{\hat{x}}$, from which it follows that

$$\langle \zeta_x, \zeta_x \rangle = \hat{\zeta}_{\hat{x}}^T G_{\hat{x}} \hat{\zeta}_{\hat{x}} = y_{\hat{x}} G_{\hat{x}}^{-1} y_{\hat{x}} = \|y_{\hat{x}}\|^2 (1 + O(\hat{x})) \qquad (4.20)$$

since we have assumed that G_0 is the identity. It follows from $y_{\hat{x}} = H_0 \hat{x} + O(\hat{x}^2)$ and $\hat{f}(\hat{x}) = \hat{f}(0) + \frac{1}{2} \hat{x}^T H_0 \hat{x} + O(\hat{x}^3)$ that, given $\epsilon \in (0, \lambda_{H,\min})$,

$$\hat{f}(\hat{x}) - \hat{f}(0) = \frac{1}{2} y_{\hat{x}}^T H_0^{-1} y_{\hat{x}} + O(\hat{x}^3) \le \frac{1}{2} \frac{1}{\lambda_{H,\min} - \epsilon} \|y_{\hat{x}}\|^2 \qquad (4.21)$$

holds for all $\hat{x}$ sufficiently close to 0. From (4.20) and (4.21), we conclude that, given $\epsilon \in (0, \lambda_{H,\min})$,

$$f(x) - f(x_*) \le \frac{1}{2} \frac{1}{\lambda_{H,\min} - \epsilon} \langle \zeta_x, \zeta_x \rangle, \qquad (4.22)$$

which is the desired lower bound on $\langle \zeta_x, \zeta_x \rangle$. Using (4.22) in (4.19) yields

$$f(R_x(t^A \zeta_x)) - f(x_*) \le (1 - 2(\lambda_{H,\min} - \epsilon) \sigma t^A)(f(x) - f(x_*)). \qquad (4.23)$$

We now turn to finding a lower bound on the Armijo step t^A. We use the notation

$$\gamma_{\hat{x},u}(t) := \hat{f}(\hat{R}_{\hat{x}}(tu))$$

and

$$h_x(t) = f(R_x(-t\zeta_x)).$$

Notice that $h_x(t) = \gamma_{\hat{x}, -\hat{\zeta}_{\hat{x}}}(t)$ and that $\dot{h}_x(0) = -\langle \zeta_x, \zeta_x \rangle = \dot{\gamma}_{\hat{x}, -\hat{\zeta}_{\hat{x}}}(0)$, from which it follows that the Armijo condition (4.19) reads

$$h_x(t^A) \le h_x(0) - \sigma t^A \dot{h}_x(0). \qquad (4.24)$$

We want to find a lower bound on t^A. From a Taylor expansion of h_x with the residual in Lagrange form (see Appendix A.6), it follows that the t's at which the left- and right-hand sides of (4.24) are equal satisfy

$$t = \frac{-2(1 - \sigma)\dot{h}_x(0)}{\ddot{h}_x(\tau)},$$

where $\tau \in (0, t)$. In view of the definition of the Armijo point, we conclude that

$$t^A \geq \min\left(\bar{\alpha}, \frac{-2\beta(1-\sigma)\dot{h}_x(0)}{\max_{\tau \in (0,\bar{\alpha})} \ddot{h}_x(\tau)}\right). \qquad (4.25)$$

Let $B_\delta := \{\hat{x} : \|\hat{x}\| < \delta\}$ and

$$M := \sup_{\hat{x} \in B_\delta, \|u\|=1, t \in (0,\bar{\alpha}\|\hat{\zeta}_{\hat{x}}\|)} \ddot{\gamma}_{\hat{x},u}(t).$$

Then $\max_{\tau \in (0,\bar{\alpha})} \ddot{h}_x(\tau) \leq M\|\hat{\zeta}_{\hat{x}}\|^2$. Notice also that $\ddot{\gamma}_{\hat{x},u}(0) = u^T H_0 u \leq \lambda_{H,\max}\|u\|^2$, so that $M \to \lambda_{H,\max}$ as $\delta \to 0$. Finally, notice that $\dot{h}_x(0) = -\hat{\zeta}_{\hat{x}}^T G_{\hat{x}}\hat{\zeta}_{\hat{x}} = \|\hat{\zeta}_{\hat{x}}\|^2(1 + O(\hat{x}))$. Using these results in (4.25) yields that, given $\epsilon > 0$,

$$t^A \geq \min\left(\bar{\alpha}, \frac{2\beta(1-\sigma)}{\lambda_{H,\max} + \epsilon}\right) \qquad (4.26)$$

holds for all x sufficiently close to x_*.

We can now combine (4.26) and (4.23) to obtain a suitable upper bound on $f(R_x(t^A \zeta_x))$:

$$f(R_x(t^A \zeta_x)) - f(x_*) \leq c_1(f(x) - f(x_*)) \qquad (4.27)$$

with

$$c_1 = 1 - \sigma \min\left(\bar{\alpha}, \frac{2\beta(1-\sigma)}{\lambda_{H,\max} + \epsilon}\right) 2(\lambda_{H,\min} - \epsilon).$$

Finally, the bound (4.27), along with the bound (4.12) imposed on the value of f at the next iterate, yields

$$\begin{aligned}
f(x_{k+1}) - f(x_*) &= f(x_{k+1}) - f(x_k) + f(x_k) - f(x_*) \\
&\leq -c(f(x_k) - f(R_{x_k}(t_k^A \zeta_{x_k}))) + f(x_k) - f(x_*) \\
&= (1-c)(f(x_k) - f(x_*)) + c(f(R_{x_k}(t_k^A \zeta_{x_k})) - f(x_*)) \\
&\leq (1 - c + c\, c_1)(f(x_k) - f(x_*)) \\
&= (c_1 + (1-c_1)(1-c))(f(x_k) - f(x_*)),
\end{aligned}$$

where $c \in (0,1)$ is the constant in the bound (4.12). $\square$

4.6 RAYLEIGH QUOTIENT MINIMIZATION ON THE SPHERE

In this section we apply algorithms of the class described by Algorithm 1 to the problem of finding a minimizer of

$$f : S^{n-1} \to \mathbb{R} : x \mapsto x^T A x, \qquad (4.28)$$

the Rayleigh quotient on the sphere. The matrix A is assumed to be symmetric $(A = A^T)$ but not necessarily positive-definite. We let λ_1 denote the smallest eigenvalue of A and v_1 denote an associated unit-norm eigenvector.

4.6.1 Cost function and gradient calculation

Consider the function

$$\overline{f} : \mathbb{R}^n \to \mathbb{R} : x \mapsto x^T A x,$$

whose restriction to the unit sphere S^{n-1} yields (4.28).

We view S^{n-1} as a Riemannian submanifold of the Euclidean space $\mathbb{R}^n$ endowed with the canonical Riemannian metric

$$\overline{g}(\xi, \zeta) = \xi^T \zeta.$$

Given $x \in S^{n-1}$, we have

$$\mathrm{D}\overline{f}(x)[\zeta] = \zeta^T A x + x^T A \zeta = 2\zeta^T A x$$

for all $\zeta \in T_x \mathbb{R}^n \simeq \mathbb{R}^n$, from which it follows, recalling the definition (3.31) of the gradient, that

$$\mathrm{grad}\,\overline{f}(x) = 2Ax.$$

The tangent space to S^{n-1}, viewed as a subspace of $T_x \mathbb{R}^n \simeq \mathbb{R}^n$, is

$$T_x S^{n-1} = \{\xi \in \mathbb{R}^n : x^T \xi = 0\}.$$

The normal space is

$$(T_x S^{n-1})^{\perp} = \{x\alpha : \alpha \in \mathbb{R}\}.$$

The orthogonal projections onto the tangent and the normal space are

$$\mathrm{P}_x \xi = \xi - xx^T \xi, \qquad \mathrm{P}_x^{\perp} \xi = xx^T \xi.$$

It follows from the identity (3.37), relating the gradient on a submanifold to the gradient on the embedding manifold, that

$$\mathrm{grad}\,f(x) = 2\mathrm{P}_x(Ax) = 2(Ax - xx^T A x). \tag{4.29}$$

The formulas above are summarized in Table 4.1.

4.6.2 Critical points of the Rayleigh quotient

To analyze an algorithm based on the Rayleigh quotient cost on the sphere, the first step is to characterize the critical points.

Proposition 4.6.1 *Let $A = A^T$ be an $n \times n$ symmetric matrix. A unit-norm vector $x \in \mathbb{R}^n$ is an eigenvector of A if and only if it is a critical point of the Rayleigh quotient (4.28).*

Proof. Let x be a critical point of (4.28), i.e., $\mathrm{grad}\,f(x) = 0$ with $x \in S^{n-1}$. From the expression (4.29) of $\mathrm{grad}\,f(x)$, it follows that x statisfies $Ax = (x^T A x)x$, where $x^T A x$ is a scalar. Conversely, if x is a unit-norm eigenvector of A, i.e., $Ax = \lambda x$ for some scalar λ, then a left multiplication by x^T yields $\lambda = x^T A x$ and thus $Ax = (x^T A x)x$, hence $\mathrm{grad}\,f(x) = 0$ in view of (4.29). $\square$

We already know from Proposition 2.1.1 that the two points $\pm v_1$ corresponding to the "leftmost" eigendirection are the global minima of the Rayleigh quotient (4.28). Moreover, the other eigenvectors are not local minima:

Table 4.1 Rayleigh quotient on the unit sphere.

	Manifold (S^{n-1})	Embedding space ($\mathbb{R}^n$)
cost	$f(x) = x^T A x,\ x \in S^{n-1}$	$\overline{f}(x) = x^T A x,\ x \in \mathbb{R}^n$
metric	induced metric	$\overline{g}(\xi, \zeta) = \xi^T \zeta$
tangent space	$\xi \in \mathbb{R}^n : x^T \xi = 0$	$\mathbb{R}^n$
normal space	$\xi \in \mathbb{R}^n : \xi = \alpha x$	$\emptyset$
projection onto tangent space	$\mathrm{P}_x \xi = (I - x x^T) \xi$	identity
gradient	$\mathrm{grad}\, f(x) = \mathrm{P}_x \mathrm{grad}\, \overline{f}(x)$	$\mathrm{grad}\, \overline{f}(x) = 2 A x$
retraction	$R_x(\xi) = \mathrm{qf}(x + \xi)$	$R_x(\xi) = x + \xi$

Proposition 4.6.2 *Let $A = A^T$ be an $n \times n$ symmetric matrix with eigen-values $\lambda_1 \leq \cdots \leq \lambda_n$ and associated orthonormal eigenvectors $v_1, \ldots, v_n$. Then*

(i) *$\pm v_1$ are local and global minimizers of the Rayleigh quotient (4.28); if the eigenvalue λ_1 is simple, then they are the only minimizers.*

(ii) *$\pm v_n$ are local and global maximizers of (4.28); if the eigenvalue λ_n is simple, then they are the only maximizers.*

(iii) *$\pm v_q$ corresponding to interior eigenvalues (i.e., strictly larger than λ_1 and strictly smaller than λ_n) are saddle points of (4.28).*

Proof. Point (i) follows from Proposition 2.1.1. Point (ii) follows from the same proposition by noticing that replacing A by $-A$ exchanges maxima with minima and leftmost eigenvectors with rightmost eigenvectors. For point (iii), let v_q be an eigenvector corresponding to an interior eigenvalue λ_q and consider the curve $\gamma : t \mapsto (v_q + t v_1)/\|v_q + t v_1\|$. Simple calculus shows that

$$\frac{\mathrm{d}^2}{\mathrm{d}t^2}(f(\gamma(t))|_{t=0} = \lambda_1 - \lambda_q < 0.$$

Likewise, for the curve $\gamma : t \mapsto (v_q + t v_n)/\|v_q + t v_n\|$, we have

$$\frac{\mathrm{d}^2}{\mathrm{d}t^2}(f(\gamma(t))|_{t=0} = \lambda_n - \lambda_q > 0.$$

It follows that v_q is a saddle point of the Rayleigh quotient f.   □

It follows from Proposition 4.6.1 and the global convergence analysis of line-search methods (Proposition 4.3.1) that all methods within the class of Algorithm 1 produce iterates that converge to the set of eigenvectors of A. Furthermore, in view of Proposition 4.6.1, and since we are considering descent methods, it follows that, if λ_1 is simple, convergence is stable to $\pm v_1$ and unstable to all other eigenvectors.

Hereafter we consider the instances of Algorithm 1 where

$$\eta_k := -\operatorname{grad} f(x_k) = 2(Ax_k - x_k x_k^T Ax_k).$$

It is clear that this choice of search direction is gradient-related. Next we have to pick a retraction. A reasonable choice is (see Example 4.1.1)

$$R_x(\xi) := \frac{x + \xi}{\|x + \xi\|}, \tag{4.30}$$

where $\| \cdot \|$ denotes the Euclidean norm in $\mathbb{R}^n$, $\|y\| := \sqrt{y^T y}$. Another possibility is

$$R_x(\xi) := x \cos \|\xi\| + \frac{\xi}{\|\xi\|} \sin \|\xi\|, \tag{4.31}$$

for which the curve $t \mapsto R_x(t\xi)$ is a big circle on the sphere. (The second retraction corresponds to the exponential mapping defined in Section 5.4.)

4.6.3 Armijo line search

We now have all the necessary ingredients to apply a simple backtracking instance of Algorithm 1 to the problem of minimizing the Rayleigh quotient on the sphere S^{n-1}. This yields the matrix algorithm displayed in Algorithm 2. Note that with the retraction R defined in (4.30), the function $f(R_{x_k}(t\eta_k))$ is a quadratic rational function in t. Therefore, the Armijo step size is easily computed as an expression of the reals $\eta_k^T \eta_k$, $\eta_k^T A\eta_k$, $x_k^T A\eta_k$, and $x_k^T Ax_k$.

Algorithm 2 Armijo line search for the Rayleigh quotient on S^{n-1}

Require: Symmetric matrix A, scalars $\bar{\alpha} > 0$, $\beta, \sigma \in (0, 1)$.
Input: Initial iterate x_0, $\|x_0\| = 1$.
Output: Sequence of iterates $\{x_k\}$.
  1: **for** $k = 0, 1, 2, \ldots$ **do**
  2:    Compute $\eta_k = -2(Ax_k - x_k x_k^T Ax_k)$.
  3:    Find the smallest integer $m \geq 0$ such that

$$f\left(R_{x_k}(\bar{\alpha}\beta^m \eta_k)\right) \leq f(x_k) - \sigma\bar{\alpha}\beta^m \eta_k^T \eta_k,$$

   with f defined in (4.28) and R defined in (4.30).
  4:    Set

$$x_{k+1} = R_{x_k}(\bar{\alpha}\beta^m \eta_k).$$

  5: **end for**

Numerical results for Algorithm 2 are presented in Figure 4.3 for the case $A = \operatorname{diag}(1, 2, \ldots, 100)$, $\sigma = 0.5$, $\bar{\alpha} = 1$, $\beta = 0.5$. The initial point x_0 is chosen from a uniform distribution on the sphere. (The point x_0 is obtained by normalizing a vector whose entries are selected from a normal distribution).

Let us evaluate the upper bound r_* on the linear convergence factor given by Theorem 4.5.6. The extreme eigenvalues $\lambda_{H,\min}$ and $\lambda_{H,\max}$ of the Hessian at the solution v_1 can be obtained as

$$\lambda_{H,\min} = \min_{v_1^T u = 0, u^T u = 1} \left. \frac{\mathrm{d}^2(f(\gamma_{v_1,u}(t)))}{\mathrm{d}t^2} \right|_{t=0}$$

$$\lambda_{H,\max} = \max_{v_1^T u = 0, u^T u = 1} \left. \frac{\mathrm{d}^2(f(\gamma_{v_1,u}(t)))}{\mathrm{d}t^2} \right|_{t=0},$$

where

$$\gamma_{v_1,u}(t) := R_{v_1}(tu) = \frac{v_1 + tu}{\|v_1 + tu\|}.$$

This yields

$$\left. \frac{\mathrm{d}^2(f(\gamma_{v_1,u}(t)))}{\mathrm{d}t^2} \right|_{t=0} = 2(u^T A u - \lambda_1)$$

and thus

$$\lambda_{H,\min} = \lambda_2 - \lambda_1, \quad \lambda_{H,\max} = \lambda_n - \lambda_1.$$

For the considered numerical example, it follows that the upper bound on the linear convergence factor given by Theorem 4.5.6 is $r_* = 0.9949....$ The convergence factor estimated from the experimental result is below 0.97, which is in accordance with Theorem 4.5.6. This poor convergence factor, very close to 1, is due to the small value of the ratio

$$\frac{\lambda_{H,\min}}{\lambda_{H,\max}} = \frac{\lambda_2 - \lambda_1}{\lambda_n - \lambda_1} \approx 0.01.$$

The convergence analysis of Algorithm 2 is summarized as follows.

Theorem 4.6.3 *Let $\{x_k\}$ be an infinite sequence of iterates generated by Algorithm 2. Let $\lambda_1 \leq \cdots \leq \lambda_n$ denote the eigenvalues of A.*

(i) *The sequence $\{x_k\}$ converges to the eigenspace of A associated to some eigenvalue.*

(ii) *The eigenspace related to λ_1 is an attractor of the iteration defined by Algorithm 2. The other eigenspaces are unstable.*

(iii) *Assuming that the eigenvalue λ_1 is simple, the linear convergence factor to the eigenvector $\pm v_1$ associated with λ_1 is smaller or equal to*

$$r_* = 1 - 2\sigma(\lambda_2 - \lambda_1) \min\left(\overline{\alpha}, \frac{2\beta(1 - \sigma)}{\lambda_n - \lambda_1}\right).$$

Proof. Points (i) and (iii) follow directly from the convergence analysis of the general Algorithm 1 (Theorems 4.3.1 and 4.5.6). For (ii), let $\mathcal{S}_1 := \{x \in S^{n-1} : Ax = \lambda_1 x\}$ denote the eigenspace related to λ_1. Any neighborhood of $\mathcal{S}_1$ contains a sublevel set $\mathcal{L}$ of f such that the only critical points of f in $\mathcal{L}$ are the points of $\mathcal{S}_1$. Any sequence of Algorithm 2 starting in $\mathcal{L}$ converges to $\mathcal{S}_1$. The second part follows from Theorem 4.4.1. $\square$

4.6.4 Exact line search

In this version of Algorithm 1, x_{k+1} is selected as $R_{x_k}(t_k\eta_k)$, where

$$t_k := \arg\min_{t>0} f(R_{x_k}(t\eta_k)).$$

We consider the case of the projected retraction (4.30), and we define again $\eta_k := -\operatorname{grad} f(x_k)$. It is assumed that $\operatorname{grad} f(x_k) \neq 0$, from which it also follows that $\eta_k^T A x_k \neq 0$. An analysis of the function $t \mapsto f(R_{x_k}(t\eta_k))$ reveals that it admits one and only one minimizer $t_k > 0$. This minimizer is the positive solution of a quadratic equation. In view of the particular choice of the retraction, the points $\pm R_{x_k}(t_k\eta_k)$ can also be expressed as

$$\arg\min_{x\in S^{n-1}, x\in\operatorname{span}\{x_k,\eta_k\}} f(x),$$

which are also equal to

$$\pm Xw,$$

where $X := [x_k, \frac{\eta_k}{\|\eta_k\|}]$ and w is a unit-norm eigenvector associated with the smaller eigenvalue of the interaction matrix $X^T A X$.

   Numerical results are presented in Figure 4.3. Note that in this example the distance to the solution as a function of the number of iterates is slightly better with the selected Armijo method than with the exact line-search method. This may seem to be in contradiction to the fact that the exact line-search method chooses the optimal step size. However, the exact minimization only implies that if the two algorithms start at the same point x_0, then the *cost function* will be lower at the *first* iterate of the exact line-search method than at the *first* iterate of the Armijo method. This does not imply that the distance to the solution will be lower with the exact line search. Neither does it mean that the exact line search will achieve a lower cost function at subsequent iterates. (The first step of the Armijo method may well produce an iterate from which a larger decrease can be obtained.)

4.6.5 Accelerated line search: locally optimal conjugate gradient

In this version of Algorithm 1, η_k is selected as $-\operatorname{grad} f(x_k)$ and x_{k+1} is selected as $R_{x_k}(\xi_k)$, where ξ_k is a minimizer over the two-dimensional subspace of $T_{x_k}\mathcal{M}$ spanned by η_k and $R_{x_k}^{-1}(x_{k-1})$, as described in (4.13). When applied to the Rayleigh quotient on the sphere, this method reduces to the locally optimal conjugate-gradient (LOCG) algorithm of A. Knyazev. Its fast convergence (Figure 4.3) can be explained by its link with conjugate-gradient (CG) methods (see Section 8.3).

4.6.6 Links with the power method and inverse iteration

The power method,

$$x_{k+1} = \frac{Ax_k}{\|Ax_k\|},$$

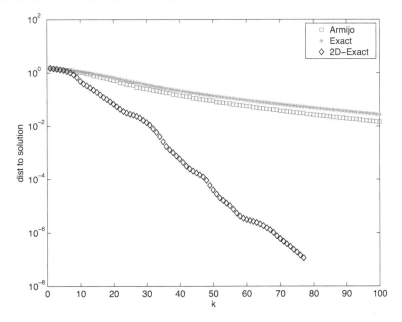

Figure 4.3 Minimization of the Rayleigh quotient of $A = \mathrm{diag}(1, 2, \ldots, n)$ on S^{n-1}, with $n = 100$. The distance to the solution is defined as the angle between the direction of the current iterate and the eigendirection associated with the smallest eigenvalue of A.

is arguably the simplest method for eigenvector computation. Let A be a symmetric matrix, assume that there is an eigenvalue λ that is simple and larger in absolute value than all the other eigenvalues, and let v denote the corresponding eigenvector. Then the power method converges to $\pm v$ for almost all initial points x_0.

We mention, as a curiosity, a relation between the power method and the steepest-descent method for the Rayleigh quotient on the sphere. Using the projective retraction (4.30), the choice $t_k = \frac{1}{2x_k^T A x_k}$ yields

$$R_{x_k}(t_k \operatorname{grad} f(x_k)) = \frac{A x_k}{\|A x_k\|},$$

i.e., the power method.

There is no such relation for the inverse iteration

$$x_{k+1} = \frac{A^{-1} x_k}{\|A^{-1} x_k\|}.$$

In fact, inverse iteration is in general much more expensive computationally than the power method since the former requires solving a linear system of size n at each iteration while the latter requires only a matrix-vector multiplication. A comparison between inverse iteration and the previous direct methods in terms of the number of iterations is not informative since an iteration of inverse iteration is expected to be computationally more demanding than an iteration of the other methods.

4.7 REFINING EIGENVECTOR ESTIMATES

All the critical points of the Rayleigh quotient correspond to eigenvectors of A, but only the extreme eigenvectors correspond to extrema of the cost function. For a given cost function f, it is, however, possible to define a new cost function that transforms *all* critical points of f into (local) minimizers. The new cost function is simply defined by

$$\tilde{f}(x) := \|\operatorname{grad} f(x)\|^2.$$

In the particular case of the Rayleigh quotient (4.28), one obtains

$$\tilde{f} : S^{n-1} \to \mathbb{R} : x \mapsto \|\mathrm{P}_x Ax\|^2 = x^T A \mathrm{P}_x Ax = x^T A^2 x - (x^T Ax)^2,$$

where $\mathrm{P}_x = (I - xx^T)$ is the orthogonal projector onto the tangent space $T_x S^{n-1} = \{\xi \in \mathbb{R}^n : x^T \xi = 0\}$. Following again the development in Section 3.6.1, we define the function

$$\overline{f} : \mathbb{R}^n \to \mathbb{R} : x \mapsto x^T A^2 x - (x^T Ax)^2$$

whose restriction to S^{n-1} is $\tilde{f}$. We obtain

$$\operatorname{grad} \overline{f}(x) = 2(A^2 x - 2Axx^T Ax),$$

hence

$$\operatorname{grad} \tilde{f}(x) = \mathrm{P}_x(\operatorname{grad} \overline{f}(x)) = 2\mathrm{P}_x(AAx - 2Axx^T Ax).$$

Applying a line-search method to the cost function $\tilde{f}$ provides a descent algorithm that (locally) converges to any eigenvector of A.

4.8 BROCKETT COST FUNCTION ON THE STIEFEL MANIFOLD

Following up on the study of descent algorithms for the Rayleigh quotient on the sphere, we now consider a cost function defined as a weighted sum $\sum_i \mu_i x_{(i)}^T Ax_{(i)}$ of Rayleigh quotients on the sphere under an orthogonality constraint, $x_{(i)}^T x_{(j)} = \delta_{ij}$.

4.8.1 Cost function and search direction

The cost function admits a more friendly expression in matrix form:

$$f : \mathrm{St}(p, n) \to \mathbb{R} : X \mapsto \operatorname{tr}(X^T AXN), \tag{4.32}$$

where $N = \operatorname{diag}(\mu_1, \cdots, \mu_p)$, with $0 \le \mu_1 \le \ldots \le \mu_p$, and $\mathrm{St}(p, n)$ denotes the orthogonal Stiefel manifold

$$\mathrm{St}(p, n) = \{X \in \mathbb{R}^{n \times p} : X^T X = I_p\}.$$

As in Section 3.3.2, we view $\mathrm{St}(p, n)$ as an embedded submanifold of the Euclidean space $\mathbb{R}^{n \times p}$. The tangent space is (see Section 3.5.7)

$$T_X \mathrm{St}(p, n) = \{Z \in \mathbb{R}^{n \times p} : X^T Z + Z^T X = 0\}$$
$$= \{X\Omega + X_\perp K : \Omega^T = -\Omega, \ K \in \mathbb{R}^{(n-p) \times p}\}.$$

We further consider $\text{St}(p, n)$ as a Riemannian submanifold of $\mathbb{R}^{n \times p}$ endowed with the canonical inner product

$$\langle Z_1, Z_2 \rangle := \text{tr}\left(Z_1^T Z_2\right).$$

It follows that the normal space to $\text{St}(p, n)$ at a point X is

$$(T_X \text{St}(p, n))^{\perp} = \{XS : \; S^T = S\}.$$

The orthogonal projection P_X onto $T_X \text{St}(p, n)$ is given by

$$\text{P}_X Z = Z - X \, \text{sym}(X^T Z) = (I - XX^T)Z + X \, \text{skew}(X^T Z),$$

where

$$\text{sym}(M) := \tfrac{1}{2}(M + M^T), \qquad \text{skew}(M) = \tfrac{1}{2}(M - M^T)$$

denote the symmetric part and the skew-symmetric part of the decomposition of M into a symmetric and a skew-symmetric term.

Consider the function

$$\overline{f} : \mathbb{R}^{n \times p} \to \mathbb{R} : X \mapsto \text{tr}(X^T AXN),$$

so that $f = \overline{f}|_{\text{St}(p,n)}$. We have

$$\text{D}\overline{f}(X)[Z] = 2 \, \text{tr}\left(Z^T AXN\right),$$

hence

$$\text{grad}\,\overline{f}(X) = 2AXN$$

and

$$\begin{aligned}
\text{grad}\,f(X) &= \text{P}_X \, \text{grad}\,\overline{f}(X) \\
&= 2AXN - 2X \, \text{sym}(X^T AXN) \\
&= 2AXN - XX^T AXN - XNX^T AX.
\end{aligned}$$

It remains to select a retraction. Choices are proposed in Section 4.1.1, such as

$$R_X(\xi) := \text{qf}(X + \xi).$$

This is all we need to turn various versions of the general Algorithm 1 into practical matrix algorithms for minimizing the cost fuction (4.32) on the orthogonal Stiefel manifold.

4.8.2 Critical points

We now show that X is a critical point of f if and only if the columns of X are eigenvectors of A.

The gradient of f admits the expression

$$\begin{aligned}
\text{grad}\,f(X) &= 2(I - XX^T)AXN + 2X \, \text{skew}(X^T AXN) \qquad (4.33) \\
&= 2(I - XX^T)AXN + X[X^T AX, N],
\end{aligned}$$

Table 4.2 Brockett cost function on the Stiefel manifold.

	Manifold $(\mathrm{St}(p,n))$	Total space $(\mathbb{R}^{n \times p})$
cost	$\mathrm{tr}(X^T A X N),\ X^T X = I_p$	$\mathrm{tr}(X^T A X N),\ X \in \mathbb{R}^{n \times p}$
metric	induced metric	$\langle Z_1, Z_2 \rangle = \mathrm{tr}(Z_1^T Z_2)$
tangent space	$Z \in \mathbb{R}^{n \times p} : \mathrm{sym}(X^T Z) = 0$	$\mathbb{R}^{n \times p}$
normal space	$Z \in \mathbb{R}^{n \times p} : Z = XS,\ S^T = S$	$\emptyset$
projection onto tangent space	$\mathrm{P}_X Z = Z - X \,\mathrm{sym}(X^T Z)$	identity
gradient	$\mathrm{grad}\, f(X) = \mathrm{P}_X \,\mathrm{grad}\, \overline{f}(X)$	$\mathrm{grad}\, \overline{f}(X) = 2AXN$
retraction	$R_X(Z) = \mathrm{qf}(X + Z)$	$R_X(Z) = X + Z$

where

$$[A, B] := AB - BA$$

denotes the (matrix) commutator of A and B. Since the columns of the first term in the expression of the gradient belong to the orthogonal complement of $\mathrm{span}(X)$, while the columns of the second term belong to $\mathrm{span}(X)$, it follows that $\mathrm{grad}\, f(X)$ vanishes if and only if

$$(I - XX^T)AXN = 0 \tag{4.34}$$

and

$$[X^T AX, N] = 0. \tag{4.35}$$

Since N is assumed to be invertible, equation (4.34) yields

$$(I - XX^T)AX = 0,$$

which means that

$$AX = XM \tag{4.36}$$

for some M. In other words, $\mathrm{span}(X)$ is an invariant subspace of A. Next, in view of the specific form of N, equation (4.35) implies that $X^T AX$ is diagonal which, used in (4.36), implies that M is diagonal, hence the columns of X are eigenvectors of A. Showing conversely that any such X is a critical point of f is straightfoward.

In the case $p = n$, $\mathrm{St}(n,n) = O_n$, and critical points of the Brockett cost function are orthogonal matrices that diagonalize A. (Note that $I - XX^T = 0$, so the first term in (4.33) trivially vanishes.) This is equivalent to saying that the columns of X are eigenvectors of A.

4.9 RAYLEIGH QUOTIENT MINIMIZATION ON THE GRASSMANN MANIFOLD

Finally, we consider a generalized Rayleigh quotient cost function on the Grassmann manifold. The Grassmann manifold is viewed as a Riemannian quotient manifold of $\mathbb{R}_*^{n \times p}$, which allows us to exploit the machinery for steepest-descent methods on quotient manifolds (see, in particular, Sections 3.4, 3.5.8, 3.6.2, and 4.1.2).

4.9.1 Cost function and gradient calculation

We start with a review of the Riemannian quotient manifold structure of the Grassmann manifold (Section 3.6.2). Let the structure space $\overline{\mathcal{M}}$ be the noncompact Stiefel manifold $\mathbb{R}_*^{n \times p} = \{Y \in \mathbb{R}^{n \times p} : Y \text{ full rank}\}$. We consider on $\overline{\mathcal{M}}$ the equivalence relation

$$X \sim Y \quad \Leftrightarrow \quad \exists M \in \mathbb{R}_*^{n \times p} : Y = XM.$$

In other words, two elements of $\mathbb{R}_*^{n \times p}$ belong to the same equivalence class if and only if they have the same column space. There is thus a one-to-one correspondence between $\mathbb{R}_*^{n \times p} / \sim$ and the set of p-dimensional subspaces of $\mathbb{R}^n$. The set $\mathbb{R}_*^{n \times p} / \sim$ has been shown (Proposition 3.4.6) to admit a unique structure of quotient manifold, called the Grassmann manifold and denoted by $\mathrm{Grass}(p, n)$ or $\mathbb{R}_*^{n \times p} / \mathrm{GL}_p$. Moreover, $\mathbb{R}_*^{n \times p} / \mathrm{GL}_p$ has been shown (Section 3.6.2) to have a structure of Riemannian quotient manifold when $\mathbb{R}_*^{n \times p}$ is endowed with the Riemannian metric

$$\overline{g}_Y(Z_1, Z_2) = \mathrm{tr}\left((Y^T Y)^{-1} Z_1^T Z_2\right).$$

The vertical space at Y is by definition the tangent space to the equivalence class of $\pi^{-1}(\pi(Y)) = \{YM : M \in \mathbb{R}_*^{p \times p}\}$, which yields

$$\mathcal{V}_Y = \{YM : M \in \mathbb{R}^{p \times p}\}.$$

The horizontal space at Y is defined as the orthogonal complement of the vertical space with respect to the metric $\overline{g}$, which yields

$$\mathcal{H}_Y = \{Z \in \mathbb{R}^{n \times p} : Y^T Z = 0\}.$$

Given $\xi \in T_{\mathrm{span}(Y)} \mathrm{Grass}(p, n)$, there exists a unique horizontal lift $\overline{\xi}_Y \in T_Y \mathbb{R}_*^{n \times p}$ satisfying

$$D\pi(Y)[\overline{\xi}_Y] = \xi.$$

Since

$$\overline{g}(\overline{\xi}_{YM}, \overline{\zeta}_{YM}) = \overline{g}(\overline{\xi}_Y, \overline{\zeta}_Y)$$

for all $M \in \mathbb{R}_*^{p \times p}$, it follows that $(\mathrm{Grass}(p, n), g)$ is a Riemannian quotient manifold of $(\mathbb{R}_*^{n \times p}, \overline{g})$ with

$$g_{\mathrm{span}(Y)}(\xi, \zeta) := \overline{g}_Y(\overline{\xi}_Y, \overline{\zeta}_Y).$$

In other words, the canonical projection π is a Riemannian submersion from $(\mathbb{R}_*^{n\times p}, \bar{g})$ to $(\mathrm{Grass}(p,n), g)$.

Let A be an $n \times n$ symmetric matrix, not necessarily positive-definite. Consider the cost function on the total space $\mathbb{R}_*^{n\times p}$ defined by

$$\bar{f} : \mathbb{R}_*^{n\times p} \to \mathbb{R} : Y \mapsto \mathrm{tr}\left((Y^T Y)^{-1} Y^T A Y\right). \tag{4.37}$$

Since $\bar{f}(YM) = \bar{f}(Y)$ whenever $M \in \mathbb{R}_*^{p\times p}$, it follows that $\bar{f}$ induces a function f on the quotient $\mathrm{Grass}(p,n)$ such that $\bar{f} = f \circ \pi$. The function f can be described as

$$f : \mathrm{Grass}(p,n) \to \mathbb{R} : \mathrm{span}(Y) \mapsto \mathrm{tr}\left((Y^T Y)^{-1} Y^T A Y\right). \tag{4.38}$$

This function can be thought of as a generalized Rayleigh quotient. Since $\bar{f}$ is smooth on $\mathbb{R}_*^{n\times p}$, it follows from Proposition 3.4.5 that f is a smooth cost function on the quotient $\mathrm{Grass}(p,n)$.

In order to obtain an expression for the gradient of f, we will make use of the trace identities (A.1) and of the formula (A.3) for the derivative of the inverse of a matrix. For all $Z \in \mathbb{R}^{n\times p}$, we have

$$\mathrm{D}\bar{f}\,(Y)\,[Z] = \mathrm{tr}\left(-(Y^T Y)^{-1}(Z^T Y + Y^T Z)(Y^T Y)^{-1} Y^T A Y\right)$$
$$+ \mathrm{tr}\left((Y^T Y)^{-1} Z^T A Y\right) + \mathrm{tr}\left((Y^T Y)^{-1} Y^T A Z\right). \tag{4.39}$$

For the last term, we have, using the two properties (A.1) of the trace,

$$\mathrm{tr}\left((Y^T Y)^{-1} Y^T A Z\right) = \mathrm{tr}\left(Z^T A Y (Y^T Y)^{-1}\right) = \mathrm{tr}\left((Y^T Y)^{-1} Z^T A Y\right).$$

Using the same properties, the first term can be rewritten as

$$-2\,\mathrm{tr}\left((Y^T Y)^{-1} Z^T Y (Y^T Y)^{-1} Y^T A Y\right).$$

Replacing these results in (4.39) yields

$$\mathrm{D}\bar{f}\,(Y)\,[Z] = \mathrm{tr}\left((Y^T Y)^{-1} Z^T\, 2(A Y - Y(Y^T Y)^{-1} Y^T A Y)\right)$$
$$= \bar{g}_Y(Z, 2(A Y - Y(Y^T Y)^{-1} Y^T A Y)).$$

It follows that

$$\mathrm{grad}\,\bar{f}(Y) = 2\left(A Y - Y(Y^T Y)^{-1} Y^T A Y\right) = \mathrm{P}_Y^h(2AY),$$

where

$$\mathrm{P}_Y^h = (I - Y(Y^T Y)^{-1} Y^T)$$

is the orthogonal projection onto the horizontal space. Note that, in accordance with the theory in Section 3.6.2, $\mathrm{grad}\,\bar{f}(Y)$ belongs to the horizontal space. It follows from the material in Section 3.6.2, in particular (3.39), that

$$\overline{\mathrm{grad}\,f}_Y = 2\mathrm{P}_Y^h A Y = 2\left(A Y - Y(Y^T Y)^{-1} Y^T A Y\right).$$

Table 4.3 Rayleigh quotient cost function on the Grassmann manifold.

	$\mathrm{Grass}(p, n)$	Total space $\mathbb{R}_*^{n \times p}$
cost	$\mathrm{span}(Y) \mapsto \overline{f}(Y)$	$\overline{f}(Y) = \mathrm{tr}((Y^T Y)^{-1} Y^T A Y)$
metric	$g_{\mathrm{span}(Y)}(\xi, \zeta) = \overline{g}_Y(\overline{\xi}_Y, \overline{\zeta}_Y)$	$\overline{g}_Y(Z_1, Z_2)$
		$= \mathrm{tr}((Y^T Y)^{-1} Z_1^T Z_2)$
horizontal space	$Z \in \mathbb{R}^{n \times p} : Y^T Z = 0$	/
projection onto horizontal space	$\mathrm{P}_Y^h Z = Z - Y(Y^T Y)^{-1} Y^T Z$	/
gradient	$\overline{\mathrm{grad}\, f}_Y = \mathrm{grad}\, \overline{f}(Y)$	$\mathrm{grad}\, \overline{f}(Y) = \mathrm{P}_Y^h(2AY)$
retraction	$R_{\mathrm{span}(Y)}(\xi) = \mathrm{span}(Y + \overline{\xi}_Y)$	$R_Y(Z) = Y + Z$

4.9.2 Line-search algorithm

In order to obtain a line-search algorithm for the Rayleigh quotient on the Grassmann manifold, it remains to pick a retraction. According to Section 4.1.2, a natural choice is

$$R_{\mathrm{span}(Y)}(\xi) = \mathrm{span}(Y + \overline{\xi}_Y). \tag{4.40}$$

In other words, $(Y + \overline{\xi}_Y)M$ is a matrix representation of $R_{\mathrm{span}(Y)}(\xi)$ for any $M \in \mathbb{R}_*^{p \times p}$. The matrix M can be viewed as a normalization factor that can be used to prevent the iterates from becoming ill-conditioned, the best-conditioned form being orthonormal matrices. We now have all the necessary elements (see the summary in Table 4.3) to write down explicitly a line-search method for the Rayleigh quotient (4.38).

The matrix algorithm obtained by applying the Armijo line-search version of Algorithm 1 to the problem of minimizing the generalized Rayleigh quotient (4.38) is stated in Algorithm 3.

The following convergence results follow from the convergence analysis of the general line-search Algorithm 1 (Theorems 4.3.1 and 4.5.6).

Theorem 4.9.1 *Let $\{Y_k\}$ be an infinite sequence of iterates generated by Algorithm 3. Let $\lambda_1 \leq \cdots \leq \lambda_n$ denote the eigenvalues of A.*

(i) *The sequence $\{\mathrm{span}(Y_k)\}$ converges to the set of p-dimensional invariant subspaces of A.*

(ii) *Assuming that the eigenvalue λ_p is simple, the (unique) invariant subspace associated with $(\lambda_1, \ldots, \lambda_p)$ is asymptotically stable for the iteration defined by Algorithm 3, and the convergence is linear with a factor smaller than or equal to*

$$r_* = 1 - 2\sigma(\lambda_{p+1} - \lambda_p) \min\left(\bar{\alpha}, \frac{2\beta(1 - \sigma)}{\lambda_n - \lambda_1}\right).$$

Algorithm 3 Armijo line search for the Rayleigh quotient on $\mathrm{Grass}(p, n)$

Require: Symmetric matrix A, scalars $\bar{\alpha} > 0$, $\beta, \sigma \in (0, 1)$.
Input: Initial iterate $Y_0 \in \mathbb{R}^{n \times p}$, Y_0 full rank.
Output: Sequence of iterates $\{Y_k\}$.
  1: **for** $k = 0, 1, 2, \ldots$ **do**
  2:    Compute $\eta_k = -2(AY - Y(Y^T Y)^{-1} AY)$.
  3:    Find the smallest integer $m \geq 0$ such that

$$\bar{f}(Y_k + \bar{\alpha}\beta^m \eta_k) \leq \bar{f}(Y_k) - \sigma \bar{\alpha}\beta^m \, \mathrm{tr}(\eta_k^T \eta_k),$$

      with $\bar{f}$ defined in (4.37).
  4:    Select $Y_{k+1} := (Y_k + \bar{\alpha}\beta^m \eta_k)M$, with some invertible $p \times p$ matrix
      M chosen to preserve good conditioning. (For example, select Y_{k+1} as
      the Q factor of the QR decomposition of $Y_k + \bar{\alpha}\beta^m \eta_k$.)
  5: **end for**

The other invariant subspaces are unstable.

Numerical results are presented in Figure 4.4.

4.10 NOTES AND REFERENCES

Classical references on numerical optimization include Bertsekas [Ber95], Dennis and Schnabel [DS83], Fletcher [Fle01], Luenberger [Lue73], Nash and Sofer [NS96], Polak [Pol71], and Nocedal and Wright [NW99].

The choice of the qualification *complete* for Riemannian manifolds is not accidental: it can be shown that a Riemannian manifold $\mathcal{M}$ is complete (i.e., the domain of the exponential is the whole $T\mathcal{M}$) if and only if $\mathcal{M}$, endowed with the Riemannian distance, is a complete metric space; see, e.g., O'Neill [O'N83].

The idea of using computationally efficient alternatives to the Riemannian exponential was advocated by Manton [Man02, § IX] and was also touched on in earlier works [MMH94, Smi94, EAS98]. Retraction mappings are common in the field of algebraic topology [Hir76]. The definition of retraction used in this book comes from Shub [Shu86]; see also Adler *et al.* [ADM$^+$02]. Most of the material about retractions on the orthogonal group comes from [ADM$^+$02].

Selecting a computationally efficient retraction is a crucial step in developing a competitive algorithm on a manifold. This problem is linked to the question of approximating the exponential in such a way that the approximation resides on the manifold. This is a major research topic in computational mathematics, with important recent contributions; see, e.g., [CI01, OM01, IZ05, DN05] and references therein.

The concept of a locally smooth family of parameterizations was introduced by Hüper and Trumpf [HT04].

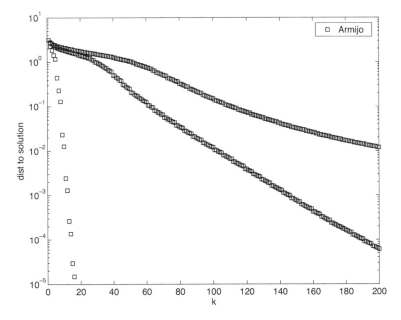

Figure 4.4 Rayleigh quotient minimization on the Grassmann manifold of p-planes in $\mathbb{R}^n$, with $p = 5$ and $n = 100$. Upper curve: $A = \text{diag}(1, 2, \ldots, 100)$. Middle curve: $A = \text{diag}(1, 102, 103, \ldots, 200)$. Lower curve: $A = \text{diag}(1, \ldots, 5, 106, 107, \ldots, 200)$.

Details on the QR and polar decompositions and algorithms to compute them can be found in Golub and Van Loan [GVL96]; the differentiability of the qf mapping is studied in Dehane [Deh95], Dieci and Eirola [DE99], and Chern and Dieci [CD00]. Formulas for the differential of qf and other smooth matrix functions can be found in Dehaene [Deh95].

Definition 4.2.1, on gradient-related sequences, is adapted from [Ber95]. Armijo's backtracking procedure was proposed in [Arm66] (or see [NW99, Ber95] for details).

Several key ideas for line-search methods on manifolds date back to Luenberger [Lue73, Ch. 11]. Luenberger proposed to use a search direction obtained by projecting the gradient in $\mathbb{R}^n$ onto the tangent space of the constraint set and mentioned the idea of performing a line search along the geodesic, "which we would use if it were computationally feasible (which it definitely is not)". He also proposed an alternative to following the geodesic that corresponds to retracting orthogonally to the tangent space. Other early contributions to optimization on manifolds can be found in Gabay [Gab82]. Line-search methods on manifolds are also proposed and analyzed in Udriṣte [Udr94]. Recently, Yang [Yan07] proposed an Armijo line-search strategy along geodesics. Exact and approximate line-search methods were proposed for matrix manifolds in a burst of research in the early 1990s [MMH94, Mah94, Bro93, Smi94]. Algorithm 1 comes from [AG05].

Many refinements exist for choosing the step length in line-search methods. For example, the backtracking parameter β can be adapted during the backtracking procedure. We refer to Dennis and Schnabel [DS83, §6.3.2] and Ortega and Rheinboldt [OR70].

The non-Hausdorff example given in Section 4.3.2 was inspired by Brickell and Clark [BC70, Ex. 3.2.1], which refers to Haefliger and Reeb [HR57].

For a local convergence analysis of classical line-search methods, see, e.g., Luenberger [Lue73] or Bertsekas [Ber95]. The proof of Theorem 4.3.1 (the global convergence of line-search methods) is a generalization of the proof of [Ber95, Prop. 1.2.1]. In Section 4.4, it is pointed out that convergence to critical points that are not local minima cannot be ruled out. Another undesirable behavior that cannot be ruled out in general is the existence of several (even infinitely many) accumulation points. Details can be found in Absil *et al.* [AMA05]; see also [GDS05]. Nevertheless, such algorithms do converge to single accumulation points, and the gap between theory and practice should not prevent one from utilizing the most computationally effective algorithm.

The notions of stability of fixed points have counterparts in dynamical systems theory; see, e.g., Vidyasagar [Vid02] or Guckenheimer and Holmes [GH83]. In fact, iterations $x_{k+1} = F(x_k)$ can be thought of as discrete-time dynamical systems.

Further information on Lojasiewicz's gradient inequality can be found in Lojasiewicz [Loj93]. The concept of Theorem 4.4.2 (the capture theorem) is borrowed from Bertsekas [Ber95]. A coordinate-free proof of our Theorem 4.5.6 (local convergence of line-search methods) is given by Smith [Smi94] in the particular case where the next iterate is obtained via an exact line search minimization along geodesics. Optimization algorithms on the Grassmann manifold can be found in Smith [Smi93], Helmke and Moore [HM94], Edelman *et al.* [EAS98], Lippert and Edelman [LE00], Manton [Man02], Manton *et al.* [MMH03], Absil *et al.* [AMS04], and Liu *et al.* [LSG04].

Gradient-descent algorithms for the Rayleigh quotient were considered as early as 1951 by Hestenes and Karush [HK51]. A detailed account is given in Faddeev and Faddeeva [FF63, §74, p. 430]. There has been limited investigation of line-search descent algorithms as numerical methods for linear algebra problems since it is clear that such algorithms are not competitive with existing numerical linear algebra algorithms. At the end of his paper on the design of gradient systems, Brockett [Bro93] provides a discrete-time analog, with an analytic step-size selection method, for a specific class of problems. In independent work, Moore *et al.* [MMH94] (see also [HM94, p. 68]) consider the symmetric eigenvalue problem directly. Chu [Chu92] proposes numerical methods for the inverse singular value problem. Smith *et al.* [Smi93, Smi94, EAS98] consider line-search and conjugate gradient updates to eigenspace tracking problems. Mahony *et al.* [Mah94, MHM96] proposes gradient flows and considers discrete updates for principal component analysis. A related approach is to consider explicit integration of the gradient flow dynamical system with a numerical integration technique that preserves the underlying matrix constraint. Moser and Veselov [MV91] use this approach directly in building numerical algorithms for matrix factorizations. The literature on structure-preserving integration algorithms is closely linked to work on the integration of Hamiltonian systems. This field is too vast to cover here, but we mention the excellent review by Iserles *et al.* [IMKNZ00] and an earlier review by Sanz-Serna [SS92].

The locally optimal conjugate gradient algorithm for the symmetric eigenvalue problem is described in Knyazev [Kny01]; see Hetmaniuk and Lehoucq [HL06] for recent developments. The connection between the power method and line-search methods for the Rayleigh quotient was studied in Mahony *et al.* [MHM96].

More information on the eigenvalue problem can be found in Golub and van der Vorst [GvdV00], Golub and Van Loan [GVL96], Parlett [Par80], Saad [Saa92], Stewart [Ste01], Sorensen [Sor02], and Bai *et al.* [BDDR00].

Linearly convergent iterative numerical methods for eigenvalue and subspace problems are not competitive with the classical numerical linear algebra techniques for one-off matrix factorization problems. However, a domain in which linear methods are commonly employed is in tracking the principal subspace of a covariance matrix associated with observations of a noisy signal. Let $\{x_1, x_2, \ldots\}$ be a sequence of elements of vectors in $\mathbb{R}^n$ and define

$$E_k^N = \frac{1}{N} \sum_{i=k+1}^{k+N} x_i x_i^T \in \mathbb{R}^{n \times n}, \qquad A_k^N = \begin{bmatrix} x_{k+1} & \cdots & x_N \end{bmatrix} \in \mathbb{R}^{n \times N}.$$

The signal subspace tracking problem is either to track a principal subspace of the covariance matrix E_k^N (a Hermitian eigenspace problem) or to directly track a signal subspace of the signal array A_k^N (a singular value problem). Common and Golub [CG90] studied classical numerical linear algebra techniques for this problem with linear update complexity. More recent review material is provided in DeGroat *et al.* [DDL99]. Most (if not all) high-accuracy linear complexity algorithms belong to a family of power-based algorithms [HXC+99]. This includes the Oja algorithm [Oja89], the PAST algorithm [Yan95], the NIC algorithm [MH98b], and the Bi-SVD algorithm [Str97], as well as gradient-based updates [FD95, EAS98]. Research in this field is extremely active at this time, with the focus on reduced-complexity updates [OH05, BDR05]. We also refer the reader to the Bayesian geometric approach followed in [Sri00, SK04].

In line-search algorithms, the limit case where the step size goes to zero corresponds to a continuous-time dynamical system of the form $\dot{x} = \eta_x$, where $\eta_x \in T_x \mathcal{M}$ denotes the search direction at $x \in \mathcal{M}$. There is a vast literature on continuous-time systems that solve computational problems, spanning several areas of computational science, including, but not limited to, linear programming [BL89a, BL89b, Bro91, Fay91b, Hel93b], continuous nonlinear optimization [Fay91a, LW00], discrete optimization [Hop84, HT85,

Vid95, AS04], signal processing [AC98, Dou00, CG03], balanced realization of linear systems [Hel93a, GL93], model reduction [HM94, YL99], and automatic control [HM94, MH98a, GS01]. Applications in linear algebra, and especially in eigenvalue and singular value problems, are particularly abundant. Important advances in the area have come from the work on isospectral flows in the early 1980s. We refer the reader to Helmke and Moore [HM94] as the seminal monograph in this area and the thesis of Dehaene [Deh95] for more information; see also [Chu94, DMV99, CG02, Prz03, MA03, BI04, CDLP05, MHM05] and the many references therein.

Chapter Five

Matrix Manifolds: Second-Order Geometry

Many optimization algorithms make use of second-order information about the cost function. The archetypal second-order optimization algorithm is Newton's method. This method is an iterative method that seeks a critical point of the cost function f (i.e., a zero of grad f) by selecting the update vector at x_k as the vector along which the directional derivative of grad f is equal to $-\text{grad } f(x_k)$. The second-order information on the cost function is incorporated through the directional derivative of the gradient.

For a quadratic cost function in $\mathbb{R}^n$, Newton's method identifies a zero of the gradient in one step. For general cost functions, the method is not expected to converge in one step and may not even converge at all. However, the use of second-order information ensures that algorithms based on the Newton step display superlinear convergence (when they do converge) compared to the linear convergence obtained for algorithms that use only first-order information (see Section 4.5).

A Newton method on Riemannian manifolds will be defined and analyzed in Chapter 6. However, to provide motivation for the somewhat abstract theory that follows in this chapter, we begin by briefly recapping Newton's method in $\mathbb{R}^n$ and identify the blocks to generalizing the iteration to a manifold setting. An important step in the development is to provide a meaningful definition of the derivative of the gradient and, more generally, of vector fields; this issue is addressed in Section 5.2 by introducing the notion of an affine connection. An affine connection also makes it possible to define parallel translation, geodesics, and exponentials (Section 5.4). These tools are not mandatory in defining a Newton method on a manifold, but they are fundamental objects of Riemannian geometry, and we will make use of them in later chapters. On a Riemannian manifold, there is one preferred affine connection, termed the *Riemannian connection*, that admits elegant specialization to Riemannian submanifolds and Riemannian quotient manifolds (Section 5.3). The chapter concludes with a discussion of the concept of a Hessian on a manifold (Sections 5.5 and 5.6).

5.1 NEWTON'S METHOD IN $\mathbb{R}^N$

In its simplest formulation, Newton's method is an iterative method for finding a solution of an equation in one unknown. Let F be a smooth function from $\mathbb{R}$ to $\mathbb{R}$ and let x_* be a *zero* (or *root*) of F, i.e., $F(x_*) = 0$. From an initial point x_0 in $\mathbb{R}$, Newton's method constructs a sequence of iterates according to

$$x_{k+1} = x_k - \frac{F(x_k)}{F'(x_k)}, \tag{5.1}$$

where F' denotes the derivative of F. Graphically, x_{k+1} corresponds to the intersection of the tangent to the graph of F at x_k with the horizontal axis (see Figure 5.1). In other words, x_{k+1} is the zero of the first-order Taylor expansion of F around x_k. This is clearly seen when (5.1) is rewritten as

$$F(x_k) + F'(x_k)(x_{k+1} - x_k) = 0. \tag{5.2}$$

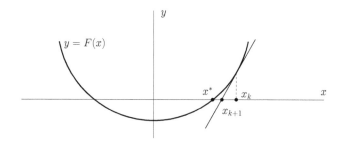

Figure 5.1 Newton's method in $\mathbb{R}$.

Let $G : \mathbb{R}^n \to \mathbb{R}^n : G(x) := x - F(x)/F'(x)$ be the iteration map from (5.1) and note that x_* is a fixed point of G. For a generic fixed point where $F(x_*) = 0$ and $F'(x_*) \neq 0$, the derivative

$$G'(x_*) = 1 - \frac{F'(x_*)}{F'(x_*)} + \frac{F(x_*)F''(x_*)}{(F'(x_*))^2} = 0,$$

and it follows that Newton's method is locally quadratically convergent to x_* (see Theorem 4.5.3).

Newton's method can be generalized to functions F from $\mathbb{R}^n$ to $\mathbb{R}^n$. Equation (5.2) becomes

$$F(x_k) + \mathrm{D}F(x_k)[x_{k+1} - x_k] = 0, \tag{5.3}$$

where $\mathrm{D}F(x)[z]$ denotes the *directional derivative* of F along z, defined by

$$\mathrm{D}F(x)[z] := \lim_{t \to 0} \frac{1}{t}(F(x + tz) - F(x)).$$

A generalization of the argument given above shows that Newton's method locally quadratically converges to isolated roots of F for which $DF(x_*)$ is full rank.

Newton's method is readily adapted to the problem of computing a critical point of a cost function f on $\mathbb{R}^n$. Simply take $F := \operatorname{grad} f$, where

$$\operatorname{grad} f(x) = (\partial_1 f(x), \dots, \partial_n f(x))^T$$

is the Euclidean gradient of f. The iterates of Newton's method then converge locally quadratically to the isolated zeros of grad f, which are the isolated critical points of f. Newton's equation then reads

$$\operatorname{grad} f(x_k) + \mathrm{D}(\operatorname{grad} f)(x_k)[x_{k+1} - x_k] = 0.$$

To generalize this approach to manifolds, we must find geometric analogs to the various components of the formula that defines the Newton iterate on $\mathbb{R}^n$. When f is a cost function an abstract Riemannian manifold, the Euclidean gradient naturally becomes the Riemannian gradient grad f defined in Section 3.6. The zeros of grad f are still the critical points of f. The difference $x_{k+1} - x_k$, which is no longer defined since the iterates x_{k+1} and x_k belong to the abstract manifold, is replaced by a tangent vector η_{x_k} in the tangent space at x_k. The new iterate x_{k+1} is obtained from η_{x_k} as $x_{k+1} = R_{x_k}(\eta_{x_k})$, where R is a retraction; see Section 4.1 for the notion of retraction. It remains to provide a meaningful definition for "$\mathrm{D}(\operatorname{grad} f)(x_k)[\eta_{x_k}]$".

More generally, for finding a zero of a tangent vector field ξ on a manifold, Newton's method takes the form

$$\xi_{x_k} + \text{"}\mathrm{D}\xi(x_k)[\eta_{x_k}]\text{"} = 0,$$
$$x_{k+1} = R_{x_k}(\eta_{x_k}).$$

The only remaining task is to provide a geometric analog of the directional derivative of a vector field.

Recall that tangent vectors are defined as derivations of real functions: given a scalar function f and a tangent vector η at x, the real $\mathrm{D}f(x)[\eta]$ is defined as $\left.\frac{\mathrm{d}(f(\gamma(t)))}{\mathrm{d}t}\right|_{t=0}$, where γ is a curve representing η; see Section 3.5. If we try to apply the same concept to vector fields instead of scalar fields, we obtain

$$\left.\frac{\mathrm{d}\,\xi_{\gamma(t)}}{\mathrm{d}t}\right|_{t=0} = \lim_{t \to 0} \frac{\xi_{\gamma(t)} - \xi_{\gamma(0)}}{t}.$$

The catch is that the two vectors $\xi_{\gamma(t)}$ and $\xi_{\gamma(0)}$ belong to two different vector spaces $T_{\gamma(t)}\mathcal{M}$ and $T_{\gamma(0)}\mathcal{M}$, and there is in general no predefined correspondence between the vector spaces that allows us to compute the difference. Such a correspondence can be introduced by means of affine connections.

5.2 AFFINE CONNECTIONS

The definition of an affine connection on a manifold is one of the most fundamental concepts in differential geometry. An affine connection is an additional structure to the differentiable structure. Any manifold admits infinitely many different affine connections. Certain affine connections, however, may have particular properties that single them out as being the most appropriate for geometric analysis. In this section we introduce the concept of an affine connection from an abstract perspective and show how it generalizes the concept of a directional derivative of a vector field.

Let $\mathfrak{X}(\mathcal{M})$ denote the set of smooth vector fields on $\mathcal{M}$. An *affine connection* ∇ (pronounced "del" or "nabla") on a manifold $\mathcal{M}$ is a mapping

$$\nabla : \mathfrak{X}(\mathcal{M}) \times \mathfrak{X}(\mathcal{M}) \to \mathfrak{X}(\mathcal{M}),$$

which is denoted by $(\eta, \xi) \xrightarrow{\nabla} \nabla_\eta \xi$ and satisfies the following properties:

i)$\mathfrak{F}(\mathcal{M})$-linearity in η: $\nabla_{f\eta+g\chi}\xi = f\nabla_\eta \xi + g\nabla_\chi \xi,$
ii)$\mathbb{R}$-linearity in ξ: $\nabla_\eta(a\xi + b\zeta) = a\nabla_\eta \xi + b\nabla_\eta \zeta,$
iii)Product rule (Leibniz' law): $\nabla_\eta(f\xi) = (\eta f)\xi + f\nabla_\eta \xi,$

in which $\eta, \chi, \xi, \zeta \in \mathfrak{X}(\mathcal{M})$, $f, g \in \mathfrak{F}(\mathcal{M})$, and $a, b \in \mathbb{R}$. (Notice that ηf denotes the application of the vector field η to the function f, as defined in Section 3.5.4.) The vector field $\nabla_\eta \xi$ is called the *covariant derivative* of ξ with respect to η for the affine connection ∇.

In $\mathbb{R}^n$, the classical directional derivative defines an affine connection,

$$(\nabla_\eta \xi)_x = \lim_{t\to 0} \frac{\xi_{x+t\eta_x} - \xi_x}{t}, \tag{5.4}$$

called the *canonical (Euclidean) connection*. (This expression is well defined in view of the canonical identification $T_x\mathcal{E} \simeq \mathcal{E}$ discussed in Section 3.5.2, and it is readily checked that (5.4) satisfies all the properties of affine connections.) This fact, along with several properties discussed below, suggests that the covariant derivatives are a suitable generalization of the classical directional derivative.

Proposition 5.2.1 *Every (second-countable Hausdorff) manifold admits an affine connection.*

In fact, every manifold admits infinitely many affine connections, some of which may be computationally more tractable than others.

We first characterize all the possible affine connections on the linear manifold $\mathbb{R}^n$. Let $(e_1, \dots, e_n)$ be the canonical basis of $\mathbb{R}^n$. If ∇ is a connection on $\mathbb{R}^n$, we have

$$\nabla_\eta \xi = \nabla_{\sum_i \eta^i e_i}\left(\sum_j \xi^j e_j\right) = \sum_i \eta^i \nabla_{e_i}\left(\sum_j \xi^j e_j\right)$$
$$= \sum_{i,j}\left(\eta^i \xi^j \nabla_{e_i}e_j + \eta^i \partial_i \xi^j e_j\right),$$

where $\eta, \xi, e_i, \nabla_\eta \xi, \nabla_{e_i}e_j$ are all vector fields on $\mathbb{R}^n$. To define ∇, it suffices to specify the n^2 vector fields $\nabla_{e_i}e_j$, $i = 1, \dots, n$, $j = 1, \dots, n$. By convention, the kth component of $\nabla_{e_i}e_j$ in the basis $(e_1, \dots, e_n)$ is denoted by Γ_{ij}^k. The n^3 real-valued functions Γ_{ij}^k on $\mathbb{R}^n$ are called *Christoffel symbols*. Each choice of smooth functions Γ_{ij}^k defines a different affine connection on $\mathbb{R}^n$. The Euclidean connection corresponds to the choice $\Gamma_{ij}^k \equiv 0$.

On an n-dimensional manifold $\mathcal{M}$, locally around any point x, a similar development can be based on a coordinate chart $(\mathcal{U}, \varphi)$. The following coordinate-based development shows how an affine connection can be defined on $\mathcal{U}$, at least in theory (in practice, the use of coordinates to define an affine connection can be cumbersome). The canonical vector e_i is replaced by the ith coordinate vector field E_i of $(\mathcal{U}, \varphi)$ which, at a point y of $\mathcal{U}$, is represented by the curve $t \mapsto \varphi^{-1}(\varphi(y) + te_i)$; in other words, given a real-valued function f defined on $\mathcal{U}$, $E_i f = \partial_i(f \circ \varphi^{-1})$. Thus, one has $D\varphi(y)[(E_i)_y] = e_i$. We will also use the notation $\partial_i f$ for $E_i f$. A vector field ξ can be decomposed as $\xi = \sum_j \xi^j E_j$, where ξ^i, $i = 1, \ldots, d$, are real-valued functions on $\mathcal{U}$, i.e., elements of $\mathfrak{F}(\mathcal{U})$. Using the characteristic properties of affine connections, we obtain

$$\nabla_\eta \xi = \nabla_{\sum_i \eta^i E_i} \left(\sum_j \xi^j E_j \right) = \sum_i \eta^i \nabla_{E_i} \left(\sum_j \xi^j E_j \right)$$

$$= \sum_{i,j} \left(\eta^i \xi^j \nabla_{E_i} E_j + \eta^i \partial_i \xi^j E_j \right). \tag{5.5}$$

It follows that the affine connection is fully specified once the n^2 vector fields $\nabla_{E_i} E_j$ are selected. We again use the Christoffel symbol Γ_{ij}^k to denote the kth component of $\nabla_{E_i} E_j$ in the basis $(E_1, \ldots, E_n)$; in other words,

$$\nabla_{E_i} E_j = \sum_k \Gamma_{ij}^k E_k.$$

The Christoffel symbols Γ_{ij}^k at a point x can be thought of as a table of n^3 real numbers that depend both on the point x in $\mathcal{M}$ and on the choice of the chart φ (for the same affine connection, different charts produce different Christoffel symbols). We thus have

$$\nabla_\eta \xi = \sum_{i,j,k} \left(\eta^i \xi^j \Gamma_{ij}^k E_k + \eta^i \partial_i \xi^j E_j \right).$$

A simple renaming of indices yields

$$\nabla_\eta \xi = \sum_{i,j,k} \eta^j \left(\xi^k \Gamma_{jk}^i + \partial_j \xi^i \right) E_i. \tag{5.6}$$

We also obtain a matrix expression as follows. Letting hat quantities denote the (column) vectors of components in the chart $(\mathcal{U}, \phi)$, we have

$$\widehat{\nabla_{\eta_x} \xi} = \hat{\Gamma}_{\hat{x}, \hat{\xi}} \hat{\eta}_{\hat{x}} + D\hat{\xi}(\hat{x}) [\hat{\eta}_{\hat{x}}], \tag{5.7}$$

where $\hat{\Gamma}_{\hat{x}, \hat{\xi}}$ denotes the matrix whose (i, j) element is the real-valued function

$$\sum_k \left(\xi^k \Gamma_{jk}^i \right) \tag{5.8}$$

evaluated at x.

From the coordinate expression (5.5), one can deduce the following properties of affine connections.

1. Dependence on η_x. The vector field $\nabla_\eta \xi$ at a point x depends only on the value η_x of η at x. Thus, an affine connection at x is a mapping $T_x \mathcal{M} \times \mathfrak{X}(x) \to \mathfrak{X}(x) : (\eta_x, \xi) \mapsto \nabla_{\eta_x} \xi$, where $\mathfrak{X}(x)$ denotes the set of vector fields on $\mathcal{M}$ whose domain includes x.

2. Local dependence on ξ. In contrast, ξ_x does not provide enough information about the vector field ξ to compute $\nabla_\eta \xi$ at x. However, if the vector fields ξ and ζ agree on some neighborhood of x, then $\nabla_\eta \xi$ and $\nabla_\eta \zeta$ coincide at x. Moreover, given two affine connections ∇ and $\tilde{\nabla}$, $\nabla_\eta \xi - \tilde{\nabla}_\eta \xi$ at x depends only on the value ξ_x of ξ at x.

3. Uniqueness at zeros. Let ∇ and $\tilde{\nabla}$ be two affine connections on $\mathcal{M}$ and let ξ and η be vector fields on $\mathcal{M}$. Then, as a corollary of the previous property,

$$\left(\nabla_\eta \xi \right)_x = \left(\tilde{\nabla}_\eta \xi \right)_x \quad \text{if } \xi_x = 0.$$

This final property is particularly important in the convergence analysis of optimization algorithms around critical points of a cost function.

5.3 RIEMANNIAN CONNECTION

On an arbitrary (second-countable Hausdorff) manifold, there are infinitely many affine connections, and *a priori*, no one is better than the others. In contrast, on a vector space $\mathcal{E}$ there is a preferred affine connection, the canonical connection (5.4), which is simple to calculate and preserves the linear structure of the vector space. On an arbitrary Riemannian manifold, there is also a preferred affine connection, called the Riemannian or the Levi-Civita connection. This connection satisfies two properties (symmetry, and invariance of the Riemannian metric) that have a crucial importance, notably in relation to the notion of Riemannian Hessian. Moreover, the Riemannian connection on Riemannian submanifolds and Riemannian quotient manifolds admits a remarkable formulation in terms of the Riemannian connection in the structure space that makes it particularly suitable in the context of numerical algorithms. Furthermore, on a Euclidean space, the Riemannian connection reduces to the canonical connection—the classical directional derivative.

5.3.1 Symmetric connections

An affine connection is symmetric if its Christoffel symbols satisfy the symmetry property $\Gamma^k_{ij} = \Gamma^k_{ji}$. This definition is equivalent to a more abstract coordinate-free approach to symmetry that provides more insight into the underlying structure of the space.

To define symmetry of an affine connection in a coordinate-free manner, we will require the concept of a *Lie bracket* of two vector fields. Let ξ and ζ be vector fields on $\mathcal{M}$ whose domains meet on an open set $\mathcal{U}$. Recall that

$\mathfrak{F}(\mathcal{U})$ denotes the set of smooth real-valued functions whose domains include $\mathcal{U}$. Let $[\xi, \eta]$ denote the function from $\mathfrak{F}(\mathcal{U})$ into itself defined by

$$[\xi, \zeta]f := \xi(\zeta f) - \zeta(\xi f). \tag{5.9}$$

It is easy to show that $[\xi, \zeta]$ is $\mathbb{R}$-linear,

$$[\xi, \eta](af + bg) = a[\xi, \eta]f + b[\xi, \eta]g,$$

and satisfies the product rule (Leibniz' law),

$$[\xi, \eta](fg) = f([\xi, \eta]g) + ([\xi, \eta]f)g.$$

Therefore, $[\xi, \zeta]$ is a derivation and defines a tangent vector field, called the *Lie bracket* of ξ and ζ.

An affine connection ∇ on a manifold $\mathcal{M}$ is said to be *symmetric* when

$$\nabla_\eta \xi - \nabla_\xi \eta = [\eta, \xi] \tag{5.10}$$

for all $\eta, \xi \in \mathfrak{X}(\mathcal{M})$.

Given a chart $(\mathcal{U}, \varphi)$, denoting by E_i the ith coordinate vector field, we have, for a symmetric connection ∇,

$$\nabla_{E_i} E_j - \nabla_{E_j} E_i = [E_i, E_j] = 0$$

since $[E_i, E_j]f = \partial_i \partial_j f - \partial_j \partial_i f = 0$ for all $f \in \mathfrak{F}(\mathcal{M})$. It follows that $\Gamma_{ij}^k = \Gamma_{ji}^k$ for every symmetric connection. Conversely, it is easy to show that connections satisfying $\Gamma_{ij}^k = \Gamma_{ji}^k$ are symmetric in the sense of (5.10) by expanding in local coordinates.

5.3.2 Definition of the Riemannian connection

The following result is sometimes referred to as the fundamental theorem of Riemannian geometry. Let $\langle \cdot, \cdot \rangle$ denote the Riemannian metric.

Theorem 5.3.1 (Levi-Civita) *On a Riemannian manifold $\mathcal{M}$ there exists a unique affine connection ∇ that satisfies*

(i) $\nabla_\eta \xi - \nabla_\xi \eta = [\eta, \xi]$ *(symmetry), and*
(ii) $\chi \langle \eta, \xi \rangle = \langle \nabla_\chi \eta, \xi \rangle + \langle \eta, \nabla_\chi \xi \rangle$ *(compatibility with the Riemannian metric),*

for all $\chi, \eta, \xi \in \mathfrak{X}(\mathcal{M})$. This affine connection ∇, called the Levi-Civita *connection or the* Riemannian connection *of $\mathcal{M}$, is characterized by the Koszul formula*

$$2\langle \nabla_\chi \eta, \xi \rangle = \chi \langle \eta, \xi \rangle + \eta \langle \xi, \chi \rangle - \xi \langle \chi, \eta \rangle - \langle \chi, [\eta, \xi] \rangle + \langle \eta, [\xi, \chi] \rangle + \langle \xi, [\chi, \eta] \rangle. \tag{5.11}$$

Recall that for vector fields $\eta, \xi, \chi \in \mathfrak{X}(\mathcal{M})$, $\langle \eta, \xi \rangle$ is a real-valued function on $\mathcal{M}$ and $\chi \langle \eta, \xi \rangle$ is the real-valued function given by the application of the vector field (i.e., derivation) χ to $\langle \eta, \xi \rangle$.)

Since the Riemannian connection is symmetric, it follows that the Christoffel symbols of the Riemannian connection satisfy $\Gamma_{ij}^k = \Gamma_{ji}^k$. Moreover, it follows from the Koszul formula (5.11) that the Christoffel symbols for the Riemannian connection are related to the coefficients of the metric by the formula

$$\Gamma^k_{ij} = \frac{1}{2} \sum_\ell g^{k\ell} \left(\partial_i g_{\ell j} + \partial_j g_{\ell i} - \partial_\ell g_{ij} \right), \qquad (5.12)$$

where $g^{k\ell}$ denotes the matrix inverse of $g_{k\ell}$, i.e., $\sum_i g^{ki} g_{i\ell} = \delta^k_\ell$. In theory, the formula (5.12) provides a means to compute the Riemannian connection. However, working in coordinates can be cumbersome in practice, and we will use a variety of tricks to avoid using (5.12) as a computational formula.

Note that on a Euclidean space, the Riemannian connection reduces to the canonical connection (5.4). A way to see this is that, in view of (5.12), the Christoffel symbols vanish since the metric is constant.

5.3.3 Riemannian connection on Riemannian submanifolds

Let $\mathcal{M}$ be a Riemannian submanifold of a Riemannian manifold $\overline{\mathcal{M}}$. By definition, the Riemannian metric on the submanifold $\mathcal{M}$ is obtained by restricting to $\mathcal{M}$ the Riemannian metric on $\overline{\mathcal{M}}$; therefore we use the same notation $\langle \cdot, \cdot \rangle$ for both. Let ∇ denote the Riemannian connection of $\mathcal{M}$, and $\overline{\nabla}$ the Riemannian connection of $\overline{\mathcal{M}}$. Let $\mathfrak{X}(\mathcal{M})$ denote the set of vector fields on $\mathcal{M}$, and $\mathfrak{X}(\overline{\mathcal{M}})$ the set of vector fields on $\overline{\mathcal{M}}$.

Given $\eta_x \in T_x\mathcal{M}$ and $\xi \in \mathfrak{X}(\mathcal{M})$, we begin by defining the object $\overline{\nabla}_{\eta_x} \xi$. To this end, since $T_x\mathcal{M}$ is a subspace of $T_x\overline{\mathcal{M}}$, let $\overline{\eta}_x$ be η_x viewed as an element of $T_x\overline{\mathcal{M}}$; moreover, let $\overline{\xi}$ be a smooth local extension of ξ over a coordinate neighborhood $\mathcal{U}$ of x in $\overline{\mathcal{M}}$. Then define

$$\overline{\nabla}_{\eta_x} \xi := \overline{\nabla}_{\overline{\eta}_x} \overline{\xi}. \qquad (5.13)$$

This expression does not depend on the local extension of ξ. However, in general, $\overline{\nabla}_{\eta_x} \xi$ does not lie in $T_x\mathcal{M}$, as illustrated in Figure 5.2. Hence the restriction of $\overline{\nabla}$ to $\mathcal{M}$, as defined in (5.13), does not qualify as a connection on $\mathcal{M}$.

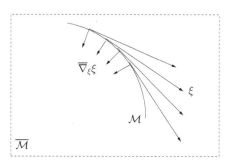

Figure 5.2 Riemannian connection $\overline{\nabla}$ in a Euclidean space $\overline{\mathcal{M}}$ applied to a tangent vector field ξ to a circle. We observe that $\overline{\nabla}_\xi \xi$ is not tangent to the circle.

Recall from Section 3.6.1 that, using the Riemannian metric on $\overline{\mathcal{M}}$, each tangent space $T_x\overline{\mathcal{M}}$ can be decomposed as the direct sum of $T_x\mathcal{M}$ and its orthogonal complement $(T_x\mathcal{M})^\perp$, called the normal space to the Riemannian submanifold $\mathcal{M}$ at x. Every vector $\xi_x \in T_x\overline{\mathcal{M}}$, $x \in \mathcal{M}$, has a unique decomposition

$$\xi_x = \mathrm{P}_x\xi_x + \mathrm{P}_x^\perp\xi_x,$$

where $\mathrm{P}_x\xi_x$ belongs to $T_x\mathcal{M}$ and $\mathrm{P}_x^\perp\xi_x$ belongs to $(T_x\mathcal{M})^\perp$. We have the following fundamental result.

Proposition 5.3.2 *Let $\mathcal{M}$ be a Riemannian submanifold of a Riemannian manifold $\overline{\mathcal{M}}$ and let ∇ and $\overline{\nabla}$ denote the Riemannian connections on $\mathcal{M}$ and $\overline{\mathcal{M}}$. Then*

$$\nabla_{\eta_x}\xi = \mathrm{P}_x\overline{\nabla}_{\eta_x}\xi \tag{5.14}$$

for all $\eta_x \in T_x\mathcal{M}$ and $\xi \in \mathfrak{X}(\mathcal{M})$.

This result is particularly useful when $\mathcal{M}$ is a Riemannian submanifold of a Euclidean space; then (5.14) reads

$$\nabla_{\eta_x}\xi = \mathrm{P}_x\left(\mathrm{D}\xi\left(x\right)\left[\eta_x\right]\right), \tag{5.15}$$

i.e., a classical directional derivative followed by an orthogonal projection.

Example 5.3.1   *The sphere* S^{n-1}
   On the sphere S^{n-1} viewed as a Riemannian submanifold of the Euclidean space $\mathbb{R}^n$, the projection P_x is given by

$$\mathrm{P}_x\xi = (I - xx^T)\xi$$

and the Riemannian connection is given by

$$\nabla_{\eta_x}\xi = (I - xx^T)\,\mathrm{D}\xi\left(x\right)\left[\eta_x\right] \tag{5.16}$$

for all $x \in S^{n-1}$, $\eta_x \in T_xS^{n-1}$, and $\xi \in \mathfrak{X}(S^{n-1})$. A practical application of this formula is presented in Section 6.4.1.

Example 5.3.2   *The orthogonal Stiefel manifold* $\mathrm{St}(p,n)$
   On the Stiefel manifold $\mathrm{St}(p,n)$ viewed as a Riemannian submanifold of the Euclidean space $\mathbb{R}^{n\times p}$, the projection P_X is given by

$$\mathrm{P}_X\xi = (I - XX^T)\xi + X\,\mathrm{skew}(X^T\xi)$$

and the Riemannian connection is given by

$$\nabla_{\eta_X}\xi = \mathrm{P}_X(\mathrm{D}\xi\left(x\right)\left[\eta_X\right]) \tag{5.17}$$

for all $X \in \mathrm{St}(p,n)$, $\eta_X \in T_X\,\mathrm{St}(p,n)$, and $\xi \in \mathfrak{X}(\mathrm{St}(p,n))$.

5.3.4 Riemannian connection on quotient manifolds

Let $\overline{\mathcal{M}}$ be a Riemannian manifold with a Riemannian metric $\overline{g}$ and let $\mathcal{M} = \overline{\mathcal{M}}/\sim$ be a Riemannian quotient manifold of $\overline{\mathcal{M}}$, i.e., $\mathcal{M}$ is endowed with a manifold structure and a Riemannian metric g that turn the natural projection $\pi : \overline{\mathcal{M}} \to \mathcal{M}$ into a Riemannian submersion. As in Section 3.6.2, the horizontal space $\mathcal{H}_y$ at a point $y \in \overline{\mathcal{M}}$ is defined as the orthogonal complement of the vertical space, and $\overline{\xi}$ denotes the horizontal lift of a tangent vector ξ.

Proposition 5.3.3 *Let $\mathcal{M} = \overline{\mathcal{M}}/\sim$ be a Riemannian quotient manifold and let ∇ and $\overline{\nabla}$ denote the Riemannian connections on $\mathcal{M}$ and $\overline{\mathcal{M}}$. Then*

$$\overline{\nabla_\eta \xi} = \mathrm{P}^h\left(\overline{\nabla}_{\overline{\eta}}\overline{\xi}\right) \tag{5.18}$$

for all vector fields ξ and η on $\mathcal{M}$, where P^h denotes the orthogonal projection onto the horizontal space.

This is a very useful result, as it provides a practical way to compute covariant derivatives in the quotient space. The result states that the horizontal lift of the covariant derivative of ξ with respect to η is given by the horizontal projection of the covariant derivative of the horizontal lift of ξ with respect to the horizontal lift of η.

If the structure space $\overline{\mathcal{M}}$ is (an open subset of) a Euclidean space, then formula (5.18) simply becomes

$$\overline{\nabla_\eta \xi} = \mathrm{P}^h\left(\mathrm{D}\overline{\xi}[\overline{\eta}]\right).$$

In some practical cases, $\overline{\mathcal{M}}$ is a vector space endowed with a Riemannian metric $\overline{g}$ that is not constant (hence $\overline{\mathcal{M}}$ is not a Euclidean space) but that is nevertheless *horizontally invariant*, namely,

$$\mathrm{D}\left(\overline{g}(\nu, \lambda)\right)(y)[\eta_y] = \overline{g}(\mathrm{D}\nu\,(y)[\eta_y], \lambda_y) + \overline{g}(\nu_y, \mathrm{D}\lambda\,(y)[\eta_y])$$

for all $y \in \overline{\mathcal{M}}$, all $\eta_y \in \mathcal{H}_y$, and all horizontal vector fields ν, λ on $\overline{\mathcal{M}}$. In this case, the next proposition states that the Riemannian connection on the quotient is still a classical directional derivative followed by a projection.

Proposition 5.3.4 *Let $\mathcal{M}$ be a Riemannian quotient manifold of a vector space $\overline{\mathcal{M}}$ endowed with a horizontally invariant Riemannian metric and let ∇ denote the Riemannian connection on $\mathcal{M}$. Then*

$$\overline{\nabla_\eta \xi} = \mathrm{P}^h\left(\mathrm{D}\overline{\xi}[\overline{\eta}]\right)$$

for all vector fields ξ and η on $\mathcal{M}$.

Proof. Let $\overline{g}(\cdot, \cdot) = \langle \cdot, \cdot \rangle$ denote the Riemannian metric on $\overline{\mathcal{M}}$ and let $\overline{\nabla}$ denote the Riemannian connection of $\overline{\mathcal{M}}$. Let χ, ν, λ be horizontal vector fields on $\overline{\mathcal{M}}$. Notice that since $\overline{\mathcal{M}}$ is a vector space, one has $[\nu, \lambda] = \mathrm{D}\lambda[\nu] - \mathrm{D}\nu[\lambda]$, and likewise for permutations between χ, ν, and λ. Moreover, since it is assumed that $\overline{g}$ is horizontally invariant, it follows that

$\mathrm{D}\overline{g}(\nu, \lambda)[\chi] = \overline{g}(\mathrm{D}\nu[\chi], \lambda) + \overline{g}(\nu, \mathrm{D}\lambda[\chi])$; and likewise for permutations. Using these identities, it follows from Koszul's formula (5.11) that

$$2\langle \overline{\nabla}_\chi \nu, \lambda \rangle = \chi\langle \nu, \lambda \rangle + \nu\langle \lambda, \chi \rangle - \lambda\langle \chi, \nu \rangle + \langle \lambda, [\chi, \nu] \rangle + \langle \nu, [\lambda, \chi] \rangle - \langle \chi, [\nu, \lambda] \rangle$$
$$= 2\overline{g}(\mathrm{D}\nu[\chi], \lambda),$$

hence $\mathrm{P}^h(\overline{\nabla}_\chi \nu) = \mathrm{P}^h(\mathrm{D}\nu[\chi])$. The result follows from Proposition 5.3.3. □

Example 5.3.3 *The Grassmann manifold*
We follow up on the example in Section 3.6.2. Recall that the Grassmann manifold $\mathrm{Grass}(p, n)$ *was viewed as a Riemannian quotient manifold of* $(\mathbb{R}^{n \times p}_*, \overline{g})$ *with*

$$\overline{g}_Y(Z_1, Z_2) = \mathrm{tr}\left((Y^T Y)^{-1} Z_1^T Z_2 \right). \tag{5.19}$$

The horizontal distribution is

$$\mathcal{H}_Y = \{ Z \in \mathbb{R}^{n \times p} : Y^T Z = 0 \} \tag{5.20}$$

and the projection onto the horizontal space is given by

$$\mathrm{P}^h_Y Z = (I - Y(Y^T Y)^{-1} Y^T)Z. \tag{5.21}$$

It is readily checked that, for all horizontal vectors $Z \in \mathcal{H}_Y$, *it holds that*

$$\mathrm{D}\overline{g}(\overline{\xi}, \overline{\zeta})(Y)[Z] = \mathrm{D}_Y(\mathrm{tr}((Y^T Y)^{-1}(\overline{\xi}_Y)^T \overline{\zeta}_Y))(Y)[Z]$$
$$= \overline{g}(\mathrm{D}\overline{\xi}(Y)[Z], \overline{\zeta}_Y) + \overline{g}(\overline{\xi}_Y, \mathrm{D}\overline{\zeta}(Y)[Z])$$

since $Y^T Z = 0$ *for all* $Z \in \mathcal{H}_Y$. *The Riemannian metric* $\overline{g}$ *is thus horizontally invariant. Consequently, we can apply the formula for the Riemannian connection on a Riemannian quotient of a manifold with a horizontally invariant metric (Proposition 5.3.4) and obtain*

$$\overline{\nabla_\eta \xi} = \mathrm{P}^h_Y \left(\mathrm{D}\overline{\xi}(Y)[\overline{\eta}_Y] \right). \tag{5.22}$$

We refer the reader to Section 6.4.2 for a practical application of this formula.

5.4 GEODESICS, EXPONENTIAL MAPPING, AND PARALLEL TRANSLATION

Geodesics on manifolds generalize the concept of straight lines in $\mathbb{R}^n$. A geometric definition of a straight line in $\mathbb{R}^n$ is that it is the image of a curve γ with zero acceleration; i.e.,

$$\frac{\mathrm{d}^2}{\mathrm{d}t^2} \gamma(t) = 0$$

for all t.

On manifolds, we have already introduced the notion of a tangent vector $\dot{\gamma}(t)$, which can be interpreted as the *velocity* of the curve γ at t. The mapping $t \mapsto \dot{\gamma}(t)$ defines the *velocity vector field* along γ. Next we define the acceleration vector field along γ.

Let $\mathcal{M}$ be a manifold equipped with an affine connection ∇ and let γ be a curve in $\mathcal{M}$ with domain $I \subseteq \mathbb{R}$. A *vector field on the curve* γ smoothly assigns to each $t \in I$ a tangent vector to $\mathcal{M}$ at $\gamma(t)$. For example, given any vector field ξ on $\mathcal{M}$, the mapping $t \mapsto \xi_{\gamma(t)}$ is a vector field on γ. The velocity vector field $t \mapsto \dot{\gamma}(t)$ is also a vector field on γ. The set of all (smooth) vector fields on γ is denoted by $\mathfrak{X}(\gamma)$. It can be shown that there is a unique function $\xi \mapsto \frac{D}{dt}\xi$ from $\mathfrak{X}(\gamma)$ to $\mathfrak{X}(\gamma)$ such that

1. $\frac{D}{dt}(a\xi + b\zeta) = a\frac{D}{dt}\xi + b\frac{D}{dt}\zeta$ $(a, b \in \mathbb{R})$,
2. $\frac{D}{dt}(f\xi) = f'\xi + f\frac{D}{dt}\xi$ $(f \in \mathfrak{F}(I))$,
3. $\frac{D}{dt}(\eta \circ \gamma)(t) = \nabla_{\dot{\gamma}(t)}\eta$ $(t \in I,\ \eta \in \mathfrak{X}(\mathcal{M}))$.

The *acceleration vector field* $\frac{D^2}{dt^2}\gamma$ on γ is defined by

$$\frac{D^2}{dt^2}\gamma := \frac{D}{dt}\dot{\gamma}. \qquad (5.23)$$

Note that the acceleration depends on the choice of the affine connection, while the velocity $\dot{\gamma}$ does not. Specifically, in a coordinate chart $(\mathcal{U}, \varphi)$, using the notation $(x^1(t), \ldots, x^n(t)) := \varphi(\gamma(t))$, the velocity $\dot{\gamma}$ simply reads $\frac{d}{dt}x^k$, which does not depend on the Christoffel symbol; on the other hand, the acceleration $\frac{D^2}{dt^2}\gamma$ reads

$$\frac{d^2}{dt^2}x^k + \sum_{i,j}\Gamma_{ij}^k(\gamma)\frac{d}{dt}x^i\frac{d}{dt}x^j,$$

where $\Gamma_{ij}^k(\gamma(t))$ are the Christoffel symbols, evaluated at the point $\gamma(t)$, of the affine connection in the chart $(\mathcal{U}, \varphi)$.

A *geodesic* γ on a manifold $\mathcal{M}$ endowed with an affine connection ∇ is a curve with zero acceleration:

$$\frac{D^2}{dt^2}\gamma(t) = 0 \qquad (5.24)$$

for all t in the domain of γ. Note that different affine connections produce different geodesics.

For every $\xi \in T_x\mathcal{M}$, there exists an interval I about 0 and a unique geodesic $\gamma(t; x, \xi) : I \to \mathcal{M}$ such that $\gamma(0) = x$ and $\dot{\gamma}(0) = \xi$. Moreover, we have the homogeneity property $\gamma(t; x, a\xi) = \gamma(at; x, \xi)$. The mapping

$$\text{Exp}_x : T_x\mathcal{M} \to \mathcal{M} : \xi \mapsto \text{Exp}_x\xi = \gamma(1; x, \xi)$$

is called the *exponential map at* x. When the domain of definition of Exp_x is the whole $T_x\mathcal{M}$ for all $x \in \mathcal{M}$, the manifold $\mathcal{M}$ (endowed with the affine connection ∇) is termed *(geodesically) complete*.

It can be shown that Exp_x defines a diffeomorphism (smooth bijection) of a neighborhood $\widehat{\mathcal{U}}$ of the origin $0_x \in T_x\mathcal{M}$ onto a neighborhood $\mathcal{U}$ of $x \in \mathcal{M}$. If, moreover, $\widehat{\mathcal{U}}$ is *star-shaped* (i.e., $\xi \in \widehat{\mathcal{U}}$ implies $t\xi \in \widehat{\mathcal{U}}$ for all $0 \leq t \leq 1$), then $\mathcal{U}$ is called a *normal neighborhood* of x.

We can further define

$$\text{Exp} : T\mathcal{M} \to \mathcal{M} : \xi \mapsto \text{Exp}_x\xi,$$

where x is the foot of ξ. The mapping Exp is differentiable, and $\text{Exp}_x 0_x = x$ for all $x \in \mathcal{M}$. Further, it can be shown that $\text{DExp}_x (0_x) [\xi] = \xi$ (with the canonical identification $T_{0_x} T_x \mathcal{M} \simeq T_x \mathcal{M}$). This yields the following result.

Proposition 5.4.1 *Let $\mathcal{M}$ be a manifold endowed with an affine connection ∇. The exponential map on $\mathcal{M}$ induced by ∇ is a retraction, termed the* exponential retraction.

The exponential mapping is an important object in differential geometry, and it has featured heavily in previously published geometric optimization algorithms on manifolds. It generalizes the concept of moving "straight" in the direction of a tangent vector and is a natural way to update an iterate given a search direction in the tangent space. However, computing the exponential is, in general, a computationally daunting task. Computing the exponential amounts to evaluating the $t = 1$ point on the curve defined by the second-order ordinary differential equation (5.24). In a coordinate chart $(\mathcal{U}, \varphi)$, (5.24) reads

$$\frac{\mathrm{d}^2}{\mathrm{d}t^2} x^k + \sum_{i,j} \Gamma_{ij}^k(\gamma) \frac{\mathrm{d}}{\mathrm{d}t} x^i \frac{\mathrm{d}}{\mathrm{d}t} x^j = 0, \quad k = 1, \dots, n,$$

where $(x^1(t), \dots, x^n(t)) := \varphi(\gamma(t))$ and Γ_{ij}^k are the Christoffel symbols of the affine connection in the chart $(\mathcal{U}, \varphi)$. In general, such a differential equation does not admit a closed-form solution, and numerically computing the geodesic involves computing an approximation to the Christoffel symbols if they are not given in closed form and then approximating the geodesic using a numerical integration scheme. The theory of general retractions is introduced to provide an alternative to the exponential in the design of numerical algorithms that retains the key properties that ensure convergence results.

Assume that a basis is given for the vector space $T_y \mathcal{M}$ and let $\mathcal{U}$ be a normal neighborhood of y. Then a chart can be defined that maps $x \in \mathcal{U}$ to the components of the vector $\xi \in T_y \mathcal{M}$ satisfying $\text{Exp}_y \xi = x$. The coordinates defined by this mapping are called *normal coordinates*.

We also point out the following fundamental result of differential geometry: if $\mathcal{M}$ is a Riemannian manifold, a curve with minimal length between two points of $\mathcal{M}$ is always a monotone reparameterization of a geodesic relative to the Riemannian connection. These curves are called *minimizing geodesics*.

Example 5.4.1 *Sphere*
  Consider the unit sphere S^{n-1} endowed with the Riemannian metric (3.33) obtained by embedding S^{n-1} in $\mathbb{R}^n$ and with the associated Riemannian connection (5.16). Geodesics $t \mapsto x(t)$ are expressed as a function of $x(0) \in S^{n-1}$ and $\dot{x}(0) \in T_{x(0)} S^{n-1}$ as follows (using the canonical inclusion of $T_{x_0} S^{n-1}$ in $\mathbb{R}^n$):

$$x(t) = x(0) \cos(\|\dot{x}(0)\| t) + \dot{x}(0) \frac{1}{\|\dot{x}(0)\|} \sin(\|\dot{x}(0)\| t). \quad (5.25)$$

(Indeed, it is readily checked that $\frac{\mathrm{D}^2}{\mathrm{d}t^2} x(t) = (I - x(t)x(t)^T) \frac{\mathrm{d}^2}{\mathrm{d}t^2} x(t) = -(I - x(t)x(t)^T) \|\dot{x}(0)\|^2 x(t) = 0.)

Example 5.4.2 *Orthogonal Stiefel manifold*

Consider the orthogonal Stiefel manifold $\mathrm{St}(p, n)$ endowed with its Riemannian metric (3.34) inherited from the embedding in $\mathbb{R}^{n \times p}$ and with the corresponding Riemannian connection ∇. Geodesics $t \mapsto X(t)$ are expressed as a function of $X(0) \in \mathrm{St}(p, n)$ and $\dot{X}(0) \in T_{X(0)} \mathrm{St}(p, n)$ as follows (using again the canonical inclusion of $T_{X(0)} \mathrm{St}(p, n)$ in $\mathbb{R}^{n \times p}$):

$$X(t) = \begin{bmatrix} X(0) & \dot{X}(0) \end{bmatrix} \exp \left(t \begin{bmatrix} A(0) & -S(0) \\ I & A(0) \end{bmatrix} \right) \begin{bmatrix} I \\ 0 \end{bmatrix} \exp(-A(0)t), \quad (5.26)$$

where $A(t) := X^T(t)\dot{X}(t)$ and $S(t) := \dot{X}^T(t)\dot{X}(t)$. It can be shown that A is an invariant of the trajectory, i.e., $A(t) = A(0)$ for all t, and that $S(t) = e^{At}S(0)e^{-At}$.

Example 5.4.3 *Grassmann manifold*

Consider the Grassmann manifold $\mathrm{Grass}(p, n)$ viewed as a Riemannian quotient manifold of $\mathbb{R}_*^{n \times p}$ with the associated Riemannian connection (5.22). Then

$$\mathcal{Y}(t) = \mathrm{span}(Y_0(Y_0^T Y_0)^{-1/2} V \cos(\Sigma t) + U \sin(\Sigma t)) \quad (5.27)$$

is the geodesic satisfying $\mathcal{Y}(0) = \mathrm{span}(Y_0)$ and $\overline{\dot{\mathcal{Y}}(0)}_{Y_0} = U \Sigma V^T$, where $U \Sigma V^T$ is a thin singular value decomposition, i.e., U is $n \times p$ orthonormal, V is $p \times p$ orthonormal, and Σ is $p \times p$ diagonal with nonnegative elements. Note that choosing Y_0 orthonormal simplifies the expression (5.27).

Let $\mathcal{M}$ be a manifold endowed with an affine connection ∇. A vector field ξ on a curve γ satisfying $\frac{\mathrm{D}}{\mathrm{d}t} \xi = 0$ is called *parallel*. Given $a \in \mathbb{R}$ in the domain of γ and $\xi_{\gamma(a)} \in T_{\gamma(a)} \mathcal{M}$, there is a unique parallel vector field ξ on γ such that $\xi(a) = \xi_{\gamma(a)}$. The operator $P_\gamma^{b \leftarrow a}$ sending $\xi(a)$ to $\xi(b)$ is called *parallel translation along* γ. In other words, we have

$$\frac{\mathrm{D}}{\mathrm{d}t} \left(P_\gamma^{t \leftarrow a} \xi(a) \right) = 0.$$

If $\mathcal{M}$ is a Riemannian manifold and ∇ is the Riemannian connection, then the parallel translation induced by ∇ is an isometry.

Much like the exponential mapping is a particular retraction, the parallel translation is a particular instance of a more general concept termed vector transport, introduced in Section 8.1. More information on vector transport by parallel translation, including formulas for parallel translation on special manifolds, can be found in Section 8.1.1. The machinery of retraction (to replace geodesic interpolation) and vector transport (to replace parallel translation) are two of the key insights in obtaining competitive numerical algorithms based on a geometric approach.

5.5 RIEMANNIAN HESSIAN OPERATOR

We conclude this chapter with a discussion of the notion of a Hessian. The Hessian matrix of a real-valued function f on $\mathbb{R}^n$ at a point $x \in \mathbb{R}^n$ is classically defined as the matrix whose (i, j) element (ith row and jth column) is given by $\partial_{ij}^2 f(x) = \frac{\partial^2}{\partial_i \partial_j} f(x)$. To formalize this concept on a manifold we need to think of the Hessian as an operator acting on geometric objects and returning geometric objects. For a real-valued function f on an abstract Euclidean space $\mathcal{E}$, the Hessian operator at x is the (linear) operator from $\mathcal{E}$ to $\mathcal{E}$ defined by

$$\operatorname{Hess} f(x)[z] := \sum_{ij} \partial_{ij}^2 \hat{f}(x^1, \dots, x^n) z^j e_i, \qquad (5.28)$$

where $(e_i)_{i=1,\dots,n}$ is an orthonormal basis of $\mathcal{E}$, $z = \sum_j z^j e_j$ and $\hat{f}$ is the function on $\mathbb{R}^n$ defined by $\hat{f}(x^1, \dots, x^n) = f(x^1 e_1 + \cdots + x^n e_n)$. It is a standard real analysis exercise to show that the definition does not depend on the choice of the orthonormal basis. Equivalently, the Hessian operator of f at x can be defined as the operator from $\mathcal{E}$ to $\mathcal{E}$ that satisfies, for all $y, z \in \mathcal{E}$,

1. $\langle \operatorname{Hess} f(x)[y], y \rangle = \mathrm{D}^2 f(x)[y, y] := \frac{\mathrm{d}^2}{\mathrm{d}t^2} f(x + ty) \big|_{t=0}$,
2. $\langle \operatorname{Hess} f(x)[y], z \rangle = \langle y, \operatorname{Hess} f(x)[z] \rangle$ (symmetry).

On an arbitrary Riemannian manifold, the Hessian operator is generalized as follows.

Definition 5.5.1 *Given a real-valued function f on a Riemannian manifold $\mathcal{M}$, the* Riemannian Hessian *of f at a point x in $\mathcal{M}$ is the linear mapping $\operatorname{Hess} f(x)$ of $T_x \mathcal{M}$ into itself defined by*

$$\operatorname{Hess} f(x)[\xi_x] = \nabla_{\xi_x} \operatorname{grad} f$$

for all ξ_x in $T_x \mathcal{M}$, where ∇ is the Riemannian connection on $\mathcal{M}$.

If $\mathcal{M}$ is a Euclidean space, this definition reduces to (5.28). (A justification for the name "Riemannian Hessian" is that the function $m_x(y) := f(x) + \langle \operatorname{grad} f(x), \operatorname{Exp}_x^{-1}(y) \rangle_x + \frac{1}{2} \langle \operatorname{Hess} f(x)[\operatorname{Exp}_x^{-1}(y)], \operatorname{Exp}_x^{-1}(y) \rangle$ is a second-order model of f around x; see Section 7.1.)

Proposition 5.5.2 *The Riemannian Hessian satisfies the formula*

$$\langle \operatorname{Hess} f[\xi], \eta \rangle = \xi(\eta f) - (\nabla_\xi \eta)f \qquad (5.29)$$

for all $\xi, \eta \in \mathfrak{X}(\mathcal{M})$.

Proof. We have $\langle \operatorname{Hess} f[\xi], \eta \rangle = \langle \nabla_\xi \operatorname{grad} f, \eta \rangle$. Since the Riemannian connection leaves the Riemannian metric invariant, this is equal to $\xi \langle \operatorname{grad} f, \eta \rangle - \langle \operatorname{grad} f, \nabla_\xi \eta \rangle$. By definition of the gradient, this yields $\xi(\eta f) - (\nabla_\xi \eta)f$. $\square$

Proposition 5.5.3 *The Riemannian Hessian is symmetric (in the sense of the Riemannian metric). That is,*

$$\langle \operatorname{Hess} f[\xi], \eta \rangle = \langle \xi, \operatorname{Hess} f[\eta] \rangle$$

for all $\xi, \eta \in \mathfrak{X}(\mathcal{M})$.

Proof. By the previous proposition, the left-hand side is equal to $\xi(\eta f) - (\nabla_\xi \eta)f$ and the right-hand side is equal to $\langle \text{Hess}\, f(x)[\eta], \xi \rangle = \eta(\xi f) - (\nabla_\eta \xi)f$. Using the symmetry property (5.10) of the Riemannian connection on the latter expression, we obtain $\eta(\xi f) - (\nabla_\eta \xi)f = \eta(\xi f) - [\eta, \xi]f - (\nabla_\xi \eta)f = \xi(\eta f) - (\nabla_\xi \eta)f$, and the result is proved. □

The following result shows that the Riemannian Hessian of a function f at a point x coincides with the Euclidean Hessian of the function $f \circ \text{Exp}_x$ at the origin $0_x \in T_x\mathcal{M}$. Note that $f \circ \text{Exp}_x$ is a real-valued function on the Euclidean space $T_x\mathcal{M}$.

Proposition 5.5.4 *Let $\mathcal{M}$ be a Riemannian manifold and let f be a real-valued function on $\mathcal{M}$. Then*

$$\text{Hess}\, f(x) = \text{Hess}\,(f \circ \text{Exp}_x)(0_x) \qquad (5.30)$$

for all $x \in \mathcal{M}$, where $\text{Hess}\, f(x)$ denotes the Riemannian Hessian of f : $\mathcal{M} \to \mathbb{R}$ at x and $\text{Hess}\,(f \circ \text{Exp}_x)(0_x)$ denotes the Euclidean Hessian of $f \circ \text{Exp}_x : T_x\mathcal{M} \to \mathbb{R}$ at the origin of $T_x\mathcal{M}$ endowed with the inner product defined by the Riemannian structure on $\mathcal{M}$.

Proof. This result can be proven by working in normal coordinates and invoking the fact that the Christoffel symbols vanish in these coordinates. We provide an alternative proof that does not make use of index notation. We have to show that

$$\langle \text{Hess}\, f(x)[\xi], \eta \rangle = \langle \text{Hess}\,(f \circ \text{Exp}_x)(0_x)[\xi], \eta \rangle \qquad (5.31)$$

for all $\xi, \eta \in T_x\mathcal{M}$. Since both sides of (5.31) are symmetric bilinear forms in ξ and η, it is sufficient to show that

$$\langle \text{Hess}\, f(x)[\xi], \xi \rangle = \langle \text{Hess}\,(f \circ \text{Exp}_x)(0_x)[\xi], \xi \rangle \qquad (5.32)$$

for all $\xi \in T_x\mathcal{M}$. Indeed, for any symmetric linear form B, we have the *polarization identity*

$$2B(\xi, \eta) = B(\xi + \eta, \xi + \eta) - B(\xi, \xi) - B(\eta, \eta),$$

which shows that the mapping $(\xi, \eta) \mapsto B(\xi, \eta)$ is fully specified by the mapping $\xi \mapsto B(\xi, \xi)$. Since the right-hand side of (5.32) involves a classical (Euclidean) Hessian, we have

$$\langle \text{Hess}\,(f \circ \text{Exp}_x)(0_x)[\xi], \xi \rangle = \frac{\mathrm{d}^2}{\mathrm{d}t^2}(f \circ \text{Exp}_x)(t\xi)\Big|_{t=0}$$

$$= \frac{\mathrm{d}}{\mathrm{d}t}\left(\frac{\mathrm{d}}{\mathrm{d}t} f(\text{Exp}_x(t\xi))\right)\Big|_{t=0} = \frac{\mathrm{d}}{\mathrm{d}t}\left(\mathrm{D}f(\text{Exp}_x t\xi)\left[\frac{\mathrm{d}}{\mathrm{d}t}\text{Exp}_x t\xi\right]\right)\Big|_{t=0}.$$

It follows from the definition of the gradient that this last expression is equal to $\frac{\mathrm{d}}{\mathrm{d}t}\langle \text{grad}\, f(\text{Exp}_x t\xi), \frac{\mathrm{d}}{\mathrm{d}t}\text{Exp}_x t\xi \rangle\big|_{t=0}$. By the invariance property of the metric, this is equal to $\langle \frac{\mathrm{D}}{\mathrm{d}t}\text{grad}\, f(\text{Exp}_x t\xi), \xi \rangle + \langle \text{grad}\, f(x), \frac{\mathrm{D}^2}{\mathrm{d}t^2}\text{Exp}_x t\xi \rangle$. By definition of the exponential mapping, we have $\frac{\mathrm{D}^2}{\mathrm{d}t^2}\text{Exp}_x t\xi = 0$ and $\frac{\mathrm{d}}{\mathrm{d}t}\text{Exp}_x t\xi\big|_{t=0} = \xi$. Hence the right-hand side of (5.32) reduces to

$$\langle \nabla_\xi \text{grad}\, f, \xi \rangle,$$

and the proof is complete. □

The result is in fact more general. It holds whenever the retraction and the Riemannian exponential agree to the second order along all rays. This result will not be used in the convergence analyses, but it may be useful to know that various retractions yield the same Hessian operator.

Proposition 5.5.5 *Let R be a retraction and suppose in addition that*

$$\frac{D^2}{dt^2} R(t\xi)\Big|_{t=0} = 0 \quad \text{for all } \xi \in T_x\mathcal{M}, \tag{5.33}$$

where $\frac{D^2}{dt^2}\gamma$ denotes acceleration of the curve γ as defined in (5.23). Then

$$\text{Hess } f(x) = \text{Hess}\,(f \circ R_x)(0_x). \tag{5.34}$$

Proof. The proof follows the proof of Proposition 5.5.4, replacing Exp_x by R_x throughout. The first-order ridigity condition of the retraction implies that $\frac{d}{dt} R_x t\xi|_{t=0} = \xi$. Because of this and of (5.33), we conclude as in the proof of Proposition 5.5.4. □

Proposition 5.5.5 provides a way to compute the Riemannian Hessian as the Hessian of a real-valued function $f \circ R_x$ defined on the Euclidean space $T_x\mathcal{M}$. In particular, this yields a way to compute $\langle \text{Hess } f(x)[\xi], \eta \rangle$ by taking second derivatives along curves, as follows. Let R be any retraction satisfying the acceleration condition (5.33). First, observe that, for all $\xi \in T_x\mathcal{M}$,

$$\langle \text{Hess } f(x)[\xi], \xi \rangle = \langle \text{Hess}\,(f \circ R_x)(0_x)[\xi], \xi \rangle = \frac{d^2}{dt^2} f(R_x(t\xi))\Big|_{t=0}. \tag{5.35}$$

Second, in view of the symmetry of the linear operator $\text{Hess } f(x)$, we have the polarization identity

$$\langle \text{Hess } f(x)[\xi], \eta \rangle = \tfrac{1}{2}(\langle \text{Hess } f(x)[\xi + \eta], \xi + \eta \rangle$$
$$- \langle \text{Hess } f(x)[\xi], \xi \rangle - \langle \text{Hess } f(x)[\eta], \eta \rangle). \tag{5.36}$$

Equations (5.35) and (5.36) yield the identity

$$\langle \text{Hess } f(x)[\xi], \eta \rangle$$
$$= \frac{1}{2} \frac{d^2}{dt^2} \left(f(R_x(t(\xi + \eta))) - f(R_x(t\xi)) - f(R_x(t\eta)) \right)\Big|_{t=0}, \tag{5.37}$$

valid for any retraction R that satisfies the zero initial acceleration condition (5.33). This holds in particular for $R = \text{Exp}$, the exponential retraction.

Retractions that satisfy the zero initial acceleration condition (5.33) will be called *second-order retractions*. For general retractions the equality of the Hessians stated in (5.34) does not hold. Nevertheless, none of our quadratic convergence results will require the retraction to be second order. The fundamental reason can be traced in the following property.

Proposition 5.5.6 *Let R be a retraction and let v be a critical point of a real-valued function f (i.e., $\text{grad } f(v) = 0$). Then*

$$\text{Hess } f(v) = \text{Hess}(f \circ R_v)(0_v).$$

Proof. We show that $\langle \operatorname{Hess} f(v)[\xi_v], \eta_v \rangle = \langle \operatorname{Hess}(f \circ R_v)(0_v)[\xi_v], \eta_v \rangle$ for all $\xi, \eta \in \mathfrak{X}(\mathcal{M})$. From Proposition 5.5.2, we have $\langle \operatorname{Hess} f(v)[\xi_v], \eta_v \rangle = \xi_v(\eta f) - (\nabla_{\xi_v} \eta) f$. The second term is an element of $T_v \mathcal{M}$ applied to f; since v is a critical point of f, this term vanishes, and we are left with $\langle \operatorname{Hess} f(v)[\xi_v], \eta_v \rangle = \xi_v(\eta f)$. Fix a basis $(e_1, \ldots, e_n)$ of $T_x \mathcal{M}$ and consider the coordinate chart φ defined by $\varphi^{-1}(y^1, \ldots, y^n) = R_v(y^1 e_1 + \cdots + y^n e_n)$. Let η^i and ξ^i denote the coordinates of η and ξ in this chart. Since v is a critical point of f, $\partial_i(f \circ \varphi^{-1})$ vanishes at 0, and we obtain $\xi_v(\eta f) = \sum_i \xi_v^i \partial_i(\sum_j \eta^j \partial_j(f \circ \varphi^{-1})) = \sum_{i,j} \xi_v^i \eta_v^j \, \partial_i \partial_j(f \circ \varphi^{-1})$. Since $\mathrm{D}R_v(0_v)$ is the identity, it follows that ξ_v^i and η_v^j are the components of ξ_v and η_v in the basis $(e_1, \ldots, e_n)$; thus the latter expression is equal to $\langle \operatorname{Hess}(f \circ R_v)(0_v)[\xi], \eta \rangle$. $\square$

5.6 SECOND COVARIANT DERIVATIVE*

In the previous section, we assumed that the manifold $\mathcal{M}$ was Riemannian. This assumption made it possible to replace the differential $\mathrm{D}f(x)$ of a function f at a point x by the tangent vector $\operatorname{grad} f(x)$, satisfying

$$\langle \operatorname{grad} f(x), \xi \rangle = \mathrm{D}f(x)[\xi] \quad \text{for all } \xi \in T_x \mathcal{M}.$$

This led to the definition of $\operatorname{Hess} f(x) : \xi_x \mapsto \nabla_{\xi_x} \operatorname{grad} f$ as a linear operator of $T_x \mathcal{M}$ into itself. This formulation has several advantages: eigenvalues and eigenvectors of the Hessian are well defined and, as we will see in Chapter 6, the definition leads to a streamlined formulation (6.4) for the Newton equation. However, on an arbitrary manifold equipped with an affine connection, it is equally possible to define a Hessian as a second covariant derivative that applies bilinearly to two tangent vectors and returns a scalar. This second covariant derivative is often called "Hessian" in the literature, but we will reserve this term for the operator $\xi_x \mapsto \nabla_{\xi_x} \operatorname{grad} f$.

To develop the theory of second covariant derivative, we will require the concept of a covector. Let $T_x^* \mathcal{M}$ denote the dual space of $T_x M$, i.e., the set of linear functionals (linear maps) $\mu_x : T_x M \to \mathbb{R}$. The set $T_x^* \mathcal{M}$ is termed the *cotangent space* of $\mathcal{M}$ at x, and its elements are called *covectors*. The bundle of cotangent spaces

$$T^* \mathcal{M} = \cup_{x \in \mathcal{M}} T_x^* \mathcal{M}$$

is termed the *cotangent bundle*. The cotangent bundle can be given the structure of a manifold in an analogous manner to the structure of the tangent bundle. A smooth section of the cotangent bundle is a smooth assignment $x \mapsto \mu_x \in T_x^* \mathcal{M}$. A smooth section of the cotangent bundle is termed a *covector field* or a *one-form* on $\mathcal{M}$. The name comes from the fact that a one-form field μ acts on "one" vector field $\xi \in \mathfrak{X}(\mathcal{M})$ to generate a scalar field on a manifold,

$$\mu[\xi] \in \mathfrak{F}(\mathcal{M}),$$

defined by $(\mu[\xi])|_x = \mu_x[\xi_x]$. The action of a covector field μ on a vector field ξ is often written simply as a concatenation of the two objects, $\mu\xi$. A covector is a $(0,1)$-tensor. The most common covector encountered is the differential of a smooth function $f : \mathcal{M} \to \mathbb{R}$:

$$\mu_x = \mathrm{D}f(x).$$

Note that the covector field $\mathrm{D}f$ is zero exactly at critical points of the function f. Thus, another way of solving for the critical points of f is to search for zeros of $\mathrm{D}f$.

Given a manifold $\mathcal{M}$ with an affine connection ∇, a real-valued function f on $\mathcal{M}$, a point $x \in \mathcal{M}$, and a tangent vector $\xi_x \in T_x\mathcal{M}$, the covariant derivative of the covector field $\mathrm{D}f$ along ξ_x is a covector $\nabla_{\xi_x}(\mathrm{D}f)$ defined by imposing the property

$$\mathrm{D}(\mathrm{D}f[\eta])(x)[\xi_x] = (\nabla_{\xi_x}(\mathrm{D}f))[\eta_x] + \mathrm{D}f(x)[\nabla_{\xi_x}\eta]$$

for all $\eta \in \mathfrak{X}(\mathcal{M})$. It is readily checked, using coordinate expressions, that $(\nabla_{\xi_x}(\mathrm{D}f))[\eta_x]$ defined in this manner depends only on η through η_x and that $(\nabla_{\xi_x}(\mathrm{D}f))[\eta_x]$ is a linear expression of ξ_x and η_x. The *second covariant derivative* of the real-valued function f is defined by

$$\nabla^2 f(x)[\xi_x, \eta_x] = (\nabla_{\xi_x}(\mathrm{D}f))[\eta_x].$$

(There is no risk of confusing $[\xi_x, \eta_x]$ with a Lie bracket since $\nabla^2 f(x)$ is known to apply to two vector arguments.) The notation ∇^2 rather than D^2 is used to emphasize that the second covariant derivative depends on the choice of the affine connection ∇.

With development analogous to that in the preceding section, one may show that

$$\nabla^2 f(x)[\xi_x, \eta_x] = \xi_x(\eta f) - (\nabla_{\xi_x}\eta)f.$$

The second covariant derivative is symmetric if and only if ∇ is symmetric. For any second-order retraction R, we have

$$\nabla^2 f(x) = \mathrm{D}^2 (f \circ R_x)(0_x),$$

where $\mathrm{D}^2 (f \circ R_x)(0_x)$ is the classical second-order derivative of $f \circ R_x$ at 0_x (see Section A.5). In particular,

$$\nabla^2 f(x)[\xi_x, \xi_x] = \mathrm{D}^2 (f \circ \mathrm{Exp}_x)(0_x)[\xi_x, \xi_x] = \frac{\mathrm{d}^2}{\mathrm{d}t^2} f(\mathrm{Exp}_x(t\xi))|_{t=0}.$$

When x is a critical point of f, we have

$$\nabla^2 f(x) = \mathrm{D}^2 (f \circ R_x)(0_x)$$

for *any* retraction R.

When $\mathcal{M}$ is a Riemannian manifold and ∇ is the Riemannian connection, we have

$$\nabla^2 f(x)[\xi_x, \eta_x] = \langle \mathrm{Hess}\, f(x)[\xi_x], \eta_x \rangle.$$

When F is a function on $\mathcal{M}$ into a vector space $\mathcal{E}$, it is still possible to uniquely define

$$\nabla^2 F(x)[\xi_x, \eta_x] = \sum_{i=1}^{n} (\nabla^2 F^i(x)[\xi_x, \eta_x])e_i,$$

where $(e_1, \ldots, e_n)$ is a basis of $\mathcal{E}$.

5.7 NOTES AND REFERENCES

Our main sources for this chapter are O'Neill [O'N83] and Brickell and Clark [BC70].

A proof of superlinear convergence for Newton's method in $\mathbb{R}^n$ can be found in [DS83, Th. 5.2.1]. A proof of Proposition 5.2.1 (the existence of affine connections) is given in [BC70, Prop. 9.1.4]. It relies on partitions of unity (see [BC70, Prop. 3.4.4] or [dC92, Th. 0.5.6] for details). For a proof of the existence and uniqueness of the covariant derivative along curves $\frac{D}{dt}$, we refer the reader to [O'N83, Prop. 3.18] for the Riemannian case and Helgason [Hel78, §I.5] for the general case. More details on the exponential can be found in do Carmo [dC92] for the Riemannian case, and in Helgason [Hel78] for the general case. For a proof of the minimizing property of geodesics, see [O'N83, §5.19]. The material about the Riemannian connection on Riemannian submanifolds comes from O'Neill [O'N83]. For more details on Riemannian submersions and the associated Riemannian connections, see O'Neill [O'N83, Lemma 7.45], Klingenberg [Kli82], or Cheeger and Ebin [CE75].

The equation (5.26) for the Stiefel geodesic is due to R. Lippert; see Edelman et al. [EAS98]. The formula (5.27) for the geodesics on the Grassmann manifold can be found in Absil et al. [AMS04]. Instead of considering the Stiefel manifold as a Riemannian submanifold of $\mathbb{R}^{n \times p}$, it is also possible to view the Stiefel manifold as a certain Riemannian quotient manifold of the orthogonal group. This quotient approach yields a different Riemannian metric on the Stiefel manifold, called the *canonical metric* in [EAS98]. The Riemannian connection, geodesics, and parallel translation associated with the canonical metric are different from those associated with the Riemannian metric (3.34) inherited from the embedding of $\mathrm{St}(p, n)$ in $\mathbb{R}^{n \times p}$. We refer the reader to Edelman et al. [EAS98] for more information on the geodesics and parallel translations on the Stiefel manifold.

The geometric Hessian is not a standard topic in differential geometry. Some results can be found in [O'N83, dC92, Sak96, Lan99]. The Hessian is often defined as a tensor of type $(0, 2)$—it applies to two vectors and returns a scalar—using formula (5.29). This does not require a Riemannian metric. Such a tensor varies under changes of coordinates via a congruence transformation. In this book, as in do Carmo [dC92], we define the Hessian as a tensor of type $(1, 1)$, which can thus be viewed as a linear transformation of the tangent space. It transforms via a similarity transformation, therefore its eigenvalues are well defined (they do not depend on the chart).

Chapter Six

Newton's Method

This chapter provides a detailed development of the archetypal second-order optimization method, Newton's method, as an iteration on manifolds. We propose a formulation of Newton's method for computing the zeros of a vector field on a manifold equipped with an affine connection and a retraction. In particular, when the manifold is Riemannian, this geometric Newton method can be used to compute critical points of a cost function by seeking the zeros of its gradient vector field. In the case where the underlying space is Euclidean, the proposed algorithm reduces to the classical Newton method. Although the algorithm formulation is provided in a general framework, the applications of interest in this book are those that have a matrix manifold structure (see Chapter 3). We provide several example applications of the geometric Newton method for principal subspace problems.

6.1 NEWTON'S METHOD ON MANIFOLDS

In Chapter 5 we began a discussion of the Newton method and the issues involved in generalizing such an algorithm on an arbitrary manifold. Section 5.1 identified the task as computing a zero of a vector field ξ on a Riemannian manifold $\mathcal{M}$ equipped with a retraction R. The strategy proposed was to obtain a new iterate x_{k+1} from a current iterate x_k by the following process.

1. Find a tangent vector $\eta_k \in T_{x_k}\mathcal{M}$ such that the "directional derivative" of ξ along η_k is equal to $-\xi$.
2. Retract η_k to obtain x_{k+1}.

In Section 5.1 we were unable to progress further without providing a generalized definition of the directional derivative of ξ along η_k. The notion of an affine connection, developed in Section 5.2, is now available to play such a role, and we have all the tools necessary to propose Algorithm 4, a geometric Newton method on a general manifold equipped with an affine connection and a retraction.

By analogy with the classical case, the operator

$$J(x) : T_x\mathcal{M} \to T_x\mathcal{M} : \eta \mapsto \nabla_\eta \xi$$

involved in (6.1) is called the *Jacobian of ξ at x*. Equation (6.1) is called the *Newton equation*, and its solution $\eta_k \in T_{x_k}\mathcal{M}$ is called the *Newton vector*.

Algorithm 4 Geometric Newton method for vector fields

Require: Manifold $\mathcal{M}$; retraction R on $\mathcal{M}$; affine connection ∇ on $\mathcal{M}$; vector field ξ on $\mathcal{M}$.

Goal: Find a zero of ξ, i.e., $x \in \mathcal{M}$ such that $\xi_x = 0$.

Input: Initial iterate $x_0 \in \mathcal{M}$.

Output: Sequence of iterates $\{x_k\}$.

1: **for** $k = 0, 1, 2, \ldots$ **do**
2:     Solve the Newton equation

$$J(x_k)\eta_k = -\xi_{x_k} \tag{6.1}$$

    for the unknown $\eta_k \in T_{x_k}\mathcal{M}$, where $J(x_k)\eta_k := \nabla_{\eta_k}\xi$.
3:     Set

$$x_{k+1} := R_{x_k}(\eta_k).$$

4: **end for**

In Algorithm 4, the choice of the retraction R and the affine connection ∇ is not prescribed. This freedom is justified by the fact that superlinear convergence holds for every retraction R and every affine connection ∇ (see forthcoming Theorem 6.3.2). Nevertheless, if $\mathcal{M}$ is a Riemannian manifold, there is a natural connection—the Riemannian connection—and a natural retraction—the exponential mapping. From a computational viewpoint, choosing ∇ as the Riemannian connection is generally a good choice, notably because it admits simple formulas on Riemannian submanifolds and on Riemannian quotient manifolds (Sections 5.3.3 and 5.3.4). In contrast, instead of choosing R as the exponential mapping, it is usually desirable to consider alternative retractions that are computationally more efficient; examples are given in Section 4.1.

When $\mathcal{M}$ is a Riemannian manifold, it is often advantageous to wrap Algorithm 4 in a line-search strategy using the framework of Algorithm 1. At the current iterate x_k, the search direction η_k is computed as the solution of the Newton equation (6.1), and x_{k+1} is computed to satisfy the descent condition (4.12) in which the cost function f is defined as

$$f := \langle \xi, \xi \rangle.$$

Note that the global minimizers of f are the zeros of the vector field ξ. Moreover, if ∇ is the Riemannian connection, then, in view of the compatibility with the Riemannian metric (Theorem 5.3.1.ii), we have

$$\mathrm{D}\langle \xi, \xi \rangle \, (x_k) \, [\eta_k] = \langle \nabla_{\eta_k}\xi, \xi \rangle + \langle \xi, \nabla_{\eta_k}\xi \rangle = -2\langle \xi, \xi \rangle_{x_k} < 0$$

whenever $\xi \neq 0$. It follows that the Newton vector η_k is a descent direction for f, although $\{\eta_k\}$ is not necessarily gradient-related. This perspective provides another motivation for choosing ∇ in Algorithm 4 as the Riemannian connection.

Note that an analytical expression of the Jacobian $J(x)$ in the Newton equation (6.1) may not be available. The Jacobian may also be singular or ill-conditioned, in which case the Newton equation cannot be reliably solved for η_k. Remedies to these difficulties are provided by the quasi-Newton approaches presented in Section 8.2.

6.2 RIEMANNIAN NEWTON METHOD FOR REAL-VALUED FUNCTIONS

We now discuss the case $\xi = \operatorname{grad} f$, where f is a cost function on a Riemannian manifold $\mathcal{M}$. The Newton equation (6.1) becomes

$$\operatorname{Hess} f(x_k)\eta_k = -\operatorname{grad} f(x_k), \qquad (6.2)$$

where

$$\operatorname{Hess} f(x) : T_x\mathcal{M} \to T_x\mathcal{M} : \eta \mapsto \nabla_\eta \operatorname{grad} f \qquad (6.3)$$

is the Hessian of f at x for the affine connection ∇. We formalize the method in Algorithm 5 for later reference. Note that Algorithm 5 is a particular case of Algorithm 4.

Algorithm 5 Riemannian Newton method for real-valued functions

---

Require: Riemannian manifold $\mathcal{M}$; retraction R on $\mathcal{M}$; affine connection ∇ on $\mathcal{M}$; real-valued function f on $\mathcal{M}$.
Goal: Find a critical point of f, i.e., $x \in \mathcal{M}$ such that $\operatorname{grad} f(x) = 0$.
Input: Initial iterate $x_0 \in \mathcal{M}$.
Output: Sequence of iterates $\{x_k\}$.
  1: **for** $k = 0, 1, 2, \ldots$ **do**
  2:    Solve the Newton equation

$$\operatorname{Hess} f(x_k)\eta_k = -\operatorname{grad} f(x_k) \qquad (6.4)$$

   for the unknown $\eta_k \in T_{x_k}\mathcal{M}$, where $\operatorname{Hess} f(x_k)\eta_k := \nabla_{\eta_k} \operatorname{grad} f$.
  3:    Set

$$x_{k+1} := R_{x_k}(\eta_k).$$

  4: **end for**

In general, the Newton vector η_k, solution of (6.2), is not necessarily a descent direction of f. Indeed, we have

$$\mathrm{D} f(x_k)[\eta_k] = \langle \operatorname{grad} f(x_k), \eta_k \rangle = -\langle \operatorname{grad} f(x_k), (\operatorname{Hess} f(x_k))^{-1} \operatorname{grad} f(x_k) \rangle, \qquad (6.5)$$

which is not guaranteed to be negative without additional assumptions on the operator $\operatorname{Hess} f(x_k)$. A sufficient condition for η_k to be a descent direction is that $\operatorname{Hess} f(x_k)$ be *positive-definite* (i.e., $\langle \xi, \operatorname{Hess} f(x_k)[\xi] \rangle > 0$ for all $\xi \neq 0_{x_k}$). When ∇ is a symmetric affine connection (such as the Riemannian connection), $\operatorname{Hess} f(x_k)$ is positive-definite if and only if all its eigenvalues are strictly positive.

In order to obtain practical convergence results, quasi-Newton methods have been proposed that select an update vector η_k as the solution of

$$(\text{Hess } f(x_k) + E_k)\eta_k = -\text{grad } f(x_k), \qquad (6.6)$$

where the operator E_k is chosen so as to make the operator $(\text{Hess } f(x_k) + E_k)$ positive-definite. For a suitable choice of operator E_k this guarantees that the sequence $\{\eta_k\}$ is gradient-related, thereby fulfilling the main hypothesis in the global convergence result of Algorithm 1 (Theorem 4.3.1). Care should be taken that the operator E_k does not destroy the superlinear convergence properties of the pure Newton iteration when the desired (local minimum) critical point is reached.

6.3 LOCAL CONVERGENCE

In this section, we study the local convergence of the plain geometric Newton method as defined in Algorithm 4. Well-conceived globally convergent modifications of Algorithm 4 should not affect its superlinear local convergence.

The convergence result below (Theorem 6.3.2) shows quadratic convergence of Algorithm 4. Recall from Section 4.3 that the notion of quadratic convergence on a manifold $\mathcal{M}$ does not require any further structure on $\mathcal{M}$, such as a Riemannian structure. Accordingly, we make no such assumption. Note also that since Algorithm 5 is a particular case of Algorithm 4, the convergence analysis of Algorithm 4 applies to Algorithm 5 as well.

We first need the following lemma.

Lemma 6.3.1 *Let* $\| \cdot \|$ *be any consistent norm on* $\mathbb{R}^{n \times n}$ *such that* $\|I\| = 1$. *If* $\|E\| < 1$, *then* $(I - E)^{-1}$ *exists and*

$$\|(I - E)^{-1}\| \leq \frac{1}{1 - \|E\|}.$$

If A *is nonsingular and* $\|A^{-1}(B - A)\| < 1$, *then* B *is nonsingular and*

$$\|B^{-1}\| \leq \frac{\|A^{-1}\|}{1 - \|A^{-1}(B - A)\|}.$$

Theorem 6.3.2 (local convergence of Newton's method) *Under the requirements and notation of Algorithm 4, assume that there exists* $x_* \in \mathcal{M}$ *such that* $\xi_{x_*} = 0$ *and* $J(x_*)^{-1}$ *exists. Then there exists a neighborhood* $\mathcal{U}$ *of* x_* *in* $\mathcal{M}$ *such that, for all* $x_0 \in \mathcal{U}$, *Algorithm 4 generates an infinite sequence* $\{x_k\}$ *converging superlinearly (at least quadratically) to* x_*.

Proof. Let $(\mathcal{U}, \varphi)$, $x_* \in \mathcal{U}$, be a coordinate chart. According to Section 4.3, it is sufficient to show that the sequence $\{\varphi(x_k)\}$ in $\mathbb{R}^d$ converges quadratically to $\varphi(x_*)$. To simplify the notation, coordinate expressions are denoted by hat quantities. In particular, $\hat{x}_k = \varphi(x_k)$, $\hat{\xi}_{\hat{x}_k} = \mathrm{D}\varphi(x_k)[\xi]$, $\hat{J}(\hat{x}_k) = (\mathrm{D}\varphi(x_k)) \circ J(x_k) \circ (\mathrm{D}\varphi(x_k))^{-1}$, $\hat{R}_{\hat{x}}\hat{\zeta} = \varphi(R_x\zeta)$. Note that $\hat{J}(\hat{x}_k)$ is a linear operator from $\mathbb{R}^d$ to $\mathbb{R}^d$, i.e., a $d \times d$ matrix. Note also that $\hat{R}_{\hat{x}}$ is a function from $\mathbb{R}^d$ to $\mathbb{R}^d$ whose differential at zero is the identity.

The iteration defined by Algorithm 4 reads

$$\hat{x}_{k+1} = \hat{R}_{\hat{x}_k}(-\hat{J}(\hat{x}_k)^{-1}\hat{\xi}_{\hat{x}_k}), \tag{6.7}$$

whereas the classical Newton method applied to the function $\hat{\xi} : \mathbb{R}^d \mapsto \mathbb{R}^d$ would yield

$$\hat{x}_{k+1} = \hat{x}_k + (-D\hat{\xi}(\hat{x}_k)^{-1}\hat{\xi}_{\hat{x}_k}). \tag{6.8}$$

The strategy in this proof is to show that (6.7) is sufficiently close to (6.8) that the superlinear convergence result of the classical Newton method is preserved. We are going to prove more than is strictly needed, in order to obtain information about the multiplicative constant in the quadratic convergence result.

Let $\beta := \|\hat{J}(\hat{x}_*)^{-1}\|$. Since ξ is a (smooth) vector field, it follows that $\hat{J}$ is smooth, too, and therefore there exists $r_J > 0$ and $\gamma_J > 0$ such that

$$\|\hat{J}(\hat{x}) - \hat{J}(\hat{y})\| \le \gamma_J\|\hat{x} - \hat{y}\|$$

for all $\hat{x}, \hat{y} \in B_{r_J}(\hat{x}_*) := \{\hat{x} \in \mathbb{R}^d : \|\hat{x} - \hat{x}_*\| < r_J\}$. Let

$$\epsilon = \min\left\{r_J, \frac{1}{2\beta\gamma_J}\right\}.$$

Assume that $\hat{x}_k \in B_\epsilon(\hat{x}_*)$. It follows that

$$\|\hat{J}(\hat{x}_*)^{-1}(\hat{J}(\hat{x}_k) - \hat{J}(\hat{x}_*))\| \le \|\hat{J}(\hat{x}_*)\|^{-1}\|\hat{J}(\hat{x}_k) - \hat{J}(\hat{x}_*)\|$$

$$\le \beta\gamma_J\|\hat{x}_k - \hat{x}_*\| \le \beta\gamma_J\epsilon \le \frac{1}{2}.$$

It follows from Lemma 6.3.1 that $\hat{J}(\hat{x}_k)$ is nonsingular and that

$$\|\hat{J}(\hat{x}_k)^{-1}\| \le \frac{\|\hat{J}(\hat{x}_*)^{-1}\|}{1 - \|\hat{J}(\hat{x}_*)^{-1}(\hat{J}(\hat{x}_k) - \hat{J}(\hat{x}_*))\|} \le 2\|\hat{J}(\hat{x}_*)^{-1}\| \le 2\beta.$$

It also follows that for all $\hat{x}_k \in B_\epsilon(\hat{x}_*)$, the Newton vector $\hat{\eta}_k := \hat{J}(\hat{x}_k)^{-1}\hat{\xi}_{\hat{x}_k}$ is well defined. Since R is a retraction (thus a smooth mapping) and $\hat{x}_*$ is a zero of $\hat{\xi}$, it follows that there exists r_R and $\gamma_R > 0$ such that

$$\|\hat{R}_{\hat{x}_k}\hat{\eta}_k - (\hat{x}_k + \hat{\eta}_k)\| \le \gamma_R\|\hat{x}_k - \hat{x}_*\|^2$$

for all $\hat{x}_k \in B_\epsilon(\hat{x}_*)$. (Indeed, since $\|\hat{J}(\hat{x}_k)^{-1}\|$ is bounded on $B_\epsilon(\hat{x}_*)$, and $\hat{\xi}$ is smooth and $\hat{\xi}_{\hat{x}_*} = 0$, we have a bound $\|\hat{\eta}_k\| \le c\|\hat{x}_k - \hat{x}_*\|$ for all x_k in a neighborhood of x_*; and in view of the local rigidity property of R, we have $\|\hat{R}_{\hat{x}_k}\hat{\eta}_k - (\hat{x}_k + \hat{\eta}_k)\| \le c\|\hat{\eta}_k\|^2$ for all x_k in a neighborhood of x_* and all η_k sufficiently small.)

Define $\hat{\Gamma}_{\hat{x},\hat{\xi}}$ by $\hat{\Gamma}_{\hat{x},\hat{\xi}}\zeta := \hat{J}(\hat{x})\zeta - D\hat{\xi}(\hat{x})\left[\zeta\right]$; see (5.7). Note that $\hat{\Gamma}_{\hat{x},\hat{\xi}}$ is a linear operator. Again by a smoothness argument, it follows that there exists r_Γ and γ_Γ such that

$$\|\hat{\Gamma}_{\hat{x},\hat{\xi}} - \hat{\Gamma}_{\hat{y},\hat{\xi}}\| \le \gamma_\Gamma\|\hat{x} - \hat{y}\|$$

for all $\hat{x}, \hat{y} \in B_{r_\Gamma}(\hat{x}_*)$. In particular, since $\hat{\xi}_{\hat{x}_*} = 0$, it follows from the uniqueness of the connection at critical points that $\hat{\Gamma}_{\hat{x}_*, \hat{\xi}} = 0$, hence

$$\|\hat{\Gamma}_{\hat{x}, \hat{\xi}}\| \le \gamma_\Gamma \|\hat{x}_k - \hat{x}_*\|$$

for all $\hat{x}_k \in B_\epsilon(\hat{x}_*)$.

We need a Lipschitz constant for $\mathrm{D}\hat{\xi}$. For all $\hat{x}, \hat{y} \in B_{\min\{r_J, r_\Gamma\}}(\hat{x}_*)$, we have

$$\|\mathrm{D}\hat{\xi}(\hat{x}) - \mathrm{D}\hat{\xi}(\hat{y})\| - \|\hat{\Gamma}_{\hat{x}, \hat{\xi}} - \hat{\Gamma}_{\hat{y}, \hat{\xi}}\|$$
$$\le \|\mathrm{D}\hat{\xi}(\hat{x}) + \hat{\Gamma}_{\hat{x}, \hat{\xi}} - \left(\mathrm{D}\hat{\xi}(\hat{y}) + \hat{\Gamma}_{\hat{y}, \hat{\xi}}\right)\| = \|\hat{J}(\hat{x}) - \hat{J}(\hat{y})\| \le \gamma_J \|\hat{x} - \hat{y}\|,$$

hence

$$\|\mathrm{D}\hat{\xi}(\hat{x}) - \mathrm{D}\hat{\xi}(\hat{y})\| \le (\gamma_J + \gamma_\Gamma)\|\hat{x} - \hat{y}\|.$$

From (6.7) we have

$$\hat{x}_{k+1} - \hat{x}_* = \hat{R}_{\hat{x}_k}(-\hat{J}(\hat{x}_k)^{-1}\hat{\xi}_{\hat{x}_k}) - \hat{x}_*.$$

Applying the bounds developed above, one obtains

$$\|\hat{x}_{k+1} - \hat{x}_*\| \le \|\hat{x}_k - \hat{J}(\hat{x}_k)^{-1}\hat{\xi}_{\hat{x}_k} - \hat{x}_*\| + \gamma_R\|\hat{x}_k - \hat{x}_*\|^2$$
$$\le \|\hat{J}(\hat{x}_k)^{-1}\left(\hat{\xi}_{\hat{x}_*} - \hat{\xi}_{\hat{x}_k} - \hat{J}(\hat{x}_k)(\hat{x}_* - \hat{x}_k)\right)\| + \gamma_R\|\hat{x}_k - \hat{x}_*\|^2$$
$$\le \|\hat{J}(\hat{x}_k)^{-1}\|\|\hat{\xi}_{\hat{x}_*} - \hat{\xi}_{\hat{x}_k} - \mathrm{D}\hat{\xi}(\hat{x}_k)[\hat{x}_* - \hat{x}_k]\|$$
$$\quad + \|\hat{J}(\hat{x}_k)^{-1}\|\|\hat{\Gamma}_{\hat{x}_k, \hat{\xi}}(\hat{x}_* - \hat{x}_k)\| + \gamma_R\|\hat{x}_k - \hat{x}_*\|^2$$
$$\le 2\,\beta\frac{1}{2}(\gamma_J + \gamma_\Gamma)\|\hat{x}_k - \hat{x}_*\|^2$$
$$\quad + 2\,\beta\,\gamma_\Gamma\|\hat{x}_k - \hat{x}_*\|^2 + \gamma_R\|\hat{x}_k - \hat{x}_*\|^2$$

whenever $\|\hat{x}_k - \hat{x}_*\| \le \min\{\epsilon, r_\Gamma, r_R\}$, where we have used Proposition A.6.1. This completes the proof. $\qquad\square$

It is interesting to note that in the classical Euclidean case, the proof holds with $\gamma_R = 0$ (because $R_x\zeta := x + \zeta$) and $\gamma_\Gamma = 0$ (because $J(x)\zeta \equiv \nabla_\zeta\xi := \mathrm{D}\xi(x)[\zeta]$).

In the case where $\mathcal{M}$ is a Riemannian metric and the Riemannian connection is used along with a second-order retraction (e.g., the exponential retraction), it is also possible to obtain a better bound. Consider normal coordinates around the point x_*. The Christoffel symbols Γ^i_{jk} vanish at $\hat{x}_*$, and the constant γ_Γ can be replaced by $O(\|\hat{x}_k - \hat{x}_*\|)$. Since we are working in normal coordinates around x_*, it follows that the Christoffel symbols at x_* vanish, hence the acceleration condition $\frac{\mathrm{D}^2}{\mathrm{dt}^2}R_{x_*}(t\zeta_{x_*})\big|_{t=0} = 0$ yields $\frac{\mathrm{d}^2}{\mathrm{dt}^2}\hat{R}_{\hat{x}_*}(t\hat{\zeta}_{\hat{x}_*})\big|_{t=0} = 0$ and, by the smoothness of R, we have $\mathrm{D}^2\hat{R}_{\hat{x}_k} = O(\|\hat{x}_k - \hat{x}_*\|)$. It follows that γ_R may be replaced by $O(\|\hat{x}_k - \hat{x}_*\|)$. Thus, the convergence bound becomes

$$\|\hat{x}_{k+1} - \hat{x}_*\| \le 2\beta\frac{1}{2}(\gamma_J + \gamma_\Gamma)\|\hat{x}_k - \hat{x}_*\|^2 + 2\beta\gamma_\Gamma\|\hat{x}_k - \hat{x}_*\|^2 + \gamma_R\|\hat{x}_k - \hat{x}_*\|^2$$
$$\le \beta\gamma_J\|\hat{x}_k - \hat{x}_*\|^2 + O(\|\hat{x}_k - \hat{x}_*\|^3).$$

In normal coordinates at x_* one has that $\mathrm{dist}(x_k, x_*) = \|\hat{x}_k - \hat{x}_*\|$.

6.3.1 Calculus approach to local convergence analysis

Theorem 6.3.2 provides a strong convergence analysis of the geometric Newton method along with explicit convergence bounds. A weaker quadratic convergence result can be obtained from a local coordinate analysis of the Newton iteration using the calculus-based convergence result of Theorem 4.5.3.

Let x_* be a critical point of a vector field ξ with a nondegenerate Jacobian
at x_*. Choose a coordinate chart around x^* and use the hat notation to
represent the coordinate expression of geometric objects. Without loss of
generality we choose $\hat{x}_* = 0$. The iteration defined in Algorithm 4 reads

$$\hat{x}_{k+1} = \hat{R}_{\hat{x}_k}(\hat{\eta}_k), \tag{6.9}$$

$$\hat{\nabla}_{\hat{\eta}_k}\hat{\xi} = -\hat{\xi}_{\hat{x}_k}. \tag{6.10}$$

Since the vector field ξ and the retraction R are smooth by assumption, this
defines a smooth iteration mapping $\hat{x}_k \mapsto \hat{x}_{k+1}(\hat{x}_k)$. Evaluating the Newton
equation (6.10) at $\hat{x}_k = \hat{x}_* = 0$ yields

$$\hat{\nabla}_{\hat{\eta}_0}\hat{\xi} = 0$$

and thus $\hat{\eta}_0 = 0$ because the Jacobian $J(x_*) : \zeta \in T_{x_*}\mathcal{M} \mapsto \nabla_\zeta\xi \in T_{x_*}\mathcal{M}$
is assumed to be nondegenerate. Since R satisfies the consistency prop-
erty $R_x(0_x) = x$ for all x, it follows that $\hat{x}_* = 0$ is a fixed point of the
iteration mapping. Recalling Theorem 4.5.3, it is sufficient to show that
$\mathrm{D}\hat{x}_{k+1}(x_*) = 0$ to prove local quadratic convergence. For clarity, we use the
notation $\hat{R}(\hat{x}, \hat{\eta})$ for $\hat{R}_{\hat{x}}(\hat{\eta})$, and we let $\mathrm{D}_1\hat{R}$ and $\mathrm{D}_2\hat{R}$ denote the differentials
with respect to the first and second arguments of the function $\hat{R}$. (Note that
$\hat{R}$ is a function from $\mathbb{R}^d \times \mathbb{R}^d$ into $\mathbb{R}^d$, where d is the dimension of the manifold $\mathcal{M}$.) Differentiating the iteration mapping $\hat{x}_k \mapsto \hat{x}_{k+1}(\hat{x}_k)$ at 0 along $\hat{\zeta}$,
one obtains

$$\mathrm{D}\hat{x}_{k+1}(0)[\hat{\zeta}] = \mathrm{D}_1\hat{R}(0,0)[\hat{\zeta}] + \mathrm{D}_2\hat{R}(0,0)[\mathrm{D}\hat{\eta}(0)[\hat{\zeta}]], \tag{6.11}$$

where $\hat{x} \mapsto \hat{\eta}(\hat{x})$ is the function implicitly defined by the Newton equation

$$\hat{\nabla}_{\hat{\eta}(\hat{x})}\hat{\xi} = -\hat{\xi}_{\hat{x}}. \tag{6.12}$$

We have $\mathrm{D}_1\hat{R}(0,0)[\hat{\zeta}] = \hat{\zeta}$ because of the consistency condition $R(0_x) =
x$. Moreover, the local rigidity condition $\mathrm{D}R_x(0_x) = \mathrm{id}_{T_x\mathcal{M}}$ (see Defini-
tion 4.1.1) ensures that $\mathrm{D}_2\hat{R}(0,0)[\mathrm{D}\hat{\eta}(0)[\hat{\zeta}]] = \mathrm{D}\hat{\eta}(0)[\hat{\zeta}]$. Hence (6.11) yields

$$\mathrm{D}\hat{x}_{k+1}(0)[\hat{\zeta}] = \hat{\zeta} + \mathrm{D}\hat{\eta}(0)[\hat{\zeta}]. \tag{6.13}$$

Using the local expression (5.7) for the affine connection, the Newton equation (6.12) reads

$$\mathrm{D}\hat{\xi}(\hat{x})[\hat{\eta}(\hat{x})] + \hat{\Gamma}_{\hat{x},\hat{\xi}_{\hat{x}}}\hat{\eta}(\hat{x}) = -\hat{\xi}_{\hat{x}}.$$

(Recall that $\hat{\Gamma}_{\hat{x},\hat{\xi}_{\hat{x}}}$ is a matrix and $\hat{\Gamma}_{\hat{x},\hat{\xi}_{\hat{x}}}\hat{\eta}(\hat{x})$ is a matrix-vector product.)
Differentiating this equation with respect to $\hat{x}$ along $\hat{\zeta}$, one obtains

$$\mathrm{D}^2\hat{\xi}(\hat{x})[\hat{\eta}(\hat{x}), \hat{\zeta}] + \mathrm{D}\hat{\xi}(\hat{x})[\mathrm{D}\hat{\eta}(\hat{x})[\hat{\zeta}]] + \mathrm{D}\hat{\Gamma}_{\cdot,\hat{\xi}}(\hat{x})[\hat{\zeta}]\hat{\eta}(\hat{x}) + \hat{\Gamma}_{\hat{x},\hat{\xi}_{\hat{x}}}\mathrm{D}\hat{\eta}(\hat{x})[\hat{\zeta}]$$
$$= -\mathrm{D}\hat{\xi}(\hat{x})[\hat{\zeta}].$$

Most of the terms in this equation vanish when evaluated at $\hat{x} = 0$ since $\hat{\xi}_0 = 0$ and $\hat{\eta}_0 = 0$. (In particular, observe that $\hat{\Gamma}_{0,0} = 0$ in view of (5.8).) This leaves us with

$$\mathrm{D}\hat{\xi}(0)[\mathrm{D}\hat{\eta}(0)[\hat{\zeta}]] = -\mathrm{D}\hat{\xi}(0)[\hat{\zeta}]. \tag{6.14}$$

Since $J(x_*)$ is nonsingular and $\hat{\Gamma}_{\hat{x}_*,\hat{\xi}_{\hat{x}_*}} = 0$, it follows in view of (5.7) that the linear operator $\mathrm{D}\hat{\xi}(0) = \hat{J}(\hat{x}_*)$ is nonsingular. Hence (6.14) reduces to

$$\mathrm{D}\hat{\eta}(0)[\hat{\zeta}] = -\hat{\zeta}.$$

Using this result in (6.13) yields $\mathrm{D}\hat{x}_{k+1}(x_*) = 0$. From Theorem 4.5.3, it follows that the iteration $\hat{x}_k \mapsto \hat{x}_{k+1}(\hat{x}_k)$ converges locally quadratically to $\hat{x}_*$. Since quadratic convergence is independent of coordinate representation, this property holds for the Newton iteration on the manifold.

6.4 RAYLEIGH QUOTIENT ALGORITHMS

In this section we show how the geometric Newton algorithms can be turned into practical numerical algorithms for the optimization of various cost functions of the Rayleigh quotient type.

6.4.1 Rayleigh quotient on the sphere

Recall the example of the Rayleigh quotient on the sphere first considered in Section 4.6. The main points were summarized in Table 4.1. The cost function is the Rayleigh quotient

$$f : S^{n-1} \to \mathbb{R} : x \mapsto x^T A x, \tag{6.15}$$

on the unit sphere S^{n-1}, viewed as a Riemannian submanifold of the Euclidean space $\mathbb{R}^n$. We also use the extension

$$\overline{f} : \mathbb{R}^n \to \mathbb{R} : x \mapsto x^T A x,$$

whose restriction to S^{n-1} is f. In Section 4.6, we obtained

$$\mathrm{grad}\, f(x) = 2\,\mathrm{P}_x(Ax) = 2\,(Ax - xx^T Ax),$$

where P_x is the orthogonal projector onto $T_x S^{n-1}$, i.e.,

$$\mathrm{P}_x z = z - xx^T z.$$

(Note that P_x can also be viewed as the matrix $(I - xx^T)$.) We also expressed a preference for the retraction

$$R_x(\xi) := \frac{x + \xi}{\|x + \xi\|}. \tag{6.16}$$

The geometric Newton method (Algorithm 4) requires an affine connection ∇. There is no reason not to pick the natural choice, the Riemannian connection. Since S^{n-1} is a Riemannian submanifold of the Euclidean space $\mathbb{R}^n$, it follows from the material in Section 5.3.3 that

$$\nabla_\eta \xi = \mathrm{P}_x \left(\mathrm{D}\xi \left(x \right) [\eta] \right)$$

for every η in the tangent space $T_x S^{n-1} = \{ z \in \mathbb{R}^n : x^T z = 0 \}$ and every vector field ξ on S^{n-1}.

We are ready to apply the geometric Newton method (Algorithm 4) to the vector field $\xi := \mathrm{grad}\, f$, where f is the Rayleigh quotient (6.15). For every $\eta \in T_{x_k} S^{n-1}$, we have

$$\begin{aligned}
\nabla_\eta \mathrm{grad}\, f(x) &= 2\mathrm{P}_x \left(\mathrm{D}\, \mathrm{grad}\, f(x)[\eta] \right) \\
&= 2\mathrm{P}_x(A\eta - \eta x^T A x) \\
&= 2(\mathrm{P}_x A \mathrm{P}_x \eta - \eta x^T A x),
\end{aligned}$$

where we took into account that $\mathrm{P}_x x = 0$ and $\mathrm{P}_x \eta = \eta$. The last expression underscores the symmetry of the Hessian operator (in the sense of Proposition 5.5.3). Consequently, the Newton equation (6.1) reads

$$\begin{cases}
\mathrm{P}_x A \mathrm{P}_x \eta - \eta x^T A x = -\mathrm{P}_x A x, \\
x^T \eta = 0,
\end{cases} \tag{6.17}$$

where the second equation is the expression of the requirement $\eta \in T_x S^{n-1}$.

In conclusion, application of the geometric Newton method to $\xi := \mathrm{grad}\, f$, where f is the Rayleigh quotient (6.15) on the sphere S^{n-1}, viewed as a Riemannian submanifold of $\mathbb{R}^n$ endowed with its Riemannian connection and with the retraction (6.16), yields the matrix algorithm displayed in Algorithm 6. Since Algorithm 6 is a particular case of the forthcoming Algorithms 7 and 8, we postpone its analysis to Section 6.5.1.

Algorithm 6 Riemannian Newton method for the Rayleigh quotient on S^{n-1}

Require: Symmetric matrix A.
Input: Initial iterate $x_0 \in \mathcal{M}$.
Output: Sequence of iterates $\{x_k\}$.
 1: **for** $k = 0, 1, 2, \ldots$ **do**
 2:    Solve the linear system (6.17), i.e.,

$$\begin{cases}
\mathrm{P}_{x_k} A \mathrm{P}_{x_k} \eta_k - \eta_k x_k^T A x_k = -\mathrm{P}_{x_k} A x_k, \\
x_k^T \eta_k = 0,
\end{cases} \tag{6.18}$$

   for the unknown $\eta_k \in \mathbb{R}^n$.
 3:    Set

$$x_{k+1} := R_{x_k} \eta_k,$$

   with R defined in (6.16).
 4: **end for**

6.4.2 Rayleigh quotient on the Grassmann manifold

Consider the cost function

$$f : \mathrm{Grass}(p, n) \to \mathbb{R} : \mathrm{span}(Y) \mapsto \mathrm{tr}\left((Y^T Y)^{-1} Y^T A Y\right). \qquad (6.19)$$

The first-order geometry of this cost function was investigated in Section 4.9 (see Table 4.3). The Grassmann manifold $\mathrm{Grass}(p, n)$ was viewed as a Riemannian quotient manifold of $(\mathbb{R}_*^{n \times p}, \bar{g})$ with

$$\bar{g}_Y(Z_1, Z_2) = \mathrm{tr}\left((Y^T Y)^{-1} Z_1^T Z_2\right). \qquad (6.20)$$

The horizontal distribution is

$$\mathcal{H}_Y = \{Z \in \mathbb{R}^{n \times p} : Y^T Z = 0\}, \qquad (6.21)$$

the projection onto the horizontal space is

$$\mathrm{P}_Y^h = (I - Y(Y^T Y)^{-1} Y^T), \qquad (6.22)$$

and we obtained

$$\overline{\mathrm{grad}\, f}_Y = 2\mathrm{P}_Y^h A Y = 2\left(A Y - Y(Y^T Y)^{-1} Y^T A Y\right).$$

It follows from this expression that if $\mathrm{grad}\, f(\mathrm{span}(Y)) = 0$, then $\mathrm{span}(Y)$ is an invariant subspace of A. Conversely, if $\mathrm{span}(Y)$ is an invariant subspace of A, then there exists an M such that $A Y = Y M$; premultiplying this equation by $(Y^T Y)^{-1} Y^T$ yields $M = (Y^T Y)^{-1} Y^T A Y$, and we obtain $\mathrm{grad}\, f(\mathrm{span}(Y)) = 0$. In conclusion, the critical points of f are the invariant subspaces of A.

In Section 5.3.4, we established the formula

$$\overline{\nabla_\eta \xi} = \mathrm{P}_Y^h\left(\mathrm{D}\bar{\xi}\,(Y)\,[\bar{\eta}_Y]\right) \qquad (6.23)$$

for the Riemannian connection on the Grassmann manifold. This yields the following expression for the Hessian of the Rayleigh quotient cost function f:

$$\overline{\nabla_\eta \mathrm{grad}\, f} = \mathrm{P}_Y^h\left(\mathrm{D}\overline{\mathrm{grad}\, f}\,(Y)\,[\bar{\eta}_Y]\right) = 2\,\mathrm{P}_Y^h\left(A\bar{\eta}_Y - \bar{\eta}_Y(Y^T Y)^{-1} Y^T A Y\right),$$

where we have utilized the identity $\mathrm{P}_Y^h(Y M) = (\mathrm{P}_Y^h Y)M = 0$. Taking the horizontal lift of the Newton equation $\nabla_\eta \mathrm{grad}\, f = -\mathrm{grad}\, f(x)$ yields the equation

$$\mathrm{P}_Y^h\left(A\bar{\eta}_Y - \bar{\eta}_Y(Y^T Y)^{-1} Y^T A Y\right) = -\mathrm{P}_Y^h A Y, \qquad \bar{\eta}_Y \in \mathcal{H}_Y,$$

whose solution $\bar{\eta}_Y$ in the horizontal space $\mathcal{H}_Y$ is the horizontal lift of the solution η of the Newton equation.

In conclusion, the geometric Newton method in Algorithm 4, for the Rayleigh quotient cost function (6.19), with the affine connection ∇ chosen as the Riemannian connection on $\mathrm{Grass}(p, n)$ seen as the Riemannian quotient manifold of $(\mathbb{R}_*^{n \times p}, \bar{g})$ with $\bar{g}$ in (6.20), and with the retraction R chosen as (4.40) yields the matrix algorithm displayed in Algorithm 7. The notation $Z_k = \bar{\eta}_{Y_k}$ is used to make the matrix expression resemble contemporary algorithms from the field of numerical linear algebra. The expression

---

Algorithm 7 Riemannian Newton method for the Rayleigh quotient on $\mathrm{Grass}(p,n)$

Require: Symmetric matrix A.
Input: Initial iterate $Y_0 \in \mathbb{R}_*^{n \times p}$.
Output: Sequence of iterates $\{Y_k\}$ in $\mathbb{R}_*^{n \times p}$.
 1: **for** $k = 0, 1, 2, \ldots$ **do**
 2:   Solve the linear system

$$\begin{cases} \mathrm{P}^h_{Y_k} \left(AZ_k - Z_k(Y_k^T Y_k)^{-1} Y_k^T AY_k \right) = -\mathrm{P}^h_{Y_k}(AY_k) \\ Y_k^T Z_k = 0 \end{cases} \tag{6.24}$$

   for the unknown Z_k, where P^h_Y is the orthogonal projector defined
   in (6.22). (The condition $Y_k^T Z_k$ expresses that Z_k belongs to the horizontal space $\mathcal{H}_{Y_k}$.)
 3:   Set

$$Y_{k+1} = (Y_k + Z_k)N_k$$

   where N_k is a nonsingular $p \times p$ matrix chosen for normalization purposes.
 4: **end for**

---

in (6.24) can be simplified since $\mathrm{P}^h_{Y_k} Z_k = Z_k$. We will tend not to simplify
such expressions in the matrix equations in order that the equations clearly
reveal the underlying geometric structure (e.g., the quantity considered belongs to the range of $\mathrm{P}^h_{Y_k}$) or to emphasize the symmetry of certain operators.

Note that Algorithm 7 is the matrix expression of an algorithm defined on
$\mathrm{Grass}(p,n)$. In other words, if $\{Y_k\}$ and $\{\check{Y}_k\}$ are two sequences generated
by Algorithm 7 (with same matrix A) and if $\mathrm{span}(Y_0) = \mathrm{span}(\check{Y}_0)$, then
$\mathrm{span}(Y_k) = \mathrm{span}(\check{Y}_k)$ for all k. Algorithm 7 thus could be written formally
as an algorithm generating a sequence on $\mathrm{Grass}(p,n)$, by taking as input an
element $\mathcal{Y}_0$ of $\mathrm{Grass}(p,n)$, picking $Y_0 \in \mathbb{R}_*^{n \times p}$ with $\mathrm{span}(Y_0) = \mathcal{Y}_0$, proceeding
as in Algorithm 7, and returning the sequence $\{\mathrm{span}(Y_k)\}$.
    Note also that when $p = 1$ and N_k (now a scalar) is chosen as $\|Y_k + Z_k\|^{-1}$,
Algorithm 7 reduces to Algorithm 6.

6.4.3 Generalized eigenvalue problem

We assume that A and B are $n \times n$ symmetric matrices with B positive-
definite, and we consider the generalized eigenvalue problem

$$Av = \lambda Bv$$

described in Section 2.1. With a view towards computing eigenspaces of a
pencil (A, B), we consider the Rayleigh quotient function

$$f(\mathrm{span}(Y)) = \mathrm{tr}((Y^T AY)(Y^T BY)^{-1}), \tag{6.25}$$

where Y is a full-rank $n \times p$ matrix and span(Y) denotes the column space of Y. It is readily checked that the right-hand side depends only on span(Y), so that f is a well-defined real-valued function on the Grassmann manifold Grass(p, n).

As in the previous section, we view Grass(p, n) as a Riemannian quotient manifold of $(\mathbb{R}_*^{n \times p}, \bar{g})$ with

$$\bar{g}_Y(Z_1, Z_2) = \mathrm{tr}((Y^T Y)^{-1} Z_1^T Z_2). \tag{6.26}$$

With a view towards applying the Riemannian Newton method given in Algorithm 5, we need formulas for the gradient and the Hessian of the Rayleigh cost function (6.25). Mimicking the calculations in Section 4.9, we obtain

$$\tfrac{1}{2} \overline{\mathrm{grad}\, f}_Y = P_{BY,Y}\, AY(Y^T BY)^{-1} Y^T Y, \tag{6.27}$$

where

$$P_{U,V} = I - U(V^T U)^{-1} V^T$$

denotes the projector parallel to the span of U onto the orthogonal complement of the span of V. Note that the projection P_Y^h onto the horizontal space $\mathcal{H}_Y$ is given by

$$\mathrm{P}_Y^h = P_{Y,Y}.$$

Using the result in Proposition 5.3.4 (on the Riemannian connection on Riemannian quotient manifolds) and the definition Hess $f[\zeta] = \nabla_\zeta \mathrm{grad}\, f$, we also have

$$\tfrac{1}{2} \overline{\mathrm{Hess}\, f[\zeta]}_Y = \tfrac{1}{2} P_{Y,Y} \mathrm{D} \overline{\mathrm{grad}\, f}\,(Y)\left[\bar{\zeta}_Y\right]. \tag{6.28}$$

Expanding this expression is possible but tedious and leads to a complicated Newton equation. Fortunately, simpler Newton equations can be obtained by exploiting the freedom in (i) the choice of the Riemannian metric $\bar{g}$; (ii) the choice of the horizontal spaces $\mathcal{H}_Y$, which need not be orthogonal to the vertical spaces $\mathcal{V}_Y$ with respect to $\bar{g}$; (iii) the choice of the affine connection ∇, which need not be the Riemannian connection induced by $\bar{g}$.

We first consider an alternative Riemannian metric. We still view the Grassmann manifold Grass(p, n) as the quotient $\mathbb{R}_*^{n \times p} / \sim$, where the equivalence classes of $\sim$ are the sets of elements of $\mathbb{R}_*^{n \times p}$ that have the same column space. However, instead of (6.26), we consider on $\mathbb{R}_*^{n \times p}$ the metric

$$\bar{g}_Y(Z_1, Z_2) = \mathrm{tr}\left((Y^T BY)^{-1} Z_1^T B Z_2\right), \tag{6.29}$$

where B is the symmetric positive-definite matrix that appears in the definition of f in (6.25). Defining again the horizontal space as the orthogonal complement—with respect to the new inner product (6.29)—of the vertical space

$$\mathcal{V}_Y := T_Y(\pi^{-1}(\pi(Y))) = \{YM : M \in \mathbb{R}^{p \times p}\},$$

we obtain

$$\mathcal{H}_Y = \{Z \in \mathbb{R}^{n \times p} : Y^T BZ = 0\}. \tag{6.30}$$

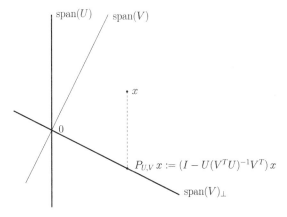

Figure 6.1 The projector $P_{U,V}$.

The orthogonal projection onto $\mathcal{H}_Y$ is given by

$$P_Y^h = P_{Y,BY} = I - Y(Y^T BY)^{-1}Y^T B.$$

The homogeneity property

$$\overline{\xi}_{YM} = \overline{\xi}_Y M$$

of Proposition 3.6.1 still holds with the new Riemannian metric (6.29). More-over,

$$\overline{g}_{YM}(\overline{\xi}_{YM}, \overline{\zeta}_{YM}) = \overline{g}_Y(\overline{\xi}_Y, \overline{\zeta}_Y).$$

Therefore, the Grassmann manifold $\mathrm{Grass}(p, n)$ admits a unique Riemannian metric

$$g(\xi, \zeta) := \overline{g}_Y(\overline{\xi}_Y, \overline{\zeta}_Y), \tag{6.31}$$

that makes $(\mathrm{Grass}(p, n), g)$ a Riemannian quotient manifold of $(\mathbb{R}_*^{n \times p}, \overline{g})$ with $\overline{g}$ defined in (6.29).

Before proceeding to obtain formulas for the gradient and the Hessian of f in $(\mathrm{Grass}(p, n), g)$, we first point out some useful properties of the projector

$$P_{U,V} = I - U(V^T U)^{-1}V^T. \tag{6.32}$$

Recall that $P_{U,V}$ is the projector that projects parallel to $\mathrm{span}(U)$ onto $\mathrm{span}(V)$; see Figure 6.1. Therefore, we have the identities

$$P_{U,V} UM = 0 \quad \text{and} \quad P_{U,V} VM = VM.$$

We also have the identity

$$P_{U,V} K = KP_{K^{-1}U, K^T V}.$$

Using the above identities and the technique of Section 3.6.2, we obtain

$$\overline{\mathrm{grad}\, f}_Y = P_{Y,BY} B^{-1}AY = B^{-1}P_{BY,Y} AY. \tag{6.33}$$

It can be checked that the new Riemannian metric (6.29) is horizontally invariant. Consequently, it follows from Proposition 5.3.4 that the Riemannian Hessian is given by

$$
\begin{aligned}
\overline{\text{Hess } f(\mathcal{Y})[\eta]}_Y &= \overline{\nabla_\eta \operatorname{grad} f}_Y \\
&= \mathrm{P}_Y^h \mathrm{D}\overline{\operatorname{grad} f}\,(Y)\,[\bar{\eta}_Y] \\
&= P_{Y,BY} B^{-1} A\bar{\eta}_Y - P_{Y,BY}\bar{\eta}_Y (Y^T BY)^{-1} Y^T AY \\
&= B^{-1} P_{BY,Y} \left(A\bar{\eta}_Y - B\bar{\eta}_Y (Y^T BY)^{-1} Y^T AY \right). \qquad (6.34)
\end{aligned}
$$

The Newton equation $\nabla_\eta \operatorname{grad} f = -\operatorname{grad} f(x)$ thus yields the equation

$$
B^{-1} P_{BY,Y} \left(A\bar{\eta}_Y - B\bar{\eta}_Y (Y^T BY)^{-1} Y^T AY \right) = -B^{-1} P_{BY,Y} AY,
$$

or equivalently,

$$
P_{BY,Y} \left(A\bar{\eta}_Y - B\bar{\eta}_Y (Y^T BY)^{-1} Y^T AY \right) = -P_{BY,Y} AY.
$$

In conclusion, the geometric Newton method in Algorithm 4, for the Rayleigh quotient cost function (6.25), with the affine connection ∇ chosen as the Riemannian connection on $\operatorname{Grass}(p,n)$ seen as the Riemannian quotient manifold of $(\mathbb{R}_*^{n\times p}, \bar{g})$ with $\bar{g}$ defined in (6.29), and with the retraction R chosen as (4.40), yields the matrix algorithm displayed in Algorithm 8. The notation $Z_k = \bar{\eta}_{Y_k}$ is used so that the algorithm resembles contemporary algorithms from the field of numerical linear algebra.

Algorithm 8 Riemannian Newton method for the Rayleigh quotient on $\operatorname{Grass}(p,n)$

Require: Symmetric matrix A, symmetric positive-definite matrix B.
Input: Initial iterate $Y_0 \in \mathbb{R}_*^{n\times p}$.
Output: Sequence of iterates $\{Y_k\}$ in $\mathbb{R}_*^{n\times p}$.
1: **for** $k = 0,1,2,\dots$ **do**
2:   Solve the linear system

$$
\begin{cases}
P_{BY_k,Y_k} \left(AZ_k - BZ_k (Y_k^T BY_k)^{-1} Y_k^T AY_k \right) = -P_{BY_k,Y_k}(AY_k) \\
Y_k^T BZ_k = 0
\end{cases}
$$

$$(6.35)$$

  for the unknown Z_k, where $P_{BY,Y} = I - BY(Y^T BY)^{-1}Y^T$. (The condition $Y_k^T BZ_k$ expresses that Z_k belongs to the horizontal space $\mathcal{H}_{Y_k}$ (6.30).)
3:   Set

$$
Y_{k+1} = (Y_k + Z_k)N_k
$$

  where N_k is a nonsingular $p \times p$ matrix chosen for normalization purposes.
4: **end for**

Algorithm 8 is related to several eigenvalues methods; see Notes and References.

A concern with the Newton equation (6.35) is that the domain $\{Z \in \mathbb{R}^{n \times p} : Y^T BZ = 0\}$ of the map

$$F : Z \mapsto P_{BY,Y} \left(AZ - BZ(Y^T BY)^{-1} Y^T AY \right)$$

differs from its range $\{Z \in \mathbb{R}^{n \times p} : Y^T Z = 0\}$. Hence, powers of F cannot be formed, and linear equation solvers based on Krylov subspaces cannot be applied directly to (6.35). A remedy based on preconditioners is discussed in Section 6.5.2. Another remedy is to exploit the freedom in the choice of the affine connection ∇, which, according to Algorithm 5, need not be the Riemannian connection. To this end, let us view $\mathrm{Grass}(p, n)$ as a Riemannian quotient manifold of $(\mathbb{R}_*^{n \times p}, \bar{g})$ with

$$\bar{g}_Y(Z_1, Z_2) = \mathrm{tr}((Y^T BY)^{-1} Z_1^T Z_2). \qquad (6.36)$$

Note that this Riemannian metric is different from the canonical Riemannian metric (6.26). The horizontal space defined as the orthogonal complement of the vertical space is still given by (6.21), but the expression of the gradient becomes

$$\tfrac{1}{2}\overline{\mathrm{grad}\, f}_Y = P_{BY,Y} AY, \qquad (6.37)$$

which is simpler than (6.27). Now, instead of choosing the affine connection ∇ as the Riemannian connection, we define ∇ by

$$\left(\overline{\nabla_\eta \xi}\right)_Y = P_{BY,Y} \mathrm{D}\bar{\xi}\,(Y)\,[\bar{\eta}_Y]. \qquad (6.38)$$

It is readily checked that (6.38) defines a horizontal lift, i.e., $\left(\overline{\nabla_\eta \xi}\right)_{YM} = \left(\overline{\nabla_\eta \xi}\right)_Y M$, and that ∇ is indeed an affine connection (see Section 5.2). With this affine connection, the horizontal lift of the Newton equation $\nabla_\eta \, \mathrm{grad}\, f = -\mathrm{grad}\, f(\mathcal{Y})$ reads

$$P_{BY,Y} \left(AZ - BZ(Y^T BY)^{-1} Y^T AY \right) = P_{BY,Y} AY, \quad Y^T Z = 0, \qquad (6.39)$$

where Z stands for $\bar{\eta}_Y$. Observe that the map

$$Z \mapsto P_{BY,Y} \left(AZ - BZ(Y^T BY)^{-1} Y^T AY \right)$$

involved in (6.39) is now from $\{Z \in \mathbb{R}^{n \times p} : Y^T Z = 0\}$ into itself. The resulting iteration is still guaranteed by Theorem 6.3.2 to converge locally at least quadratically to the spectral invariant subspaces of $B^{-1} A$ (see Section 6.5.1 for details).

Note that in this section we have always chosen the horizontal space as the orthogonal complement of the vertical space. The possibility of choosing other horizontal spaces is exploited in Section 7.5.3.

6.4.4 The nonsymmetric eigenvalue problem

The Rayleigh quotient

$$f : \mathrm{Grass}(p, n) \to \mathbb{R} : \mathrm{span}(Y) \mapsto \mathrm{tr}\left((Y^T Y)^{-1} Y^T AY\right)$$

depends only on the symmetric part of A; it is thus clear that when A is nonsymmetric, computing critical points of f in general does not produce invariant subspaces of A. A way to tackle the nonsymmetric eigenvalue problem is to consider instead the tangent vector field on the Grassmann manifold defined by

$$\overline{\xi}_Y := P_Y^h AY, \tag{6.40}$$

where P_Y^h denotes the projector (6.22) onto the horizontal space (6.21). This expression is homogeneous $(\overline{\xi}_{YM} = \overline{\xi}_Y M)$ and horizontal; therefore, as a consequence of Proposition 3.6.1, it is a well-defined horizontal lift and defines a tangent vector field given by $\xi_Y = D\pi(Y)\left[\overline{\xi}_Y\right]$ on the Grassmann manifold. Moreover, $\xi_{\mathcal{Y}} = 0$ if and only if $\mathcal{Y}$ is an invariant subspace of A. Obtaining the Newton equation (6.1) for ξ defined in (6.40) is straightforward: formula (6.23), giving the horizontal lift of the connection, leads to

$$\overline{\nabla_\eta \xi}_Y = P_Y^h \left(A\overline{\eta}_Y - \overline{\eta}_Y(Y^TY)^{-1}Y^TAY\right)$$

and the Newton equation (6.1) reads

$$\begin{cases} P_Y^h \left(A\overline{\eta}_Y - \overline{\eta}_Y(Y^TY)^{-1}Y^TAY\right) = -P_Y^h AY, \\ Y^T\overline{\eta}_Y = 0, \end{cases} \tag{6.41}$$

where the second equation expresses that $\overline{\eta}_Y$ is in the horizontal space. The resulting Newton iteration turns out to be identical to Algorithm 7, except that A is no longer required to be symmetric.

6.4.5 Newton with subspace acceleration: Jacobi-Davidson

The Jacobi-Davidson approach is a powerful technique for solving a variety of eigenproblems. It has recently become widely popular among chemists and solid-state physicists for computing a few extreme eigenpairs of large-scale eigenvalue problems. In this section, the principles of the Jacobi-Davidson approach are briefly reviewed and the method is interpreted as a Rayleigh-based Riemannian Newton method within a sequential subspace optimization scheme.

For simplicity we focus on the standard eigenvalue problem and let A be a symmetric $n \times n$ matrix. Central to the Jacobi-Davidson approach is the *Jacobi correction equation*

$$(I - x_kx_k^T)(A - \tau_kI)(I - x_kx_k^T)s_k = -(A - \tau_kI)x_k, \quad x_k^Ts_k = 0, \tag{6.42}$$

which, for the usual choice of shift $\tau_k = x_k^TAx_k$, reduces to the Newton equation for the Rayleigh quotient on the sphere (see Algorithm 6).

In the Riemannian Newton method the update vector s_k is retracted onto the manifold to produce the next iterate $x_{k+1} = R_{x_k}s_k$; for example, the choice (6.16) of the retraction R yields $x_{k+1} = (x_k + s_k)/\|x_k + s_k\|$. Instead, in the Jacobi-Davidson approach, the update vector is used to expand a low-dimensional search space on which the given eigenproblem is projected. This is the standard Rayleigh-Ritz procedure that underlies all Davidson-like methods, as well as the Lanczos and Arnoldi methods. The small projected problem is solved by standard techniques, and this leads to approximations

---

Algorithm 9 Jacobi-Davidson

Require: Symmetric matrix A.
Input: Select a set of k_0 (≥ 1) orthonomal vectors $v_1, \ldots, v_{k_0}$ and set $V_1 = [v_1 | \ldots | v_{k_0}]$.
Output: Sequence of iterates $\{x_k\}$.
  1: **for** $k = 1, 2, \ldots$ **do**
  2:   Compute the interaction matrix $H_k = V_k^T A V_k$.
  3:   Compute the leftmost eigenpair (ρ_k, y_k) of H_k (with $\|y_k\| = 1$).
  4:   Compute the Ritz vector $x_k = V_k y_k$.
  5:   If needed, shrink the search space: compute the $j_{\min}$ leftmost eigen-pairs $(\rho_k^{(j)}, y_k^{(j)})$ of H_k and reset $V_k := V_k[y_k^{(1)} | \cdots | y_k^{(j_{\min})}]$.
  6:   Obtain s_k by solving (approximately) the Jacobi equation (6.42).
  7:   Orthonormalize $[V_k | s_k]$ into V_{k+1}.
  8: **end for**

---

for the wanted eigenvector and eigenvalues of the given large problem. The procedure is described in Algorithm 9 for the case where the leftmost eigen-pair of A is sought.

Practical implementations of Algorithm 9 vary widely depending on the methods utilized to solve the Jacobi equation approximately and to reduce the search space.

Concerning the solution of the Jacobi equation, anticipating the development in Chapter 7, we point out that the solution s_k of (6.42) is the critical point of the model

$$\widehat{m}_{x_k}(s) := x_k^T A x_k + 2s^T A x_k + s^T (A - x_k^T A x_k I)s, \quad x_k^T s = 0.$$

This model is the quadratic Taylor expansion of the cost function

$$f \circ R_{x_k} : T_{x_k} S^{n-1} \to \mathbb{R} : s \mapsto \frac{(x_k + s)^T A(x_k + s)}{(x_k + s)^T (x_k + s)}$$

around the origin 0 of the Euclidean space $T_{x_k} S^{n-1}$, where f denotes the Rayleigh quotient on the sphere and R denotes the retraction (6.16). When the goal of the algorithm is to minimize the Rayleigh quotient (in order to find the leftmost eigenpair), the idea of solving the Jacobi equation, which amounts to computing the critical point s_* of the model $\widehat{m}_{x_k}(s)$, presents two drawbacks: (i) the critical point is not necessarily a minimizer of the model, it may be a saddle point or a maximizer; (ii) even when the critical point is a minimizer, it may be so far away from the origin of the Taylor expansion that there is an important mismatch between the model $\widehat{m}_{x_k}(s_*)$ and the cost function $f \circ R_{x_k}(s_*)$. The trust-region approach presented in Chapter 7 remedies these drawbacks by selecting the update vector s_k as an approximate minimizer of the model $\widehat{m}_{x_k}$, constrained to a region around $s = 0$ where its accuracy is trusted. Therefore, algorithms for approximately solving trust-region subproblems (see Section 7.3) can be fruitfully used as "intelligent" approximate solvers for the Jacobi equation that are aware of the underlying Rayleigh quotient optimization problem.

Concerning the sequential subspace approach, if the sequence of computed s_k's is gradient-related, then the Jacobi-Davidson method fits within the framework of Algorithm 1 (an accelerated line search) and the convergence analysis of Section 4.3 applies. In particular, it follows from Theorem 4.3.1 that every accumulation point of the sequence $\{x_k\}$ is a critical point of the Rayleigh quotient, and thus an eigenvector of A. A simple way to guarantee that $\{x_k\}$ stems from a gradient-related sequence is to include $\operatorname{grad} f(x_k) = Ax_k - x_k x_k^T Ax_k$ as a column of the new basis matrix V_{k+1}.

6.5 ANALYSIS OF RAYLEIGH QUOTIENT ALGORITHMS

In this section, we first formally prove quadratic convergence of the Newton algorithms for the Rayleigh quotient developed in the previous section. After this, the remainder of the section is devoted to a discussion of the numerical implementation of the proposed algorithms. Efficiently solving the Newton equations is an important step in generating numerically tractable algorithms. The structured matrix representation of the Newton equations that result from the approach taken in this book means that we can exploit the latest tools from numerical linear algebra to analyze and solve these equations.

6.5.1 Convergence analysis

For the convergence analysis, we focus on the case of Algorithm 8 (iteration on the Grassmann manifold for the generalized eigenvalue problem). The convergence analysis of Algorithm 7 (standard eigenvalue problem, on the Grassmann manifold) follows by setting $B = I$. These results also apply to Algorithm 6 (on the sphere) since it fits in the framework of Algorithm 7.

Since Algorithm 8 is a particular instance of the general geometric Newton method (Algorithm 4), the convergence analysis in Theorem 6.3.2 applies to Algorithm 8. A p-dimensional subspace span(Y_*) is a critical point of the Rayleigh quotient (6.25) if and only if span(Y_*) is an invariant subspace of the pencil (A, B). The condition in Theorem 6.3.2 that the Jacobian (here, the Hessian of f) at span(Y_*) be invertible becomes the condition that the Hessian operator

$$Z \in \mathcal{H}_{Y_*} \mapsto B^{-1} P_{BY_*,Y_*} \left(AZ - BZ(Y_*^T BY_*)^{-1} Y_*^T AY_* \right) \in \mathcal{H}_{Y_*}$$

given in (6.34) be invertible. It can be shown that this happens if and only if the invariant subspace span(Y_*) is *spectral*, i.e., for every eigenvalue λ of $B^{-1}A|_{\text{span}(Y_*)}$ the multiplicities of λ as an eigenvalue of $B^{-1}A|_{\text{span}(Y_*)}$ and as an eigenvalue of $B^{-1}A$ are identical. (To prove this, one chooses a basis where B is the identity and A is diagonal with the eigenvalues of $B^{-1}A|_{\text{span}(Y_*)}$ in the upper left block. The operator reduced to $\mathcal{H}_{Y_*}$ turns out to be diagonal with all diagonal elements different from zero.)

Theorem 6.5.1 (local convergence of Algorithm 8) *Under the requirements of Algorithm 8, assume that there is a $Y_* \in \mathbb{R}^{n \times p}$ such that* span(Y_*) *is a spectral invariant subspace of $B^{-1}A$. Then there exists a neighborhood $\mathcal{U}$ of* span(Y_*) *in* Grass(p, n) *such that, for all $Y_0 \in \mathbb{R}^{n \times p}$ with* span$(Y_0) \in \mathcal{U}$, *Algorithm 8 generates an infinite sequence $\{Y_k\}$ such that $\{$*span$(Y_k)\}$ *converges superlinearly (at least quadratically) to $\mathcal{Y}_*$ on* Grass(p, n).

Concerning the algorithm for the nonsymmetric eigenvalue problem presented in Section 6.4.4, it follows by a similar argument that the iterates of the method converge locally superlinearly to the spectral invariant subspaces of A.

6.5.2 Numerical implementation

A crucial step in a numerical implementation of the Newton algorithms lies in solving the Newton equations. We first consider the Grassmann case with $B = I$ (standard eigenvalue problem). For clarity, we drop the subscript k. The Newton equation (6.24) reads

$$\begin{cases} P_Y^h \left(AZ - Z(Y^T Y)^{-1} Y^T AY \right) = -P_Y^h(AY) \\ Y^T Z = 0, \end{cases} \tag{6.43}$$

where Z is the unknown and $P_Y^h = (I - Y(Y^T Y)^{-1} Y^T)$. In order to make this equation simpler, the first thing to do is to choose Y orthonormal, so that $Y^T Y = I$. Since $Y^T AY$ is symmetric, it is possible to further choose Y such that $Y^T AY$ is diagonal. This can be done by computing a matrix M such that $(YM)^T AYM \equiv M^T(Y^T AY)M$ is diagonal and making $Y \leftarrow YM$. This corresponds to solving a small-scale $p \times p$ eigenvalue problem. The diagonal elements $\rho_1, \ldots, \rho_p$ of the diagonal matrix $Y^T AY$ are called the *Ritz values* related to $(A, \text{span}(Y))$, and the columns of Y are the corresponding *Ritz vectors*. This decouples (6.43) into p independent systems of linear equations of the form

$$\begin{cases} P_Y^h(A - \rho_i I)P_Y^h z_i = -P_Y^h Ay_i, \\ Y^T z_i = 0, \end{cases} \tag{6.44}$$

where $z_1, \ldots, z_p \in \mathbb{R}^p$ are the columns of Z. Note that (6.44) resembles a parallel implementation of p Newton methods (6.17). However, the projection operator P_Y^h in (6.44) is equal to $(I - Y(Y^T Y)^{-1} Y^T)$, whereas the parallel implementation of (6.17) would lead to

$$\begin{cases} P_{y_i}(A - \rho_i I)P_{y_i} z_i = -P_{y_i} Ay_i, \\ y_i^T z_i = 0, \end{cases}$$

$i = 1, \ldots, p$, where $P_{y_i} = (I - y_i(y_i^T y_i)^{-1} y_i^T)$. Methods for solving (6.44) include Krylov-based methods that naturally enforce the constraint $Y^T z_i$. Another approach is to transform (6.44) into the saddle-point problem

$$\begin{bmatrix} A - \rho_i I & Y \\ Y^T & 0 \end{bmatrix} \begin{bmatrix} z_i \\ \ell \end{bmatrix} = \begin{bmatrix} -(A - f(y_i)I)y_i \\ 0 \end{bmatrix}, \tag{6.45}$$

a structured linear system for which several algorithms have been proposed in the literature.

We now look specifically into the $p = 1$ case, in the form of the Newton equation (6.17) on the sphere, repeated here for convenience:

$$\begin{cases} P_x A P_x \eta - \eta x^T A x = -P_x A x, \\ x^T \eta = 0, \end{cases}$$

where $P_x = (I - xx^T)$. The Newton equation (6.17) is a system of linear equations in the unknown $\eta_k \in \mathbb{R}^n$. It admits several equivalent formulations. For example, using the fact that $P_x x = 0$, (6.17) can be rewritten as

$$P_x(A - f(x)I)(x + \eta) = 0, \quad x^T \eta = 0,$$

where $f(x)$ still stands for the Rayleigh quotient (6.15). Equation (6.17) is also equivalent to the *saddle-point problem*

$$\begin{bmatrix} A - f(x)I & x \\ x^T & 0 \end{bmatrix} \begin{bmatrix} \eta \\ \ell \end{bmatrix} = \begin{bmatrix} -(A - f(x)I)y \\ 0 \end{bmatrix}. \tag{6.46}$$

If $A - f(x)I$ is nonsingular, then the solution η of (6.17) is given explicitly by

$$\eta = -x + (A - f(x)I)^{-1} x \frac{1}{x^T(A - f(x)I)^{-1} x}. \tag{6.47}$$

This points to an interesting link with the Rayleigh quotient iteration: with the retraction defined in (6.16), the next iterate constructed by Algorithm 6 is given by

$$\frac{x + \eta}{\|x + \eta\|} = \frac{(A - f(x)I)^{-1} x}{\|(A - f(x)I)^{-1} x\|},$$

which is the formula defining the Rayleigh quotient iteration.

With $U \in \mathbb{R}^{n \times (n-1)}$ chosen such that $[x|U]^T [x|U] = I$, (6.17) is also equivalent to

$$(U^T A U - f(x)I)s = -U^T A x, \quad \eta = Us, \tag{6.48}$$

which is a linear system in classical form in the unknown s. Note that when A is a large sparse matrix, the matrix U is large and dense, and there may not be enough memory space to store it. Moreover, the sparsity of A is in general not preserved in the reduced matrix $U^T A U$. The approach (6.48) should thus be avoided in the large sparse case. It is preferable to solve the system in the form (6.17) or (6.46) using an iterative method.

We now briefly discuss the generalized case $B \neq I$. The Newton equation (6.35) can be decoupled into p independent equations of the form

$$P_{BY,Y}(A - \rho_i I)P_{Y,BY} z_i = -P_{BY,Y} A y_i, \quad Y^T B z_i = 0. \tag{6.49}$$

The corresponding saddle-point formulation is

$$\begin{bmatrix} A - \rho_i B & BY \\ (BY)^T & 0 \end{bmatrix} \begin{bmatrix} z_i \\ \ell \end{bmatrix} = \begin{bmatrix} -(A - \rho_i B)y_i \\ 0 \end{bmatrix}, \tag{6.50}$$

where $\rho_i := y_i^T A y_i / (y_i^T B y_i)$. For the purpose of applying a Krylov-like iterative method, the Newton equations (6.49) present the difficulty that the operator on the left-hand side sends $z_i \in (\mathrm{span}(BY))^\perp$ to the subspace $(\mathrm{span}(Y))^\perp$. A remedy is to solve the equivalent equation obtained by applying the projector $P_{BY,BY}$ to (6.49).

This trick is a particular instance of a more general technique called *preconditioning*. Assume that we have a matrix K that is a reasonably good approximation of $(A - \rho_i B)$ and such that the operator K^{-1} is easily available (in other words, systems $Kx = b$ are easy to solve). Let the superscript † denote the pseudo-inverse. Then (6.49) is equivalent to

$$(P_{BY,Y} K P_{Y,BY})^\dagger P_{BY,Y} (A - \rho_i I) P_{Y,BY} z_i$$
$$= -(P_{BY,Y} K P_{Y,BY})^\dagger P_{BY,Y} A y_i, \quad Y^T B z_i = 0. \quad (6.51)$$

The advantage is that the operator acting on z_i on the left-hand side of (6.51) is close to the identity, which improves the speed of convergence of Krylov-based iterative solvers. In practice, applying this operator is made possible by the Olsen formula

$$\left(P_{\widetilde{Q},Q} K P_{U,\widetilde{U}}\right)^\dagger = P_{U,U} P_{K^{-1}\widetilde{Q},\widetilde{U}} K^{-1} P_{Q,Q} = P_{U,U} K^{-1} P_{\widetilde{Q},K^{-T}\widetilde{U}} P_{Q,Q}$$

since we have assumed that the operator K^{-1} is easily available.

6.6 NOTES AND REFERENCES

The history of Newton's method on manifolds can be traced back to Luenberger [Lue72], if not earlier. Gabay [Gab82] proposed a Newton method on embedded submanifolds of $\mathbb{R}^n$. Smith [Smi93, Smi94] and Udriște [Udr94] formulated the method on general Riemannian manifolds, and Smith [Smi93, Smi94] provided a proof of quadratic convergence. Mahony's thesis [Mah94, Mah96, MM02] develops a Newton method on Lie groups and homogeneous spaces. Related work includes [Shu86, EAS98, OW00, Man02, MM02, ADM+02, FS02, DPM03, HT04, ABM06].

Smith [Smi93, Smi94] proposes a geometric Newton method that seeks a zero of a one-form (instead of a vector field). The underlying idea is that affine connections can be extended to general tensors, and in particular to one-forms. This approach makes it possible to define a Newton method that seeks a critical point of a cost function defined on a manifold equipped with an affine connection and a retraction (cf. the requirements of Algorithm 4) but not necessarily with a Riemannian metric since the Riemannian metric is no longer needed to define the gradient vector field. (Smith nevertheless requires the manifold to be Riemannian, notably because his algorithms use the Riemannian exponential to perform the update.)

Our convergence analysis (Theorem 6.3.2) of the geometric Newton method, which was built from the $\mathbb{R}^n$ proof given by Dennis and Schnabel [DS83, Th. 5.2.1] and is strongly based on coordinates. A coordinate-free approach can be found in Smith [Smi93, Smi94] for the case of a real-valued function on a Riemannian manifold with its Riemannian connection and exponential retraction; this elegant proof exploits bounds on the second and third covariant derivatives of the cost function and on the curvature of the manifold in a neighborhood of the critical point.

The calculus-based local analysis in Section 6.3.1 was inspired by the work of Hüper; see, e.g., [HH00, Hüp02, HT04].

In general, it cannot be guaranteed that the Jacobian $J(x_k)$ in Newton's method (Algorithm 4) is nonsingular. In other words, the Newton equation (6.1) may not admit one and only one solution. Even if it does, the Jacobian operator may be poorly conditioned, so that the linear system (6.1) cannot be reliably solved. If this happens while x_k is far away from the solution, a possible remedy is to fall back to a first-order, steepest-descent-like method. Several other remedies exist that pertain to globally convergent modifications of Newton's method; see, e.g., Dennis and Schnabel [DS83] and Nocedal and Wright [NW99].

Several ways to combine Newton and line-search approaches are discussed in Dennis and Schnabel [DS83, Ch. 6]. For more information on positive-definite modifications of the Hessian, see Nocedal and Wright [NW99, §6.3].

Theorem 6.3.2 states that the sequences $\{x_k\}$ generated by Algorithm 4 (the geometric Newton method) converge to any nondegenerate zero x_* of the vector field whenever the initial point x_0 belongs to *some* neighborhood of x_*; however, Theorem 6.3.2 is silent about the *size* of this neighborhood. For Newton's method in $\mathbb{R}^n$ applied to finding a zero of a function F, Kantorovich's theorem [Kan52] (or see [Den71, DS83]) states that if the product of a Lipschitz constant for the Jacobian times a bound on the inverse of the Jacobian at x_0 times a bound on the first Newton vector is smaller than $\frac{1}{2}$, then the function F has a unique zero x_* in a ball around x_0 larger than a certain bound and the iterates of Newton's method converge to x_*. This is a very powerful result, although in applications it is often difficult to ensure that the Kantorovich condition holds. Kantorovich's theorem was generalized to the Riemannian Newton method by Ferreira and Svaiter [FS02]. Another way to obtain information about the basins of attraction for Newton's method is to use Smale's γ and α theorems, which were generalized to the Riemannian Newton method by Dedieu *et al.* [DPM03].

For the application to the computation of invariant subspaces of a matrix, Theorem 6.5.1 states that the sequences $\{\text{span}(Y_k)\}$ produced by Algorithm 7 converge locally to any p-dimensional spectral invariant subspace $\mathcal{V}$ of A provided that the initial point is in a basin of attraction that contains an open ball around $\mathcal{V}$ in the Grassmann manifold, but it does not give any information about the size of the basin of attraction. This is an important issue since a large basin of attraction means that the iteration converges to the target invariant subspace even if the initial estimate is quite imprecise. It has been shown for previously available methods that the basins of attraction are prone to deteriorate when some eigenvalues are clustered. Batterson and Smillie [BS89] have drawn the basins of attraction of the Rayleigh quotient iteration (RQI) for $n = 3$ and have shown that they deteriorate when two eigenvalues are clustered. The bounds involved in the convergence results of the methods analyzed by Demmel [Dem87] blow up when the external gap vanishes. It was shown in [ASVM04], analytically and numerically, that the Riemannian Newton method applied to the Rayleigh quotient on the Grassmann manifold suffers from a similar dependence on the eigenvalue gap. It was also shown how this drawback can be remedied by considering a Levenberg-Marquardt-like modification of the Newton algorithm. The modified algorithm depends on a real parameter whose extreme values yield the Newton method on the one hand, and the steepest-descent method for the cost function $\|\xi\|$ with ξ defined in (6.40) on the other hand. A specific choice for this parameter was proposed that significantly improves the size of the basins of attraction around each invariant subspace.

The formula for the Hessian of the Brockett cost function (4.32) on the Stiefel manifold is straightforward using formula (5.15) for the Riemannian connection on Riemannian submanifolds of Euclidean spaces. The resulting Newton equation, however, is significantly more complex than the Newton equation (6.24) for the Rayleigh quotient on the Grassmann manifold. This outcome is due to the fact that the projection (3.35) onto the tangent space to the Stiefel manifold has one more term than the projection (3.41) onto the horizontal space of the Grassmann manifold (viewed as a quotient of the noncompact Stiefel manifold). The extra complexity introduced into the Newton equation significantly complicates the evaluation of each iterate and does not significantly add to the performance of the method since the Grassmann Newton method identifies an invariant p-dimensional subspace and the numerical cost of identifying the Ritz vectors of this subspace is a negligible additional cost on top of the subspace problem.

The term *"spectral* invariant subspace" is used by Rodman *et al.* [GLR86, RR02]; Stewart [Ste01] uses the term *"simple* invariant subspace". The material in Section 6.4.4 comes from [AMS04]. There is a vast literature on saddle-point problems such as (6.46); see Benzi *et al.* [BGL05] for a survey. For the numerical computation of U in (6.48), we refer the reader to [NW99, §16.2], for example. Saad [Saa96] is an excellent reference on iterative solvers for linear systems of equations. Practical implementation issues for Newton methods applied to the Rayleigh quotient are further discussed in Absil *et al.* [ASVM04].

Several methods proposed in the literature are closely related to the eigenvalue algorithms proposed in this chapter. These methods differ on three points: (i) the matrix BY in the structured matrix involved in the Newton equation (6.50). (ii) the shifts ρ_i. (iii) the way the z_i's are used to compute the new iterate.

The modified block newton method proposed by Lösche et al. [LST98] corresponds to Algorithm 8 with $B = I$. The authors utilize formula (6.50) and prove quadratic convergence. In fact, the order of convergence is even cubic [AMS04].

The Newton method discussed by Edelman et al. [EAS98, p. 344] corresponds to Algorithm 5 applied to the Rayleigh quotient (6.19) on the Grassmann manifold with ∇ chosen as the Riemannian connection and R chosen as the exponential mapping.

Smith [Smi93, Smi94] mentions Algorithm 6 but focuses on the version where the retraction R is the exponential mapping.

The shifted Tracemin algorithm of Sameh and Wisniewski [SW82, ST00] can be viewed as a modification of Algorithm 8 where the shifts ρ_i in (6.49)— or equivalently (6.50)—are selected using a particular strategy. The simple (unshifted) version of Tracemin corresponds to $\rho_i = 0$. This algorithm is mathematically equivalent to a direct subspace iteration with matrix $A^{-1}B$. The choice $\rho_i = 0$ is further discussed and exploited in the context of a trust-region method with an adaptive model in [ABGS05].

Equation (6.47) corresponds to equation (2.10) in Sameh and Tong [ST00] and to algorithm 2 in Lundström and Elden [LE02, p. 825] to some extent.

Relations between the RQI and various Newton-based approaches are mentioned in several references, e.g., [PW79, Shu86, Smi94, ADM+02, MA03]. This equivalence still holds when certain Galerkin techniques are used to approximately solve the Newton and RQI equations [SE02]. A block generalization of the RQI is proposed in [Smi97, AMSV02]. The connection with the Newton method does not hold for the block version [AMSV02].

The method proposed by Fattebert [Fat98] is connected with (6.50). The idea is to replace BY in (6.50) by thinner matrices where some columns that are not essential for the well conditioning of the linear system are omitted. This approach is thus midway between that of the Newton method and the RQI.

As discussed in [EAS98, AMSV02], the Newton method proposed by Chatelin [Cha84, Dem87] corresponds to performing a classical Newton method in a fixed coordinate chart of the Grassmann manifold. In contrast, the algorithms proposed in this chapter can be viewed as using an adaptive coordinate chart, notably because the retraction used, for example, in Step 3 of Algorithm 6 depends on the current iterate x_k.

For more information on the Jacobi-Davidson approach and related generalized Davidson methods, see Sleijpen et al. [SVdV96, SvdVM98], Morgan and Scott [MS86], Stathopoulos et al. [SS98, Sta05, SM06], Notay [Not02, Not03, Not05], van den Eshof [vdE02], Brandts [Bra03], and references therein. Methods for (approximately) solving the Jacobi (i.e., Newton) equation can also be found in these references. The Newton algorithm on the sphere for the Rayleigh quotient (Algorithm 6) is very similar to the simplified Jacobi-Davidson algorithm given in [Not02]; see the discussion in [ABG06b]. The Newton equations (6.24), (6.35), and (6.39) can be thought of as block versions of particular instances of the Jacobi correction

equations found in [SBFvdV96, HS03]. References related to sequential subspace optimization include Hager [Hag01, HP05], Absil and Gallivan [AG05],
and Narkiss and Zibulevsky [NZ05].

Chapter Seven

Trust-Region Methods

The plain Newton method discussed in Chapter 6 was shown to be locally convergent to any critical point of the cost function. The method does not distinguish among local minima, saddle points, and local maxima: all (nondegenerate) critical points are asymptotically stable fixed points of the Newton iteration. Moreover, it is possible to construct cost functions and initial conditions for which the Newton sequence does not converge. There even exist examples where the set of nonconverging initial conditions contains an open subset of search space.

To exploit the desirable superlinear local convergence properties of the Newton algorithm in the context of global optimization, it is necessary to embed the Newton update in some form of descent method. In Chapter 6 we briefly outlined how the Newton equation can be used to generate a descent direction that is used in a line-search algorithm. Such an approach requires modification of the Newton equation to ensure that the resulting sequence of search directions is gradient-related and an implementation of a standard line-search iteration. The resulting algorithm will converge to critical points of the cost function for *all* initial points. Moreover, saddle points and local maxima are rendered unstable, thus favoring convergence to local minimizers.

Trust-region methods form an alternative class of algorithms that combine desirable global convergence properties with a local superlinear rate of convergence. In addition to providing good global convergence, trust-region methods also provide a framework to relax the computational burden of the plain Newton method when the iterates are too far away from the solution for fast local convergence to set in. This is particularly important in the development of optimization algorithms on matrix manifolds, where the inverse Hessian computation can involve solving relatively complex matrix equations.

Trust-region methods can be understood as an enhancement of Newton's method. To this end, however, we need to consider this method from another viewpoint: instead of looking for an update vector along which the derivative of grad f is equal to $-\text{grad } f(x_k)$, it is equivalent to think of Newton's method (in $\mathbb{R}^n$) as the algorithm that selects the new iterate x_{k+1} to be the critical point of the quadratic Taylor expansion of the cost function f about x_k.

To this end, the chapter begins with a discussion of generalized quadratic models on manifolds (Section 7.1). Here again, a key role is played by the concept of retraction, which provides a way to pull back the cost function on the manifold to a cost function on the tangent space. It is therefore sufficient to define quadratic models on abstract vector spaces and to understand how these models correspond to the real-valued function on the manifold $\mathcal{M}$.

Once the notion of a quadratic model is established, a trust-region algorithm can be defined on a manifold (Section 7.2). It is less straightforward to show that all the desirable convergence properties of classical trust-region methods in $\mathbb{R}^n$ still hold, *mutatis mutandis*, for their manifold generalizations. The difficulty comes from the fact that trust-region methods on manifolds do not work with a single cost function but rather with a succession of cost functions whose domains are different tangent spaces. The issue of computing an (approximate but sufficiently accurate) solution of the trust-region subproblems is discussed in Section 7.3. The convergence analysis is carried out in Section 7.4. The chapter is concluded in Section 7.5 with a "checklist" of steps one has to go through in order to turn the abstract geometric trust-region schemes into practical numerical algorithms on a given manifold for a given cost function; this checklist is illustrated for several examples related to Rayleigh quotient minimization.

7.1 MODELS

Several classical optimization schemes rely on successive local minimization of quadratic models of the cost function. In this section, we review the notion of quadratic models in $\mathbb{R}^n$ and in general vector spaces. Then, making use of retractions, we extend the concept to Riemannian manifolds.

7.1.1 Models in $\mathbb{R}^n$

The fundamental mathematical tool that justifies the use of local models is Taylor's theorem (see Appendix A.6). In particular, we have the following results.

Proposition 7.1.1 *Let f be a smooth real-valued function on $\mathbb{R}^n$, $x \in \mathbb{R}^n$, $\mathcal{U}$ a bounded neighborhood of x, and H any symmetric matrix. Then there exists $c > 0$ such that, for all $(x + h) \in \mathcal{U}$,*

$$\left\| f(x + h) - \left(f(x) + \partial f(x)h + \tfrac{1}{2}h^T H h \right) \right\| \leq c\|h\|^2,$$

where $\partial f(x) := (\partial_1 f(x), \ldots, \partial_n f(x))$. If, moreover, $H_{i,j} = \partial_i \partial_j f(x)$, then there exists $c > 0$ such that, for all $(x + h) \in \mathcal{U}$,

$$\left\| f(x + h) - \left(f(x) + \partial f(x)h + \tfrac{1}{2}h^T H h \right) \right\| \leq c\|h\|^3.$$

7.1.2 Models in general Euclidean spaces

The first step towards defining quadratic models on Riemannian manifolds is to generalize the above results to (abstract) Euclidean spaces, i.e., finite-dimensional vector spaces endowed with an inner product $\langle \cdot, \cdot \rangle$. This is readily done using the results in Appendix A.6. (Note that Euclidean spaces are naturally finite-dimensional normed vector spaces for the induced norm $\|\xi\| := \sqrt{\langle\xi,\xi\rangle}.$)

Proposition 7.1.2 *Let f be a smooth real-valued function on a Euclidean space $\mathcal{E}$, $x \in \mathcal{E}$, $\mathcal{U}$ a bounded neighborhood of x, and $H : \mathcal{E} \to \mathcal{E}$ any symmetric operator. Then there exists $c > 0$ such that, for all $(x + h) \in \mathcal{U}$,*

$$\left\| f(x + h) - \left(f(x) + \langle\operatorname{grad} f(x), h\rangle + \tfrac{1}{2}\langle H[h], h\rangle\right) \right\| \le c\|h\|^2.$$

If, moreover, $H = \operatorname{Hess} f(x) : h \mapsto \mathrm{D}(\operatorname{grad} f)(x)[h]$, then there exists $c > 0$ such that, for all $(x + h) \in \mathcal{U}$,

$$\left\| f(x + h) - \left(f(x) + \langle\operatorname{grad} f(x), h\rangle + \tfrac{1}{2}\langle H[h], h\rangle\right) \right\| \le c\|h\|^3.$$

7.1.3 Models on Riemannian manifolds

Let f be a real-valued function on a Riemannian manifold $\mathcal{M}$ and let $x \in \mathcal{M}$. A model m_x of f around x is a real-valued function defined on a neighborhood of x such that (i) m_x is a "sufficiently good" approximation of f and (ii) m_x has a "simple" form that makes it easy to tackle with optimization algorithms.

The quality of the model m_x is assessed by evaluating how the discrepancy between $m_x(y)$ and $f(y)$ evolves as a function of the Riemannian distance $\operatorname{dist}(x, y)$ between x and y. The model m_x is an *order-q model*, $q > 0$, if there exists a neighborhood $\mathcal{U}$ of x in $\mathcal{M}$ and a constant $c > 0$ such that

$$|f(y) - m_x(y)| \le c \, (\operatorname{dist}(x, y))^{q+1} \quad \text{for all } y \in \mathcal{U}.$$

Note that an order-q model automatically satisfies $m_x(x) = f(x)$. The following result shows that the order of a model can be assessed using any retraction R on $\mathcal{M}$.

Proposition 7.1.3 *Let f be a real-valued function on a Riemannian manifold $\mathcal{M}$ and let $x \in \mathcal{M}$. A model m_x of f is order-q if and only if there exists a neighborhood $\mathcal{U}$ of x and a constant $c > 0$ such that*

$$|f(y) - m_x(y)| \le c\|R_x^{-1}(y)\|^{q+1} \quad \text{for all } y \in \mathcal{U}.$$

Proof. In view of the local rigidity property of retractions, it follows that $\mathrm{D}(\operatorname{Exp}^{-1} \circ R_x)(0_x) = \operatorname{id}_{T_x\mathcal{M}}$, hence $\|R_x^{-1}(y)\| = \|(R_x^{-1} \circ \operatorname{Exp}^{-1})(\operatorname{Exp}_x y)\| = \Omega(\|\operatorname{Exp}_x y\|) = \Omega(\operatorname{dist}(x, y))$; see Section A.4 for the definition of the asymptotic notation Ω. In other words, there is a neighborhood $\mathcal{U}$ of x and constants c_1 and c_2 such that

$$c_1 \operatorname{dist}(x, y) \le \|R_x^{-1}(y)\| \le c_2 \operatorname{dist}(x, y) \quad \text{for all } y \in \mathcal{U}.$$

$\square$

Proposition 7.1.3 yields a conceptually simple way to build an order-q model of f around x. Pick a retraction on $\mathcal{M}$. Consider

$$\widehat{f}_x := f \circ R_x,$$

the pullback of f to $T_x\mathcal{M}$ through R_x. Let $\widehat{m}_x$ be the order-q Taylor expansion of $\widehat{f}_x$ around the origin 0_x of $T_x\mathcal{M}$. Finally, push $\widehat{m}_x$ forward through R_x to obtain $m_x = \widehat{m}_x \circ R_x^{-1}$. This model is an order-q model of f around x. In particular, the obvious order-2 model to choose is

$$\widehat{m}_x(\xi) = \widehat{f}_x(0_x) + \mathrm{D}\widehat{f}_x(0_x)[\xi] + \tfrac{1}{2}\mathrm{D}^2\widehat{f}_x(0_x)[\xi,\xi]$$
$$= f(x) + \langle \mathrm{grad}\, f(x), \xi\rangle + \tfrac{1}{2}\langle \mathrm{Hess}\, \widehat{f}_x(0_x)[\xi], \xi\rangle.$$

Note that, since $\mathrm{D}R_x(0_x) = \mathrm{id}_{T_x\mathcal{M}}$ (see Definition 4.1.1), it follows that $\mathrm{D}f(x) = \mathrm{D}\widehat{f}_x(0_x)$, hence $\mathrm{grad}\,\widehat{f}_{x_k}(0_x) = \mathrm{grad}\, f(x)$, where $\mathrm{grad}\, f(x)$ denotes the (Riemannian) gradient of f (see Section 3.6).

In practice, the second-order differentials of the function $\widehat{f}_x$ may be difficult to compute. (The first-order differential is always straightforward since the rigidity condition of the retraction ensures $\mathrm{D}\widehat{f}_x(0_x) = \mathrm{D}f(x)$.) On the other hand, the Riemannian connection ∇ admits nice formulations on Riemannian submanifolds and Riemannian quotient manifolds (see Section 5.3). This suggests the model

$$m_x = \widehat{m}_x \circ R_x^{-1}$$

with

$$\widehat{m}_x(\xi) = f(x) + \langle \mathrm{grad}\, f(x), \xi\rangle + \tfrac{1}{2}\langle \mathrm{Hess}\, f(x)[\xi], \xi\rangle, \quad \xi \in T_x\mathcal{M}, \qquad (7.1)$$

where the quadratic term is given by the Riemannian Hessian

$$\mathrm{Hess}\, f(x)[\xi] = \nabla_\xi \mathrm{grad}\, f(x). \qquad (7.2)$$

In general, this model m_x is only order 1 because $\mathrm{Hess}\, f(x) \neq \mathrm{Hess}\, \widehat{f}_x(0_x)$. However, if R is a second-order retraction, then $\mathrm{Hess}\, f(x) = \mathrm{Hess}\, \widehat{f}_x(0_x)$ (Proposition 5.5.5) and m_x is order 2. More importantly, for any retraction, $\mathrm{Hess}\, f(x_*) = \mathrm{Hess}\, \widehat{f}_{x_*}(0_{x_*})$ when x_* is a critical point of f (Proposition 5.5.6).

The quadratic model (7.1) has a close connection to the Newton algorithm. Assuming that the Hessian is nonsingular at x, the critical point ξ_* of $\widehat{m}_x$ satisfies the Newton equation

$$\mathrm{Hess}\, f(x)[\eta_*] + \mathrm{grad}\, f(x) = 0.$$

It follows that the geometric Newton method (Algorithm 5), with retraction R and affine connection ∇, defines its next iterate as the critical point of the quadratic model (7.1).

We point out that all the models we have considered so far assume a quadratic form on $T_x\mathcal{M}$, i.e., there is a symmetric operator $H : T_x\mathcal{M} \to T_x\mathcal{M}$ such that

$$\widehat{m}_x(\xi) = f(x) + \langle \mathrm{grad}\, f(x), \xi\rangle + \tfrac{1}{2}\langle H[\xi], \xi\rangle.$$

Quadratic models are particularly interesting because the problem of minimizing a quadratic function under trust-region constraints is well understood and several algorithms are available (see Section 7.3).

7.2 TRUST-REGION METHODS

We first briefly review the principles of trust-region methods in $\mathbb{R}^n$. Extending the concept to manifolds is straightforward given the material developed in the preceding section.

7.2.1 Trust-region methods in $\mathbb{R}^n$

The basic trust-region method in $\mathbb{R}^n$ for a cost function f consists of adding to the current iterate $x \in \mathbb{R}^n$ the update vector $\eta \in \mathbb{R}^n$, solving (up to some approximation) the *trust-region subproblem*

$$\min_{\eta \in \mathbb{R}^n} m(\eta) = f(x) + \partial f(x)\eta + \tfrac{1}{2}\eta^T H\eta, \qquad \|\eta\| \leq \Delta, \qquad (7.3)$$

where H is some symmetric matrix and Δ is the trust-region radius. Clearly, a possible choice for H is the Hessian matrix $H_{i,j} = \partial_i \partial_j f(x)$; classical convergence results guarantee superlinear convergence if the chosen H is "sufficiently close" to the Hessian matrix. The algorithm used to compute an approximate minimizer η of the model within the trust region is termed the *inner iteration*. Once an η has been returned by the inner iteration, the quality of the model m is assessed by forming the quotient

$$\rho = \frac{f(x) - f(x + \eta)}{m(0) - m(\eta)}. \qquad (7.4)$$

Depending on the value of ρ, the new iterate $x + \eta$ is accepted or discarded and the trust-region radius Δ is updated. A specific procedure (in the Riemannian setting) is given in Algorithm 10, or see the textbooks mentioned in Notes and References.

7.2.2 Trust-region methods on Riemannian manifolds

We can now lay out the structure of a trust-region method on a Riemannian manifold $(\mathcal{M}, g)$ with retraction R. Given a cost function $f : \mathcal{M} \to \mathbb{R}$ and a current iterate $x_k \in \mathcal{M}$, we use R_{x_k} to locally map the minimization problem for f on $\mathcal{M}$ into a minimization problem for the *pullback* of f under R_{x_k},

$$\widehat{f}_{x_k} : T_{x_k}\mathcal{M} \to \mathbb{R} : \xi \mapsto f(R_{x_k}\xi). \qquad (7.5)$$

The Riemannian metric g turns $T_{x_k}\mathcal{M}$ into a Euclidean space endowed with the inner product $g_{x_k}(\cdot, \cdot)$—which we usually denote by $\langle \cdot, \cdot \rangle_{x_k}$—and the trust-region subproblem on $T_{x_k}\mathcal{M}$ reads

$$\min_{\eta \in T_{x_k}\mathcal{M}} \widehat{m}_{x_k}(\eta) = f(x_k) + \langle \operatorname{grad} f(x_k), \eta \rangle + \tfrac{1}{2}\langle H_k[\eta], \eta \rangle,$$
$$\text{subject to } \langle \eta, \eta \rangle_{x_k} \leq \Delta_k^2, \qquad (7.6)$$

where H_k is some symmetric operator on $T_{x_k}\mathcal{M}$. (A possible choice is $H_k := \operatorname{Hess} f(x_k)$, the Riemannian Hessian (7.2).) This is called the *trust-region subproblem*.

Next, an (approximate) solution η_k of the Euclidean trust-region subproblem (7.6) is computed using any available method (inner iteration). The candidate for the new iterate is then given by $R_{x_k}(\eta_k)$. Notice that the inner iteration has to operate on the Euclidean space $T_x\mathcal{M}$ which, in the case of embedded submanifolds and quotient manifolds of a matrix space $\mathbb{R}^{n \times p}$, is represented as a linear subspace of $\mathbb{R}^{n \times p}$. Fortunately, most (or even all?) classical algorithms for the trust-region subproblem are readily adapted to Euclidean matrix spaces (see Section 7.3 for details).

The decisions on accepting or rejecting the candidate $R_{x_k}(\eta_k)$ and on selecting the new trust-region radius Δ_{k+1} are based on the quotient

$$\rho_k = \frac{f(x_k) - f(R_{x_k}(\eta_k))}{\widehat{m}_{x_k}(0_{x_k}) - \widehat{m}_{x_k}(\eta_k)} = \frac{\widehat{f}_{x_k}(0_{x_k}) - \widehat{f}_{x_k}(\eta_k)}{\widehat{m}_{x_k}(0_{x_k}) - \widehat{m}_{x_k}(\eta_k)}. \tag{7.7}$$

If ρ_k is exceedingly small, then the model is very inaccurate: the step must be rejected, and the trust-region radius must be reduced. If ρ_k is small but less dramatically so, then the step is accepted but the trust-region radius is reduced. If ρ_k is close to 1, then there is a good agreement between the model and the function over the step, and the trust-region radius can be expanded. If $\rho_k \gg 1$, then the model is inaccurate, but the overall optimization iteration is producing a significant decrease in the cost. In this situation a possible strategy is to increase the trust region in the hope that your luck will hold and that bigger steps will result in a further decrease in the cost, regardless of the quality of the model approximation. This procedure is formalized in Algorithm 10.

Later in the chapter we sometimes drop the subscript k and denote x_{k+1} by x_+.

In general, there is no assumption on the operator H_k in (7.6) other than being a symmetric linear operator. Consequently, the choice of the retraction R does not impose any constraint on $\widehat{m}_{x_k}$. In order to achieve superlinear convergence, however, H_k must approximate the Hessian (Theorem 7.4.11). The issue of obtaining an approximate Hessian in practice is addressed in Section 7.5.1.

7.3 COMPUTING A TRUST-REGION STEP

Step 2 in Algorithm 10 computes an (approximate) solution of the trust-region subproblem (7.6),

$$\min_{\eta \in T_{x_k}\mathcal{M}} \widehat{m}_{x_k}(\eta) = f(x_k) + \langle \operatorname{grad} f(x_k), \eta \rangle + \frac{1}{2}\langle H_k[\eta], \eta \rangle,$$

$$\text{subject to } \langle \eta, \eta \rangle_{x_k} \leq \Delta_k^2.$$

Methods for solving trust-region subproblems in the $\mathbb{R}^n$ case can be roughly classified into two broad classes: (i) methods (based on the work of Moré and Sorensen) that compute nearly exact solutions of the subproblem; (ii) methods that compute an approximate solution to the trust-region subproblem

---

Algorithm 10 Riemannian trust-region (RTR) meta-algorithm

Require: Riemannian manifold $(\mathcal{M}, g)$; scalar field f on $\mathcal{M}$; retraction R from $T\mathcal{M}$ to $\mathcal{M}$ as in Definition 4.1.1.
  Parameters: $\bar{\Delta} > 0$, $\Delta_0 \in (0, \bar{\Delta})$, and $\rho' \in [0, \frac{1}{4})$.
Input: Initial iterate $x_0 \in \mathcal{M}$.
Output: Sequence of iterates $\{x_k\}$.
1: **for** $k = 0, 1, 2, \ldots$ **do**
2:   Obtain η_k by (approximately) solving (7.6);
3:   Evaluate ρ_k from (7.7);
4:   **if** $\rho_k < \frac{1}{4}$ **then**
5:     $\Delta_{k+1} = \frac{1}{4}\Delta_k$;
6:   **else if** $\rho_k > \frac{3}{4}$ and $\|\eta_k\| = \Delta_k$ **then**
7:     $\Delta_{k+1} = \min(2\Delta_k, \bar{\Delta})$;
8:   **else**
9:     $\Delta_{k+1} = \Delta_k$;
10:   **end if**
11:   **if** $\rho_k > \rho'$ **then**
12:     $x_{k+1} = R_x \eta_k$;
13:   **else**
14:     $x_{k+1} = x_k$;
15:   **end if**
16: **end for**

---

using a computationally simple iteration that achieves at least the decrease in cost obtained for the Cauchy point. For the trust-region subproblem (7.6), assuming that $\operatorname{grad} f(x_k) \neq 0$, we define the *Cauchy point* as the solution η_k^C of the one-dimensional problem

$$\eta_k^C = \arg \min_\eta \{\widehat{m}_{x_k}(\eta) : \eta = -\tau \operatorname{grad} f(x_k), \ \tau > 0, \ \|\eta\| \leq \Delta_k\}, \qquad (7.8)$$

which reduces to the classical definition of the Cauchy point when $\mathcal{M} = \mathbb{R}^n$. The *Cauchy decrease* is given by $\widehat{m}_{x_k}(0) - \widehat{m}_{x_k}(\eta_k^C)$. We present a brief outline of the first method before providing a more detailed development of a conjugate gradient algorithm for the second approach that we prefer for the later developments.

7.3.1 Computing a nearly exact solution

The following statement is a straightforward adaptation of a result of Moré and Sorensen to the case of the trust-region subproblem on $T_{x_k}\mathcal{M}$ as expressed in (7.6).

Proposition 7.3.1 *The vector η^* is a global solution of the trust-region subproblem (7.6) if and only if there exists a scalar $\mu \geq 0$ such that the* *following conditions are satisfied:*

$$(H_k + \mu\,\mathrm{id})\eta^* = -\mathrm{grad}\,f(x_k), \tag{7.9a}$$

$$\mu(\Delta_k - \|\eta^*\|) = 0, \tag{7.9b}$$

$$(H_k + \mu\,\mathrm{id}) \text{ is positive-semidefinite,} \tag{7.9c}$$

$$\|\eta^*\| \leq \Delta_k. \tag{7.9d}$$

This result suggests a strategy for computing the solution of the subproblem (7.6). Either solve (7.9) with $\mu = 0$ or define

$$\eta(\mu) := -(H_k + \mu\,\mathrm{id})^{-1}\,\mathrm{grad}\,f(x_k)$$

and adjust μ to achieve $\|\eta(\mu)\| = \Delta_k$. Several algorithms have been proposed to perform this task; see Notes and References.

7.3.2 Improving on the Cauchy point

In many applications, the dimension d of the manifold $\mathcal{M}$ is extremely large (see, for example, computation of an invariant subspace of a large matrix A in Section 7.5). In such cases, solving the linear system (7.9a) of size d or checking the positive-definiteness of a $d \times d$ matrix (7.9c) is unfeasible. Many algorithms exist that scale down the precision of the solution of the trust-region subproblem (7.6) and lighten the numerical burden.

A number of these methods start by computing the Cauchy point (7.8), and then attempt to improve on it. The improvement strategy is often designed so that, when H_k is positive-definite, and given sufficient iterations, the estimate eventually reaches the minimizer $\eta_k^N = (H_k)^{-1}\,\mathrm{grad}\,f(x_k)$ provided that the minimizer lies within the trust region. Among these strategies, the *truncated conjugate-gradient method* is one of the most popular. Algorithm 11 is a straightforward adaptation of the truncated CG method in $\mathbb{R}^n$ to the trust-region subproblem (7.6) in $T_{x_k}\mathcal{M}$. Note that we use superscripts to denote the evolution of η within the inner iteration, while subscripts are used in the outer iteration.

Several comments about Algorithm 11 are in order.

The following result will be useful in the convergence analysis of trust-region methods.

Proposition 7.3.2 *Let η^i, $i = 0, \ldots, j$, be the iterates generated by Algorithm 11 (truncated CG method). Then $\widehat{m}_{x_k}(\eta^i)$ is strictly decreasing and $\widehat{m}_{x_k}(\eta_k) \leq \widehat{m}_{x_k}(\eta^i)$, $i = 0, \ldots, j$. Further, $\|\eta^i\|$ is strictly increasing and $\|\eta_k\| > \|\eta^i\|$, $i = 0, \ldots, j$.*

The simplest stopping criterion to use in Step 14 of Algorithm 11 is to truncate after a fixed number of iterations. In order to achieve superlinear convergence (see Section 7.4.2), one may take the stopping criterion

$$\|r_{j+1}\| \leq \|r_0\| \min(\|r_0\|^\theta, \kappa), \tag{7.10}$$

where $\theta > 0$ is a real parameter chosen in advance.

Algorithm 11 Truncated CG (tCG) method for the trust-region subproblem

---

Goal: This algorithm handles Step 2 of Algorithm 10.

  1: Set $\eta^0 = 0$, $r_0 = \operatorname{grad} f(x_k)$, $\delta_0 = -r_0$; $j = 0$;

  2: **loop**

  3:     **if** $\langle \delta_j, H_k \delta_j \rangle_{x_k} \leq 0$ **then**

  4:         Compute τ such that $\eta = \eta^j + \tau \delta_j$ minimizes $\widehat{m}_{x_k}(\eta)$ in (7.6) and satisfies $\|\eta\|_{g_x} = \Delta$;

  5:         **return** $\eta_k := \eta$;

  6:     **end if**

  7:     Set $\alpha_j = \langle r_j, r_j \rangle_{x_k} / \langle \delta_j, H_k \delta_j \rangle_{x_k}$;

  8:     Set $\eta^{j+1} = \eta^j + \alpha_j \delta_j$;

  9:     **if** $\|\eta^{j+1}\|_{g_x} \geq \Delta$ **then**

10:         Compute $\tau \geq 0$ such that $\eta = \eta^j + \tau \delta_j$ satisfies $\|\eta\|_{g_x} = \Delta$;

11:         **return** $\eta_k := \eta$;

12:     **end if**

13:     Set $r_{j+1} = r_j + \alpha_j H_k \delta_j$;

14:     **if** a stopping criterion is satisfied **then**

15:         **return** $\eta_k := \eta^{j+1}$;

16:     **end if**

17:     Set $\beta_{j+1} = \langle r_{j+1}, r_{j+1} \rangle_{x_k} / \langle r_j, r_j \rangle_{x_k}$;

18:     Set $\delta_{j+1} = -r_{j+1} + \beta_{j+1} \delta_j$;

19:     Set $j = j + 1$;

20: **end loop**

In Steps 4 and 10 of Algorithm 11, τ is found by computing the positive root of the quadratic equation

$$\tau^2 \langle \delta_j, \delta_j \rangle_{x_k} + 2\tau \langle \eta^j, \delta_j \rangle_{x_k} = \Delta_k^2 - \langle \eta^j, \eta^j \rangle_{x_k}.$$

Notice that the truncated CG algorithm is "inverse-free", as it uses H_k only in the computation of $H_k[\delta_j]$.

Practical implementations of Algorithm 11 usually include several additional features to reduce the numerical burden and improve the robustness to numerical errors. For example, the value of $\langle r_{j+1}, r_{j+1} \rangle_{x_k}$ can be stored since it will be needed at the next iteration. Because the Hessian operator H_k is an operator on a vector space of dimension d, where d may be very large, it is important to implement an efficient routine for computing $H_k \delta$. In many practical cases, the tangent space $T_x \mathcal{M}$ to which the quantities η, r, and δ belong will be represented as a linear subspace of a higher-dimensional Euclidean space; to prevent numerical errors it may be useful from time to time to reproject the above quantities onto the linear subspace.

7.4 CONVERGENCE ANALYSIS

In this section, we study the global convergence properties of the RTR scheme (Algorithm 10) without any assumption on the way the trust-region subproblems are solved (Step 2), except that the approximate solution η_k must produce a decrease in the model that is at least a fixed fraction of the Cauchy decrease. Under mild additional assumptions on the retraction and on the cost function, it is shown that the sequences $\{x_k\}$ produced by Algorithm 10 converge to the set of critical points of the cost function. This result is well known in the $\mathbb{R}^n$ case; in the case of manifolds, the convergence analysis has to address the fact that a different lifted cost function $\widehat{f}_{x_k}$ is considered at each iterate x_k.

In the second part of the section we analyze the local convergence of Algorithm 10-11 around nondegenerate local minima. Algorithm 10-11 refers to the RTR framework where the trust-region subproblems are approximately solved using the truncated CG algorithm with stopping criterion (7.10). It is shown that the iterates of the algorithm converge to nondegenerate critical points with an order of convergence of at least $\min\{\theta + 1, 2\}$, where θ is the parameter chosen for the stopping condition (7.10).

7.4.1 Global convergence

The objective of this section is to show that, under appropriate assumptions, the sequence $\{x_k\}$ generated by Algorithm 10 converges to the critical set of the cost function; this generalizes a classical convergence property of trust-region methods in $\mathbb{R}^n$. In what follows, $(\mathcal{M}, g)$ is a Riemannian manifold of dimension d and R is a retraction on $\mathcal{M}$ (Definition 4.1.1). We define the pullback cost

$$\widehat{f} : T\mathcal{M} \mapsto \mathbb{R} : \xi \mapsto f(R\xi) \tag{7.11}$$

and, in accordance with (7.5), $\widehat{f}_x$ denotes the restriction of $\widehat{f}$ to $T_x\mathcal{M}$. We denote by $B_\delta(0_x) = \{\xi \in T_x\mathcal{M} : \|\xi\| < \delta\}$ the open ball in $T_x\mathcal{M}$ of radius δ centered at 0_x, and $B_\delta(x)$ stands for the set $\{y \in M : \mathrm{dist}(x, y) < \delta\}$, where dist denotes the Riemannian distance (i.e., the distance defined in terms of the Riemannian metric; see Section 3.6). We denote by $P_\gamma^{t \leftarrow t_0} v$ the vector of $T_{\gamma(t)}\mathcal{M}$ obtained by parallel translation (with respect to the Riemannian connection) of the vector $v \in T_{\gamma(t_0)}\mathcal{M}$ along a curve γ.

As in the classical $\mathbb{R}^n$ proof, we first show that at least one accumulation point of $\{x_k\}$ is a critical point of f. The convergence result requires that $\widehat{m}_{x_k}(\eta_k)$ be a sufficiently good approximation of $\widehat{f}_{x_k}(\eta_k)$. In classical proofs, this is often guaranteed by the assumption that the Hessian of the cost function is bounded. It is, however, possible to weaken this assumption, which leads us to consider the following definition.

Definition 7.4.1 (radially L-C^1 function) *Let $\widehat{f} : T\mathcal{M} \to \mathbb{R}$ be defined as in (7.11). We say that $\widehat{f}$ is* radially Lipschitz continuously differentiable *if there exist reals $\beta_{RL} > 0$ and $\delta_{RL} > 0$ such that, for all $x \in \mathcal{M}$, for all $\xi \in T_x\mathcal{M}$ with $\|\xi\| = 1$, and for all $t < \delta_{RL}$, it holds that*

$$\left| \frac{\mathrm{d}}{\mathrm{d}\tau} \widehat{f}_x(\tau\xi)|_{\tau=t} - \frac{\mathrm{d}}{\mathrm{d}\tau} \widehat{f}_x(\tau\xi)|_{\tau=0} \right| \leq \beta_{RL} t. \tag{7.12}$$

For the purpose of Algorithm 10, which is a descent algorithm, this condition needs only to be imposed for all x in the level set

$$\{x \in M : f(x) \leq f(x_0)\}. \tag{7.13}$$

A key assumption in the classical global convergence result in $\mathbb{R}^n$ is that the approximate solution η_k of the trust-region subproblem (7.6) produces at least as much decrease in the model function as a fixed fraction of the Cauchy decrease. The definition (7.8) of the Cauchy point is equivalent to the closed-form definition $\eta_k^C = -\tau_k \operatorname{grad} f(x_k)$ with

$$\tau_k = \begin{cases} \frac{\Delta_k}{\|\operatorname{grad} f(x_k)\|} & \text{if } \langle H_k[\operatorname{grad} f(x_k)], \operatorname{grad} f(x_k)\rangle_{x_k} \leq 0, \\ \frac{\Delta_k}{\|\operatorname{grad} f(x_k)\|} \min\left(\frac{\|\operatorname{grad} f(x_k)\|^3}{\Delta_k \langle H_k[\operatorname{grad} f(x_k)], \operatorname{grad} f(x_k)\rangle_{x_k}}, 1\right) & \text{otherwise.} \end{cases}$$

(Note that the definition of the Cauchy point excludes the case $\operatorname{grad} f(x_k) = 0$, for which convergence to a critical point becomes trivial.) The assumption on the decrease in f then becomes

$$\widehat{m}_{x_k}(0) - \widehat{m}_{x_k}(\eta_k) \geq c_1 \|\operatorname{grad} f(x_k)\| \min\left(\Delta_k, \frac{\|\operatorname{grad} f(x_k)\|}{\|H_k\|}\right), \tag{7.14}$$

for some constant $c_1 > 0$, where $\|H_k\|$ is defined as

$$\|H_k\| := \sup\{\|H_k\zeta\| : \zeta \in T_{x_k}\mathcal{M}, \|\zeta\| = 1\}. \tag{7.15}$$

In particular, the Cauchy point satisfies (7.14) with $c_1 = \frac{1}{2}$. Hence the tangent vector η_k returned by the truncated CG method (Algorithm 11) satisfies (7.14) with $c_1 = \frac{1}{2}$ since the truncated CG method first computes the Cauchy point and then attempts to improve the model decrease.

With these ideas in place, we can state and prove the first global convergence result. Note that this theorem is presented under weak assumptions; stronger but arguably easier to check assumptions are given in Proposition 7.4.5.

Theorem 7.4.2 *Let $\{x_k\}$ be a sequence of iterates generated by Algorithm 10 with $\rho' \in [0, \frac{1}{4})$. Suppose that f is C^1 and bounded below on the level set (7.13), that $\widehat{f}$ is radially L-C^1 (Definition 7.4.1), and that there is a constant β such that $\|H_k\| \leq \beta$ for all k. Further suppose that all η_k's obtained in Step 2 of Algorithm 10 satisfy the Cauchy decrease inequality (7.14) for some positive constant c_1. We then have*

$$\liminf_{k \to \infty} \|\operatorname{grad} f(x_k)\| = 0.$$

Proof. From the definition of the ratio ρ_k in (7.7), we have

$$|\rho_k - 1| = \left| \frac{\widehat{m}_{x_k}(\eta_k) - \widehat{f}_{x_k}(\eta_k)}{\widehat{m}_{x_k}(0) - \widehat{m}_{x_k}(\eta_k)} \right|. \tag{7.16}$$

Proposition A.6.1 (Taylor) applied to the function $t \mapsto \widehat{f}_{x_k}(t \frac{\eta_k}{\|\eta_k\|})$ yields

$$\widehat{f}_{x_k}(\eta_k) = \widehat{f}_{x_k}(0_{x_k}) + \|\eta_k\| \left. \frac{\mathrm{d}}{\mathrm{d}\tau} \widehat{f}_{x_k}(\tau \frac{\eta_k}{\|\eta_k\|}) \right|_{\tau=0} + \epsilon'$$

$$= f(x_k) + \langle \operatorname{grad} f(x_k), \eta_k \rangle_{x_k} + \epsilon',$$

where $|\epsilon'| = \frac{1}{2}\beta_{RL}\|\eta_k\|^2$ whenever $\|\eta_k\| < \delta_{RL}$ and β_{RL} and δ_{RL} are the constants in the radially L-C^1 property (7.12). Therefore, it follows from the definition (7.6) of $\widehat{m}_{x_k}$ that

$$|\widehat{m}_{x_k}(\eta_k) - \widehat{f}_{x_k}(\eta_k)| = \left| \frac{1}{2}\langle H_k \eta_k, \eta_k \rangle_{x_k} - \epsilon' \right|$$
$$\leq \frac{1}{2}\beta\|\eta_k\|^2 + \frac{1}{2}\beta_{RL}\|\eta_k\|^2 \leq \beta'\|\eta_k\|^2 \tag{7.17}$$

whenever $\|\eta_k\| < \delta_{RL}$, where $\beta' = \max(\beta, \beta_{RL})$.

Assume for contradiction that the claim does not hold; i.e., assume there exist $\epsilon > 0$ and a positive index K such that

$$\|\operatorname{grad} f(x_k)\| \geq \epsilon \quad \text{for all } k \geq K. \tag{7.18}$$

From (7.14), for $k \geq K$, we have

$$\widehat{m}_{x_k}(0) - \widehat{m}_{x_k}(\eta_k) \geq c_1 \|\operatorname{grad} f(x_k)\| \min\left(\Delta_k, \frac{\|\operatorname{grad} f(x_k)\|}{\|H_k\|} \right)$$
$$\geq c_1 \epsilon \min\left(\Delta_k, \frac{\epsilon}{\beta'} \right). \tag{7.19}$$

Substituting (7.17) and (7.19) into (7.16), we have that

$$|\rho_k - 1| \leq \frac{\beta'\|\eta_k\|^2}{c_1 \epsilon \min\left(\Delta_k, \frac{\epsilon}{\beta'} \right)} \leq \frac{\beta' \Delta_k^2}{c_1 \epsilon \min\left(\Delta_k, \frac{\epsilon}{\beta'} \right)} \tag{7.20}$$

whenever $\|\eta_k\| < \delta_{RL}$. Let $\hat{\Delta}$ be defined as $\hat{\Delta} = \min\left(\frac{c_1 \epsilon}{2\beta'}, \frac{\epsilon}{\beta'}, \delta_{RL} \right)$. If $\Delta_k \leq \hat{\Delta}$, then $\min\left(\Delta_k, \frac{\epsilon}{\beta'} \right) = \Delta_k$ and (7.20) becomes

$$|\rho_k - 1| \leq \frac{\beta' \hat{\Delta} \Delta_k}{c_1 \epsilon \min\left(\Delta_k, \frac{\epsilon}{\beta'} \right)} \leq \frac{\Delta_k}{2 \min\left(\Delta_k, \frac{\epsilon}{\beta'} \right)} = \frac{1}{2}.$$

Therefore, $\rho_k \geq \frac{1}{2} > \frac{1}{4}$ whenever $\Delta_k \leq \hat{\Delta}$, so that by the workings of Algorithm 10, it follows (from the argument above) that $\Delta_{k+1} \geq \Delta_k$ whenever $\Delta_k \leq \hat{\Delta}$. It follows that a reduction in Δ_k (by a factor of $\frac{1}{4}$) can occur in Algorithm 10 only when $\Delta_k > \hat{\Delta}$. Therefore, we conclude that

$$\Delta_k \geq \min\left(\Delta_K, \hat{\Delta}/4 \right) \quad \text{for all } k \geq K. \tag{7.21}$$

Suppose now that there is an infinite subsequence $\mathcal{K}$ such that $\rho_k \geq \frac{1}{4} > \rho'$ for $k \in \mathcal{K}$. If $k \in \mathcal{K}$ and $k \geq K$, we have from (7.19) that

$$f(x_k) - f(x_{k+1}) = f_{x_k} - \hat{f}_{x_k}(\eta_k)$$

$$\geq \frac{1}{4}(\widehat{m}_{x_k}(0) - \widehat{m}_{x_k}(\eta_k)) \geq \frac{1}{4}c_1\epsilon \min\left(\Delta_k, \frac{\epsilon}{\beta'}\right). \quad (7.22)$$

Since f is bounded below on the level set containing these iterates, it follows from this inequality that $\lim_{k \in \mathcal{K}, k \to \infty} \Delta_k = 0$, clearly contradicting (7.21). Then such an infinite subsequence as $\mathcal{K}$ cannot exist. It follows that we must have $\rho_k < \frac{1}{4}$ for all k sufficiently large so that Δ_k will be reduced by a factor of $\frac{1}{4}$ on every iteration. Then we have $\lim_{k \to \infty} \Delta_k = 0$, which again contradicts (7.21). Hence our assumption (7.18) is false, and the proof is complete. $\qquad\qquad\qquad\qquad\qquad\qquad\qquad\qquad\qquad\qquad\qquad\qquad\qquad\qquad\square$

To further show that all accumulation points of $\{x_k\}$ are critical points, we need to make an additional regularity assumption on the cost function f. The convergence result in $\mathbb{R}^n$ requires that f be Lipschitz continuously differentiable. That is, for any $x, y \in \mathbb{R}^n$,

$$\|\operatorname{grad} f(y) - \operatorname{grad} f(x)\| \leq \beta_1 \|y - x\|. \quad (7.23)$$

A key to obtaining a Riemannian counterpart of this global convergence result is to adapt the notion of being Lipschitz continuously differentiable to the Riemannian manifold $(\mathcal{M}, g)$. The expression $\|x - y\|$ on the right-hand side of (7.23) naturally becomes the Riemannian distance $\operatorname{dist}(x, y)$. For the left-hand side of (7.23), observe that the operation $\operatorname{grad} f(x) - \operatorname{grad} f(y)$ is not well defined in general on a Riemannian manifold since $\operatorname{grad} f(x)$ and $\operatorname{grad} f(y)$ belong to two different tangent spaces, namely, $T_x\mathcal{M}$ and $T_y\mathcal{M}$. However, if y belongs to a normal neighborhood of x, then there is a unique geodesic $\alpha(t) = \operatorname{Exp}_x(t \operatorname{Exp}_x^{-1} y)$ in this neighborhood such that $\alpha(0) = x$ and $\alpha(1) = y$, and we can parallel-translate $\operatorname{grad} f(y)$ along α to obtain the vector $P_\alpha^{0 \leftarrow 1} \operatorname{grad} f(y)$ in $T_x\mathcal{M}$. A lower bound on the size of the normal neighborhoods is given by the *injectivity radius*, defined as

$$i(\mathcal{M}) := \inf_{x \in \mathcal{M}} i_x,$$

where

$$i_x := \sup\{\epsilon > 0 : \operatorname{Exp}_x|_{B_\epsilon(0_x)} \text{ is a diffeomorphism for all } x \in \mathcal{M}\}.$$

This yields the following definition.

Definition 7.4.3 (Lipschitz continuously differentiable) *Assume that $(\mathcal{M}, g)$ has a positive injectivity radius. A real function f on $\mathcal{M}$ is Lipschitz continuously differentiable if it is differentiable and if there exists β_1 such that, for all x, y in $\mathcal{M}$ with $\operatorname{dist}(x, y) < i(\mathcal{M})$, it holds that*

$$\|P_\alpha^{0 \leftarrow 1} \operatorname{grad} f(y) - \operatorname{grad} f(x)\| \leq \beta_1 \operatorname{dist}(y, x), \quad (7.24)$$

where α is the unique minimizing geodesic with $\alpha(0) = x$ and $\alpha(1) = y$.

Note that (7.24) is symmetric in x and y; indeed, since the parallel transport is an isometry, it follows that

$$\|P_\alpha^{0\leftarrow 1}\operatorname{grad} f(y) - \operatorname{grad} f(x)\| = \|\operatorname{grad} f(y) - P_\alpha^{1\leftarrow 0}\operatorname{grad} f(x)\|.$$

Moreover, we place one additional requirement on the retraction R, that there exist $\mu > 0$ and $\delta_\mu > 0$ such that

$$\|\xi\| \geq \mu \operatorname{dist}(x, R_x\xi) \quad \text{for all } x \in \mathcal{M}, \text{for all } \xi \in T_x\mathcal{M}, \ \|\xi\| \leq \delta_\mu. \quad (7.25)$$

Note that for the exponential retraction, (7.25) is satisfied as an equality with $\mu = 1$. The bound is also satisfied when $\mathcal{M}$ is compact (Corollary 7.4.6).

We are now ready to show that under some additional assumptions, the gradient of the cost function converges to zero on the whole sequence of iterates. Here again we refer to Proposition 7.4.5 for a simpler (but slightly stronger) set of assumptions that yield the same result.

Theorem 7.4.4 *Let $\{x_k\}$ be a sequence of iterates generated by Algorithm 10. Suppose that all the assumptions of Theorem 7.4.2 are satisfied. Further suppose that $\rho' \in (0, \frac{1}{4})$, that f is Lipschitz continuously differentiable (Definition 7.4.3), and that (7.25) is satisfied for some $\mu > 0$, $\delta_\mu > 0$. It then follows that*

$$\lim_{k\to\infty} \operatorname{grad} f(x_k) = 0.$$

Proof. Consider any index m such that $\operatorname{grad} f(x_m) \neq 0$. Define the scalars

$$\epsilon = \frac{1}{2}\|\operatorname{grad} f(x_m)\|, \quad r = \min\left(\frac{\|\operatorname{grad} f(x_m)\|}{2\beta_1}, i(M)\right) = \min\left(\frac{\epsilon}{\beta_1}, i(M)\right).$$

In view of the Lipschitz property (7.24), we have for all $x \in B_r(x_m)$,

$$\begin{aligned}
\|\operatorname{grad} f(x)\| &= \|P_\alpha^{0\leftarrow 1}\operatorname{grad} f(x)\| \\
&= \|P_\alpha^{0\leftarrow 1}\operatorname{grad} f(x) + \operatorname{grad} f(x_m) - \operatorname{grad} f(x_m)\| \\
&\geq \|\operatorname{grad} f(x_m)\| - \|P_\alpha^{0\leftarrow 1}\operatorname{grad} f(x) - \operatorname{grad} f(x_m)\| \\
&\geq 2\epsilon - \beta_1 \operatorname{dist}(x, x_m) \\
&> 2\epsilon - \beta_1 \min\left(\frac{\|\operatorname{grad} f(x_m)\|}{2\beta_1}, i(M)\right) \\
&\geq 2\epsilon - \frac{1}{2}\|\operatorname{grad} f(x_m)\| \\
&= \epsilon.
\end{aligned}$$

If the entire sequence $\{x_k\}_{k\geq m}$ stays inside the ball $B_r(x_m)$, then we have $\|\operatorname{grad} f(x_k)\| > \epsilon$ for all $k \geq m$, a contradiction to Theorem 7.4.2. Thus the sequence eventually leaves the ball $B_r(x_m)$. Let the index $l \geq m$ be such that x_{l+1} is the first iterate after x_m outside $B_r(x_m)$. Since $\|\operatorname{grad} f(x_k)\| > \epsilon$ for

$k = m, m+1, \ldots, l$, we have, in view of the Cauchy decrease condition (7.14),

$$f(x_m) - f(x_{l+1}) = \sum_{k=m}^{l} f(x_k) - f(x_{k+1})$$

$$\geq \sum_{k=m, x_k \neq x_{k+1}}^{l} \rho'\left(\widehat{m}_{x_k}(0) - \widehat{m}_{x_k}(\eta_k)\right)$$

$$\geq \sum_{k=m, x_k \neq x_{k+1}}^{l} \rho' c_1 \|\operatorname{grad} f(x_k)\| \min\left(\Delta_k, \frac{\|\operatorname{grad} f(x_k)\|}{\|B_k\|}\right)$$

$$\geq \sum_{k=m, x_k \neq x_{k+1}}^{l} \rho' c_1 \epsilon \min\left(\Delta_k, \frac{\epsilon}{\beta}\right).$$

We distinguish two cases. If $\Delta_k > \epsilon/\beta$ in at least one of the terms of the sum, then

$$f(x_m) - f(x_{l+1}) \geq \rho' c_1 \epsilon \frac{\epsilon}{\beta}. \tag{7.26}$$

In the other case, we have

$$f(x_m) - f(x_{l+1}) \geq \rho' c_1 \epsilon \sum_{k=m, x_k \neq x_{k+1}}^{l} \Delta_k \geq \rho' c_1 \epsilon \sum_{k=m, x_k \neq x_{k+1}}^{l} \|\eta_k\|. \tag{7.27}$$

If $\|\eta_k\| > \delta_\mu$ in at least one term in the sum, then

$$f(x_m) - f(x_{l+1}) \geq \rho' c_1 \epsilon \delta_\mu. \tag{7.28}$$

Otherwise, (7.27) yields

$$f(x_m) - f(x_{l+1}) \geq \rho' c_1 \epsilon \sum_{k=m, x_k \neq x_{k+1}}^{l} \mu \operatorname{dist}(x_k, R_{x_k} \eta_k)$$

$$= \rho' c_1 \epsilon \mu \sum_{k=m, x_k \neq x_{k+1}}^{l} \operatorname{dist}(x_k, x_{k+1})$$

$$\geq \rho' c_1 \epsilon \mu r = \rho' c_1 \epsilon \mu \min\left(\frac{\epsilon}{\beta_1}, i(M)\right). \tag{7.29}$$

It follows from (7.26), (7.28), and (7.29) that

$$f(x_m) - f(x_{l+1}) \geq \rho' c_1 \epsilon \min\left(\frac{\epsilon}{\beta}, \delta_\mu, \frac{\epsilon\mu}{\beta_1}, i(M)\mu\right). \tag{7.30}$$

Because $\{f(x_k)\}_{k=0}^{\infty}$ is decreasing and bounded below, we have

$$f(x_k) \downarrow f^* \tag{7.31}$$

for some $f^* > -\infty$. It then follows from (7.30) that

$$f(x_m) - f^* \geq f(x_m) - f(x_{l+1})$$

$$\geq \rho' c_1 \epsilon \min \left(\frac{\epsilon}{\beta}, \delta_\mu, \frac{\epsilon \mu}{\beta_1}, i(M)\mu \right)$$

$$= \frac{1}{2} \rho' c_1 \|\operatorname{grad} f(x_m)\|$$

$$\min \left(\frac{\|\operatorname{grad} f(x_m)\|}{2\beta}, \delta_\mu, \frac{\|\operatorname{grad} f(x_m)\|\mu}{2\beta_1}, i(M)\mu \right).$$

Taking $m \to \infty$ in the latter expression yields $\lim_{m \to \infty} \|\operatorname{grad} f(x_m)\| = 0$.
□

Note that this theorem reduces gracefully to the classical $\mathbb{R}^n$ case, taking $\mathcal{M} = \mathbb{R}^n$ endowed with the classical inner product and $R_x \xi := x + \xi$. Then $i(M) = +\infty > 0$, R satisfies (7.25), and the Lipschitz condition (7.24) reduces to the classical expression, which subsumes the radially L-C^1 condition.

The following proposition shows that the regularity conditions on f and $\widehat{f}$ required in the previous theorems are satisfied under stronger but possibly easier to check conditions. These conditions impose a bound on the Hessian of f and on the "acceleration" along curves $t \mapsto R(t\xi)$. Note also that all these conditions need only be checked on the level set $\{x \in \mathcal{M} : f(x) \leq f(x_0)\}$.

Proposition 7.4.5 *Suppose that $\|\operatorname{grad} f(x)\| \leq \beta_g$ and that $\|\operatorname{Hess} f(x)\| \leq \beta_H$ for some constants β_g, β_H, and all $x \in \mathcal{M}$. Moreover, suppose that*

$$\left\| \frac{\mathrm{D}}{\mathrm{d}t} \frac{\mathrm{d}}{\mathrm{d}t} R(t\xi) \right\| \leq \beta_D \tag{7.32}$$

for some constant β_D, for all $\xi \in T\mathcal{M}$ with $\|\xi\| = 1$ and all $t < \delta_D$, where $\frac{\mathrm{D}}{\mathrm{d}t}$ denotes the covariant derivative along the curve $t \mapsto R(t\xi)$. Then the Lipschitz-C^1 condition on f (Definition 7.4.3) is satisfied with $\beta_L = \beta_H$; the radially Lipschitz-C^1 condition on $\widehat{f}$ (Definition 7.4.1) is satisfied for $\delta_{RL} < \delta_D$ and $\beta_{RL} = \beta_H(1 + \beta_D \delta_D) + \beta_g \beta_D$; and the condition (7.25) on R is satisfied for values of μ and δ_μ satisfying $\delta_\mu < \delta_D$ and $\frac{1}{2}\beta_D \delta_\mu < \frac{1}{\mu} - 1$.

Proof. By a standard Taylor argument (see Lemma 7.4.7), boundedness of the Hessian of f implies the Lipschitz-C^1 property of f.

For (7.25), define $u(t) = R(t\xi)$ and observe that

$$\operatorname{dist}(x, R(t\xi)) \leq \int_0^t \|u'(\tau)\| \, \mathrm{d}\tau$$

where $\int_0^t \|u'(\tau)\| \, \mathrm{d}\tau$ is the length of the curve u between 0 and t. Using the Cauchy-Schwarz inequality and the invariance of the metric by the connection, we have

$$\left| \frac{\mathrm{d}}{\mathrm{d}\tau} \|u'(\tau)\| \right| = \left| \frac{\mathrm{d}}{\mathrm{d}\tau} \sqrt{\langle u'(\tau), u'(\tau) \rangle_{u(\tau)}} \right| = \left| \frac{\langle \frac{\mathrm{D}}{\mathrm{d}t} u'(\tau), u'(\tau) \rangle_{u(\tau)}}{\|u'(\tau)\|} \right|$$

$$\leq \frac{\beta_D \|u'(\tau)\|}{\|u'(\tau)\|} \leq \beta_D$$

for all $t < \delta_D$. Therefore

$$\int_0^t \|u'(\tau)\| \, d\tau \le \int_0^t \|u'(0)\| + \beta_D \tau \, d\tau = \|\xi\| t + \tfrac{1}{2}\beta_D t^2 = t + \tfrac{1}{2}\beta_D t^2,$$

which is smaller than $\frac{t}{\mu}$ if $\frac{1}{2}\beta_D t < \frac{1}{\mu} - 1$.

For the radially Lipschitz-C^1 condition, let $u(t) = R(t\xi)$ and $h(t) = f(u(t)) = \widehat{f}(t\xi)$ with $\xi \in T_x\mathcal{M}$, $\|\xi\| = 1$. Then

$$h'(t) = \langle \operatorname{grad} f(u(t)), u'(t)\rangle_{u(t)}$$

and

$$h''(t) = \frac{D}{dt}\langle \operatorname{grad} f(u(t)), u'(t)\rangle_{u(t)}$$

$$= \langle \frac{D}{dt} \operatorname{grad} f(u(t)), u'(t)\rangle_{u(t)} + \langle \operatorname{grad} f(u(t)), \frac{D}{dt} u'(t)\rangle_{u(t)}.$$

Now, $\frac{D}{dt} \operatorname{grad} f(u(t)) = \nabla_{u'(t)} \operatorname{grad} f(u(t)) = \operatorname{Hess} f(u(t))[u'(t)]$. It follows that $|h''(t)|$ is bounded on $t \in [0, \delta_D)$ by the constant $\beta_{RL} = \beta_H(1 + \beta_D\delta_D) + \beta_g\beta_D$. Then

$$|h'(t) - h'(0)| \le \int_0^t |h''(\tau)| \, d\tau \le t\beta_{RL}.$$

$\square$

Corollary 7.4.6 (smoothness and compactness) *If the cost function f is smooth and the Riemannian manifold $\mathcal{M}$ is compact, then all the conditions in Proposition 7.4.5 are satisfied.*

The major manifolds considered in this book (the Grassmann manifold and the Stiefel manifold) are compact, and the cost functions based on the Rayleigh quotient are smooth.

7.4.2 Local convergence

We now state local convergence properties of Algorithm 10-11: local convergence to local minimizers (Theorem 7.4.10) and superlinear convergence (Theorem 7.4.11). We begin with a few preparatory lemmas.

As before, $(\mathcal{M}, g)$ is a Riemannian manifold of dimension d and R is a retraction on $\mathcal{M}$ (Definition 4.1.1). The first lemma is a first-order Taylor formula for tangent vector fields.

Lemma 7.4.7 (Taylor's theorem) *Let $x \in \mathcal{M}$, let $\mathcal{V}$ be a normal neighborhood of x, and let ζ be a C^1 tangent vector field on $\mathcal{M}$. Then, for all $y \in \mathcal{V}$,*

$$P_\gamma^{0\leftarrow 1}\zeta_y = \zeta_x + \nabla_\xi\zeta + \int_0^1 \left(P_\gamma^{0\leftarrow\tau}\nabla_{\gamma'(\tau)}\zeta - \nabla_\xi\zeta\right) d\tau, \qquad (7.33)$$

where γ is the unique minimizing geodesic satisfying $\gamma(0) = x$ and $\gamma(1) = y$, and $\xi = \operatorname{Exp}_x^{-1} y = \gamma'(0)$.

Proof. Start from

$$P_\gamma^{0\leftarrow1}\zeta_y = \zeta_x + \int_0^1 \frac{\mathrm{d}}{\mathrm{d}\tau}\, P_\gamma^{0\leftarrow\tau}\zeta \,\mathrm{d}\tau = \zeta_x + \nabla_\xi\zeta + \int_0^1 \left(\frac{\mathrm{d}}{\mathrm{d}\tau}\, P_\gamma^{0\leftarrow\tau}\zeta - \nabla_\xi\zeta\right)\mathrm{d}\tau$$

and use the formula for the connection in terms of the parallel transport (see [dC92, Ch. 2, Ex. 2]), to obtain

$$\frac{\mathrm{d}}{\mathrm{d}\tau}\, P_\gamma^{0\leftarrow\tau}\zeta = \frac{\mathrm{d}}{\mathrm{d}\epsilon}\, P_\gamma^{0\leftarrow\tau} P_\gamma^{\tau\leftarrow\tau+\epsilon}\zeta\bigg|_{\epsilon=0} = P_\gamma^{0\leftarrow\tau}\nabla_{\gamma'}\zeta.$$

$\square$

We use this lemma to show that in some neighborhood of a nondegenerate local minimizer v of f, the norm of the gradient of f can be taken as a measure of the Riemannian distance to v.

Lemma 7.4.8 *Let $v \in \mathcal{M}$ and let f be a C^2 cost function such that $\mathrm{grad}\, f(v) = 0$ and $\mathrm{Hess}\, f(v)$ is positive-definite with maximal and minimal eigenvalues $\lambda_{\max}$ and $\lambda_{\min}$. Then, given $c_0 < \lambda_{\min}$ and $c_1 > \lambda_{\max}$, there exists a neighborhood $\mathcal{V}$ of v such that, for all $x \in \mathcal{V}$, it holds that*

$$c_0 \,\mathrm{dist}(v, x) \leq \|\,\mathrm{grad}\, f(x)\| \leq c_1 \,\mathrm{dist}(v, x). \tag{7.34}$$

Proof. From Taylor (Lemma 7.4.7), it follows that

$$P_\gamma^{0\leftarrow1}\,\mathrm{grad}\, f(v) = \mathrm{Hess}\, f(v)[\gamma'(0)]$$

$$+ \int_0^1 \left(P_\gamma^{0\leftarrow\tau}\,\mathrm{Hess}\, f(\gamma(\tau))[\gamma'(\tau)] - \mathrm{Hess}\, f(v)[\gamma'(0)]\right)\mathrm{d}\tau. \tag{7.35}$$

Since f is C^2 and since $\|\gamma'(\tau)\| = \mathrm{dist}(v, x)$ for all $\tau \in [0, 1]$, we have the following bound for the integral in (7.35):

$$\left\|\int_0^1 P_\gamma^{0\leftarrow\tau}\,\mathrm{Hess}\, f(\gamma(\tau))[\gamma'(\tau)] - \mathrm{Hess}\, f(v)[\gamma'(0)]\,\mathrm{d}\tau\right\|$$

$$= \left\|\int_0^1 \left(P_\gamma^{0\leftarrow\tau} \circ \mathrm{Hess}\, f(\gamma(\tau)) \circ P_\gamma^{\tau\leftarrow0} - \mathrm{Hess}\, f(v)\right)[\gamma'(0)]\,\mathrm{d}\tau\right\|$$

$$\leq \epsilon(\mathrm{dist}(v, x))\,\mathrm{dist}(v, x),$$

where $\lim_{t\to0}\epsilon(t) = 0$. Since $\mathrm{Hess}\, f(v)$ is nonsingular, it follows that $|\lambda_{\min}| > 0$. Take $\mathcal{V}$ sufficiently small so that $\lambda_{\min} - \epsilon(\mathrm{dist}(v, x)) > c_0$ and $\lambda_{\max} + \epsilon(\mathrm{dist}(v, x)) < c_1$ for all x in $\mathcal{V}$. Then, using the fact that the parallel translation is an isometry, (7.34) follows from (7.35). $\square$

We need a relation between the gradient of f at $R_x(\xi)$ and the gradient of $\widehat{f}_x$ at ξ.

Lemma 7.4.9 *Let R be a retraction on $\mathcal{M}$ and let f be a C^1 cost function on $\mathcal{M}$. Then, given $v \in \mathcal{M}$ and $c_5 > 1$, there exists a neighborhood $\mathcal{V}$ of v and $\delta > 0$ such that*

$$\|\mathrm{grad}\, f(R\xi)\| \leq c_5\|\mathrm{grad}\, \widehat{f}(\xi)\|$$

for all $x \in \mathcal{V}$ and all $\xi \in T_x\mathcal{M}$ with $\|\xi\| \leq \delta$, where $\widehat{f}$ is as in (7.11).

Proof. Consider a parameterization of $\mathcal{M}$ at v and consider the corresponding parameterization of $T\mathcal{M}$ (see Section 3.5.3). We have

$$\partial_i \widehat{f}_x(\xi) = \sum_j \partial_j f(R\xi) A_i^j(\xi),$$

where $A(\xi)$ stands for the differential of R_x at $\xi \in T_x\mathcal{M}$. Then,

$$
\begin{aligned}
\|\operatorname{grad} \widehat{f}_x(\xi)\|^2 &= \sum_{i,j} \partial_i \widehat{f}_x(\xi) g^{ij}(x)\, \partial_j \widehat{f}_x(\xi) \\
&= \sum_{i,j,k,\ell} \partial_k f(R_x\xi) A_i^k(\xi) g^{ij}(x) A_j^\ell(\xi)\, \partial_\ell f(R_x\xi)
\end{aligned}
$$

and

$$\|\operatorname{grad} f(R_x\xi)\|^2 = \sum_{i,j} \partial_i f(R_x\xi) g^{ij}(R_x\xi)\, \partial_j f(R_x\xi).$$

The conclusion follows by a real analysis argument, invoking the smoothness properties of R and g, and the compactness of the set $\{(x,\xi) : x \in \mathcal{V},\ \xi \in T_x\mathcal{M},\ \|\xi\| \leq \delta\}$ and using $A(0_x) = \mathrm{id}$. $\qquad\square$

Finally, we will make use of Lemma 5.5.6 stating that the Hessians of f and $\widehat{f}$ coincide at critical points.

We now state and prove the local convergence results. The first result states that the nondegenerate local minima are attractors of Algorithm 10-11.

Theorem 7.4.10 (local convergence to local minima) *Consider Algorithm 10-11—i.e., the Riemannian trust-region algorithm where the trust-region subproblems (7.6) are solved using the truncated CG algorithm with stopping criterion (7.10)—with all the assumptions of Theorem 7.4.2. Let v be a nondegenerate local minimizer of f, i.e., $\operatorname{grad} f(v) = 0$ and $\operatorname{Hess} f(v)$ is positive-definite. Assume that $x \mapsto \|H_x^{-1}\|$ is bounded on a neighborhood of v and that (7.25) holds for some $\mu > 0$ and $\delta_\mu > 0$. Then there exists a neighborhood $\mathcal{V}$ of v such that, for all $x_0 \in \mathcal{V}$, the sequence $\{x_k\}$ generated by Algorithm 10-11 converges to v.*

Proof. Take $\delta_1 > 0$ with $\delta_1 < \delta_\mu$ such that $\|H_x^{-1}\|$ is bounded on $B_{\delta_1}(v)$, that $B_{\delta_1}(v)$ contains only v as critical point, and that $f(x) > f(v)$ for all $x \in \bar{B}_{\delta_1}(v)$. (In view of the assumptions, such a δ_1 exists.) Take δ_2 small enough that, for all $x \in B_{\delta_2}(v)$, it holds that $\|\eta^*(x)\| \leq \mu(\delta_1 - \delta_2)$, where η^* is the (unique) solution of $H_x\eta^* = -\operatorname{grad} f(x)$; such a δ_2 exists because of Lemma 7.4.8 and the bound on $\|H_x^{-1}\|$. Consider a level set $\mathcal{L}$ of f such that $\mathcal{V} := \mathcal{L} \cap B_{\delta_1}(v)$ is a subset of $B_{\delta_2}(v)$; invoke that $f \in C^1$ to show that such a level set exists. Let $\eta^{tCG}(x, \Delta)$ denote the tangent vector η_k returned by the truncated CG algorithm (Algorithm 11) when $x_k = x$ and $\Delta_k = \Delta$. Then, $\mathcal{V}$ is a neighborhood of v, and for all $x \in \mathcal{V}$ and all $\Delta > 0$, we have

$$\operatorname{dist}(x, x_+) \leq \frac{1}{\mu}\|\eta^{tCG}(x, \Delta)\| \leq \frac{1}{\mu}\|\eta^*\| \leq (\delta_1 - \delta_2),$$

where we used the fact that $\|\eta\|$ is increasing along the truncated CG process (Proposition 7.3.2). It follows from the equation above that x_+ is in $B_{\delta_1}(v)$. Moreover, since $f(x_+) \leq f(x)$, it follows that $x_+ \in \mathcal{V}$. Thus $\mathcal{V}$ is invariant. But the only critical point of f in $\mathcal{V}$ is v, so $\{x_k\}$ goes to v whenever x_0 is in $\mathcal{V}$.   □

Now we study the order of convergence of sequences that converge to a nondegenerate local minimizer.

Theorem 7.4.11 (order of convergence) *Consider Algorithm 10-11 with stopping criterion (7.10). Suppose that f is a C^2 cost function on $\mathcal{M}$ and that*

$$\|H_k - \mathrm{Hess}\, \widehat{f}_{x_k}(0_k)\| \leq \beta_{\mathcal{H}} \|\mathrm{grad}\, f(x_k)\|, \qquad (7.36)$$

i.e., H_k is a sufficiently good approximation of $\mathrm{Hess}\, \widehat{f}_{x_k}(0_{x_k})$. Let $v \in \mathcal{M}$ be a nondegenerate local minimizer of f (i.e., $\mathrm{grad}\, f(v) = 0$ and $\mathrm{Hess}\, f(v)$ is positive-definite). Further assume that $\mathrm{Hess}\, \widehat{f}_x$ is Lipschitz-continuous at 0_x uniformly in x in a neighborhood of v, i.e., there exist $\beta_{L2} > 0$, $\delta_1 > 0$, and $\delta_2 > 0$ such that, for all $x \in B_{\delta_1}(v)$ and all $\xi \in B_{\delta_2}(0_x)$, it holds that

$$\|\mathrm{Hess}\, \widehat{f}_x(\xi) - \mathrm{Hess}\, \widehat{f}_x(0_x)\| \leq \beta_{L2} \|\xi\|, \qquad (7.37)$$

where $\|\cdot\|$ on the left-hand side denotes the operator norm in $T_x\mathcal{M}$ defined as in (7.15). Then there exists $c > 0$ such that, for all sequences $\{x_k\}$ generated by the algorithm converging to v, there exists $K > 0$ such that for all $k > K$,

$$\mathrm{dist}(x_{k+1}, v) \leq c\, (\mathrm{dist}(x_k, v))^{\min\{\theta+1, 2\}} \qquad (7.38)$$

with $\theta > 0$ as in (7.10).

Proof. The proof relies on a set of bounds which are justified after the main result is proved. Assume that there exist $\tilde{\Delta}, c_0, c_1, c_2, c_3, c_3', c_4, c_5$ such that, for all sequences $\{x_k\}$ satisfying the conditions asserted, all $x \in \mathcal{M}$, all ξ with $\|\xi\| < \tilde{\Delta}$, and all k greater than some K, it holds that

$$c_0\, \mathrm{dist}(v, x_k) \leq \|\mathrm{grad}\, f(x_k)\| \leq c_1\, \mathrm{dist}(v, x_k), \qquad (7.39)$$

$$\|\eta_k\| \leq c_4 \|\mathrm{grad}\, \widehat{m}_{x_k}(0)\| \leq \tilde{\Delta}, \qquad (7.40)$$

$$\rho_k > \rho', \qquad (7.41)$$

$$\|\mathrm{grad}\, f(R_{x_k}\xi)\| \leq c_5 \|\mathrm{grad}\, \widehat{f}_{x_k}(\xi)\|, \qquad (7.42)$$

$$\|\mathrm{grad}\, \widehat{m}_{x_k}(\xi) - \mathrm{grad}\, \widehat{f}_{x_k}(\xi)\| \leq c_3 \|\xi\|^2 + c_3' \|\mathrm{grad}\, f(x_k)\|\, \|\xi\|, \qquad (7.43)$$

$$\|\mathrm{grad}\, \widehat{m}_{x_k}(\eta_k)\| \leq c_2 \|\mathrm{grad}\, \widehat{m}_{x_k}(0)\|^{\theta+1}, \qquad (7.44)$$

where $\{\eta_k\}$ is the sequence of update vectors corresponding to $\{x_k\}$.

Given the bounds (7.39) to (7.44), the proof proceeds as follows. For all $k > K$, it follows from (7.39) and (7.41) that

$$c_0\, \mathrm{dist}(v, x_{k+1}) \leq \|\mathrm{grad}\, f(x_{k+1})\| = \|\mathrm{grad}\, f(R_{x_k}\eta_k)\|,$$

from (7.42) and (7.40) that

$$\|\mathrm{grad}\, f(R_{x_k}\eta_k)\| \leq c_5 \|\mathrm{grad}\, \widehat{f}_{x_k}(\eta_k)\|,$$

from (7.40) and (7.43) and (7.44) that

$$\|\operatorname{grad} \widehat{f}_{x_k}(\eta_k)\| \leq \|\operatorname{grad} \widehat{m}_{x_k}(\eta_k) - \operatorname{grad} \widehat{f}_{x_k}(\eta_k)\| + \|\operatorname{grad} \widehat{m}_{x_k}(\eta_k)\|$$
$$\leq (c_3 c_4^2 + c_3' c_4)\|\operatorname{grad} \widehat{m}_{x_k}(0)\|^2 + c_2\|\operatorname{grad} \widehat{m}_{x_k}(0)\|^{1+\theta},$$

and from (7.39) that

$$\|\operatorname{grad} \widehat{m}_{x_k}(0)\| = \|\operatorname{grad} f(x_k)\| \leq c_1 \operatorname{dist}(v, x_k).$$

Consequently, taking K larger if necessary so that $\operatorname{dist}(v, x_k) < 1$ for all $k > K$, it follows that

$$c_0 \operatorname{dist}(v, x_{k+1}) \leq \quad \|\operatorname{grad} f(x_{k+1})\| \tag{7.45}$$
$$\leq \quad c_5(c_3 c_4^2 + c_3' c_4)\|\operatorname{grad} f(x_k)\|^2 + c_5 c_2\|\operatorname{grad} f(x_k)\|^{\theta+1} \tag{7.46}$$
$$\leq \quad c_5((c_3 c_4^2 + c_3' c_4)c_1^2 (\operatorname{dist}(v, x_k))^2 + c_2 c_1^{1+\theta}(\operatorname{dist}(v, x_k))^{1+\theta})$$
$$\leq \quad c_5((c_3 c_4^2 + c_3' c_4)c_1^2 + c_2 c_1^{1+\theta})(\operatorname{dist}(v, x_k))^{\min\{2, 1+\theta\}}$$

for all $k > K$, which is the desired result.

It remains to show that the bounds (7.39)–(7.44) hold under the assumptions or the theorem.

Equation (7.39) comes from Lemma 7.4.8 and is due to the fact that v is a nondegenerate critical point.

We prove (7.40). Since $\{x_k\}$ converges to the nondegenerate local minimizer v where $\operatorname{Hess} \widehat{f}_v(0_v) = \operatorname{Hess} f(v)$ (in view of Lemma 5.5.6), and since $\operatorname{Hess} f(v)$ is positive-definite with $f \in C^2$, it follows from the approximation condition (7.36) and from (7.39) that there exists $c_4 > 0$ such that, for all k greater than some K, H_k is positive-definite and $\|H_k^{-1}\| < c_4$. Given a $k > K$, let η_k^* be the solution of $H_k \eta_k^* = -\operatorname{grad} \widehat{m}_{x_k}(0)$. It follows that $\|\eta_k^*\| \leq c_4\|\operatorname{grad} \widehat{m}_{x_k}(0)\|$. Since $\{x_k\}$ converges to a critical point of f and since $\operatorname{grad} \widehat{m}_{x_k}(0) = \operatorname{grad} f(x_k)$ in view of (7.6), we obtain that $\|\eta_k^*\| \leq c_4\|\operatorname{grad} \widehat{m}_{x_k}(0)\| \leq \tilde{\Delta}$ for any given $\tilde{\Delta} > 0$ by choosing K larger if necessary. Then, since the sequence of η_k^j's constructed by the truncated CG inner iteration (Algorithm 11) is strictly increasing in norm (Proposition 7.3.2) and would reach η_k^* at $j = d$ in the absence of the stopping criterion, it follows that (7.40) holds.

We prove (7.41). Let γ_k denote $\|\operatorname{grad} f(x_k)\|$. It follows from the definition of ρ_k that

$$\rho_k - 1 = \frac{\widehat{m}_{x_k}(\eta_k) - \widehat{f}_{x_k}(\eta_k)}{\widehat{m}_{x_k}(0_{x_k}) - \widehat{m}_{x_k}(\eta_k)}. \tag{7.47}$$

From Taylor's theorem, it holds that

$$\widehat{f}_{x_k}(\eta_k) = \widehat{f}_{x_k}(0_{x_k}) + \langle \operatorname{grad} f(x_k), \eta_k \rangle_{x_k}$$
$$+ \int_0^1 \langle \operatorname{Hess} \widehat{f}_{x_k}(\tau \eta_k)[\eta_k], \eta_k \rangle_{x_k} (1 - \tau) d\tau.$$

It follows that

$$
\left| \widehat{m}_{x_k}(\eta_k) - \widehat{f}_{x_k}(\eta_k) \right|
$$

$$
= \left| \int_0^1 \left(\langle H_k[\eta_k], \eta_k \rangle_{x_k} - \langle \operatorname{Hess} \widehat{f}_{x_k}(\tau\eta_k)[\eta_k], \eta_k \rangle_{x_k} \right) (1-\tau)\, d\tau \right|
$$

$$
\leq \int_0^1 \left| \langle (H_k - \operatorname{Hess} \widehat{f}_{x_k}(0_{x_k}))[\eta_k], \eta_k \rangle_{x_k} \right| (1-\tau)\, d\tau
$$

$$
+ \int_0^1 \left| \langle (\operatorname{Hess} \widehat{f}_{x_k}(0_{x_k}) - \operatorname{Hess} \widehat{f}(\tau\eta_k))[\eta_k], \eta_k \rangle_{x_k} \right| (1-\tau)\, d\tau
$$

$$
\leq \frac{1}{2} \beta_{\mathcal{H}} \gamma_k \|\eta_k\|^2 + \frac{1}{6} \beta_{L2} \|\eta_k\|^3.
$$

It then follows from (7.47), using the Cauchy bound (7.14), that

$$
|\rho_k - 1| \leq \frac{(3\beta_{\mathcal{H}}\gamma_k + \beta_{L2}\|\eta_k\|)\,\|\eta_k\|^2}{6\gamma_k \min\{\Delta_k, \gamma_k/\beta\}},
$$

where β is an upper bound on the norm of H_k. Since $\|\eta_k\| \leq \Delta_k$ and $\|\eta_k\| \leq c_4\gamma_k$, it follows that

$$
|\rho_k - 1| \leq \frac{(3\beta_{\mathcal{H}} + \beta_{L2}c_4)\,(\min\{\Delta_k, c_4\gamma_k\})^2}{6\min\{\Delta_k, \gamma_k/\beta\}}. \tag{7.48}
$$

Either Δ_k is active in the denominator of (7.48), in which case we have

$$
|\rho_k - 1| \leq \frac{(3\beta_{\mathcal{H}} + \beta_{L2}c_4)\,\Delta_k c_4\gamma_k}{6\Delta_k} = \frac{(3\beta_{\mathcal{H}} + \beta_{L2}c_4)\,c_4}{6}\gamma_k,
$$

or γ_k/β is active in the denominator of (7.48), in which case we have

$$
|\rho_k - 1| \leq \frac{(3\beta_{\mathcal{H}} + \beta_{L2}c_4)\,(c_4\gamma_k)^2}{6\gamma_k/\beta} = \frac{(3\beta_{\mathcal{H}} + \beta_{L2}c_4)\,c_4^2\beta}{6}\gamma_k.
$$

In both cases, $\lim_{k\to\infty} \rho_k = 1$ since, in view of (7.39), $\lim_{k\to\infty} \gamma_k = 0$.

Equation (7.42) comes from Lemma 7.4.9.

We prove (7.43). It follows from Taylor's formula (Lemma 7.4.7, where the parallel translation becomes the identity since the domain of $\widehat{f}_{x_k}$ is the Euclidean space $T_{x_k}\mathcal{M}$) that

$$
\operatorname{grad} \widehat{f}_{x_k}(\xi) = \operatorname{grad} \widehat{f}_{x_k}(0_{x_k}) + \operatorname{Hess} \widehat{f}_{x_k}(0_{x_k})[\xi]
$$

$$
+ \int_0^1 \left(\operatorname{Hess} \widehat{f}_{x_k}(\tau\xi) - \operatorname{Hess} \widehat{f}_{x_k}(0_{x_k}) \right) [\xi]\, d\tau.
$$

The conclusion comes by the Lipschitz condition (7.37) and the approximation condition (7.36).

Finally, equation (7.44) comes from the stopping criterion (7.10) of the inner iteration. More precisely, the truncated CG loop (Algorithm 11) terminates if $\langle \delta_j, H_k\delta_j \rangle \leq 0$ or $\|\eta_{j+1}\| \geq \Delta$ or the criterion (7.10) is satisfied. Since $\{x_k\}$ converges to v and $\operatorname{Hess} f(v)$ is positive-definite, it follows that H_k is positive-definite for all k greater than a certain K. Therefore, for all

$k > K$, the criterion $\langle \delta_j, H_k \delta_j \rangle \leq 0$ is never satisfied. In view of (7.40) and (7.41), it can be shown that the trust region is eventually inactive. Therefore, increasing K if necessary, the criterion $\|\eta_{j+1}\| \geq \Delta$ is never satisfied for all $k > K$. In conclusion, for all $k > K$, the stopping criterion (7.10) is satisfied each time a computed η_k is returned by the truncated CG loop. Therefore, the truncated CG loop behaves as a classical linear CG method. Consequently, $\operatorname{grad} \widehat{m}_{x_k}(\eta_j) = r_j$ for all j. Choose K such that for all $k > K$, $\|\operatorname{grad} f(x_k)\| = \|\operatorname{grad} \widehat{m}_{x_k}(0)\|$ is so small—it converges to zero in view of (7.39)—that the stopping criterion (7.10) yields

$$\|\operatorname{grad} \widehat{m}_{x_k}(\eta_j)\| = \|r_j\| \leq \|r_0\|^{1+\theta} = \|\operatorname{grad} \widehat{m}_{x_k}(0)\|^{1+\theta}. \qquad (7.49)$$

This is (7.44) with $c_2 = 1$. $\qquad\qquad\qquad\qquad\qquad\qquad\qquad\qquad\qquad\qquad\quad\square$

The constants in the proof of Theorem 7.4.11 can be chosen as $c_0 < \lambda_{min}$, $c_1 > \lambda_{\max}$, $c_4 > 1/\lambda_{\min}$, $c_5 > 1$, $c_3 \geq \beta_{L2}$, $c_3' \geq \beta_{\mathcal{H}}$, $c_2 \geq 1$, where $\lambda_{\min}$ and $\lambda_{\max}$ are the smallest and largest eigenvalue of Hess $f(v)$, respectively. Consequently, the constant c in the convergence bound (7.38) can be chosen as

$$c > \frac{1}{\lambda_{\min}} \left((\beta_{L2}/\lambda_{\min}^2 + \beta_{\mathcal{H}}/\lambda_{\min}) \lambda_{\max}^2 + \lambda_{\max}^{1+\theta} \right). \qquad (7.50)$$

A nicer-looking bound holds when convergence is evaluated in terms of the norm of the gradient, as expressed in the theorem below which is a direct consequence of (7.45) and (7.46).

Theorem 7.4.12 *Under the assumptions of Theorem 7.4.11, if $\theta + 1 < 2$, then given $c_g > 1$ and $\{x_k\}$ generated by the algorithm, there exists $K > 0$ such that*

$$\|\operatorname{grad} f(x_{k+1})\| \leq c_g \|\operatorname{grad} f(x_k)\|^{\theta+1}$$

for all $k > K$.

Nevertheless, (7.50) suggests that the algorithm may not perform well when the relative gap $\lambda_{\max}/\lambda_{\min}$ is large. In spite of this, numerical experiments on eigenvalue problems have shown that the method tends to behave as well as, or even better than, other methods in the presence of a small relative gap.

7.4.3 Discussion

The main global convergence result (Theorem 7.4.4) shows that RTR-tCG method (Algorithm 10-11) converges to a set of critical points of the cost function for *all* initial conditions. This is an improvement on the pure Newton method (Algorithm 5), for which only local convergence results exist. However, the convergence theory falls short of showing that the algorithm always converges to a local minimizer. This is not surprising: since we have ruled out the possibility of checking the positive-definiteness of the Hessian of the cost function, we have no way of testing whether a critical point is a local minimizer or not. (Note as an aside that even checking the positive-definiteness of the Hessian is not always sufficient for determining if a critical point is a local minimizer or not: if the Hessian is singular and nonnegative-definite, then no conclusion can be drawn.) In fact, for the vast majority of optimization methods, only convergence to critical points can be secured unless some specific assumptions (like convexity) are made. Nevertheless, it is observed in numerical experiments with random initial conditions that the algorithm systematically converges to a local minimizer; convergence to a saddle point is observed only on specifically crafted problems, e.g., when the iteration is started on a point that is a saddle point in computer arithmetic. This is due to the fact that the algorithm is a descent method, i.e., $f(x_{k+1}) < f(x_k)$ whenever $x_{k+1} \neq x_k$. Therefore, saddle points or local maxima are unstable fixed points of the algorithm.

There are cases where the bound (7.38) holds with order $\min\{\theta + 1, 3\}$; i.e., by choosing $\theta \geq 2$, one obtains cubic convergence. This situation occurs when the Taylor expansion of the cost function around the limit point has no third-order contribution. Thus, the second-order approximation used in the algorithm becomes an effective third-order approximation, and the order of convergence benefits as expected. In practice, this condition holds more often than one might guess since any cost function that is symmetric around the local minimizer v, i.e., $f(\mathrm{Exp}_x(\xi)) = f(\mathrm{Exp}_x(-\xi))$, will have only even contributions to its Taylor expansion.

7.5 APPLICATIONS

In this section, we briefly review the essential "ingredients" necessary for applying the RTR-tCG method (Algorithm 10-11), and we present two examples in detail as an illustration. One of these examples is optimization of the Rayleigh quotient, leading to an algorithm for computing extreme invariant subpaces of symmetric matrices. Since trust-region algorithms with an exact Hessian can be thought of as enhanced Newton methods, and since the Newton equation for Rayleigh quotient optimization is equivalent to the Jacobi equation (see Notes and References in Chapter 6), it is not surprising that this algorithm has close links with the celebrated Jacobi-Davidson approach to the eigenproblem. The Riemannian trust-region approach sheds new light on the Jacobi-Davidson method. In particular, it yields new ways to deal with the Jacobi equation so as to reduce the computational burden while preserving the superlinear convergence inherited from the Newton approach and obtaining strong global convergence results supported by a detailed analysis.

7.5.1 Checklist

The following elements are required for applying the RTR method to optimizing a cost function f on a Riemannian manifold $(\mathcal{M}, g)$: (i) a tractable numerical representation for points x on $\mathcal{M}$, for tangent spaces $T_x\mathcal{M}$, and for the inner products $\langle \cdot, \cdot \rangle_x$ on $T_x\mathcal{M}$; (ii) a retraction $R_x : T_x\mathcal{M} \to \mathcal{M}$ (Definition 4.1.1); (iii) formulas for $f(x)$, $\operatorname{grad} f(x)$ and an approximate Hessian H_x that satisfies the properties required for the convergence results in Section 7.4.

Formulas for $\operatorname{grad} f(x)$ and $\operatorname{Hess} \widehat{f}_x(0_x)$ can be obtained by identification in a Taylor expansion of the lifted cost function $\widehat{f}_x$, namely
$$\widehat{f}_x(\eta) = f(x) + \langle \operatorname{grad} f(x), \eta \rangle_x + \tfrac{1}{2}\langle \operatorname{Hess} \widehat{f}_x(0_x)[\eta], \eta \rangle_x + O(\|\eta\|^3),$$
where $\operatorname{grad} f(x) \in T_x\mathcal{M}$ and $\operatorname{Hess} \widehat{f}_x(0_x)$ is a linear transformation of $T_x\mathcal{M}$. A formula for $\operatorname{Hess} \widehat{f}_x(0_x)$ is not needed, though; the convergence theory requires only an "approximate Hessian" H_x that satisfies the approximation condition (7.36). To obtain such an H_x, one can pick $H_x := \operatorname{Hess}(f \circ \widetilde{R}_x)(0_x)$, where $\widetilde{R}$ is any retraction. Then, assuming sufficient smoothness of f, the bound (7.36) follows from Lemmas 7.4.8 and 5.5.6. In particular, the choice $\widetilde{R}_x = \operatorname{Exp}_x$ yields
$$H_x := \nabla \operatorname{grad} f(x) \quad (= \operatorname{Hess} f(x)), \tag{7.51}$$
where ∇ denotes the Riemannian connection, and the model $\widehat{m}_x$ takes the form (7.1). If $\mathcal{M}$ is a Riemannian submanifold or a Riemannian quotient of a Euclidean space, then $\nabla \operatorname{grad} f(x)$ admits a simple formula; see Section 5.3.

7.5.2 Symmetric eigenvalue decomposition

Let $\mathcal{M}$ be the orthogonal group
$$\mathcal{M} = O_n = \{Q \in \mathbb{R}^{n \times n} : Q^T Q = I_n\}.$$
This manifold is an embedded submanifold of $\mathbb{R}^{n \times n}$ (Section 3.3). The tangent spaces are given by $T_Q O_n = \{Q\Omega : \Omega = -\Omega^T\}$ (Section 3.5.7). The canonical Euclidean metric $g(A, B) = \operatorname{tr}(A^T B)$ on $\mathbb{R}^{n \times n}$ induces on O_n the metric
$$\langle Q\Omega_1, Q\Omega_2 \rangle_Q = \operatorname{tr}(\Omega_1^T \Omega_2). \tag{7.52}$$
A retraction $R_Q : T_Q O_n \to O_n$ must be chosen that satisfies the properties stated in Section 7.2. Several possibilities are mentioned in Section 4.1.1.

Consider the cost function
$$f(Q) = \operatorname{tr}(Q^T A Q N),$$
where A and N are $n \times n$ symmetric matrices. For $N = \operatorname{diag}(\mu_1, \ldots, \mu_n)$, $\mu_1 < \cdots < \mu_n$, the minimum of f is realized by the orthonormal matrices of eigenvectors of A sorted in decreasing order of corresponding eigenvalue (this is a consequence of the critical points analysis in Section 4.8). Assume that the retraction R approximates the exponential at least to order 2. With the metric g defined as in (7.52), we obtain
$$\begin{aligned}
\widehat{f}_Q(Q\Omega) &:= f(R_Q(Q\Omega)) \\
&= \operatorname{tr}((I + \Omega + \tfrac{1}{2}\Omega^2 + O(\Omega^3))^T Q^T A Q(I + \Omega + \tfrac{1}{2}\Omega^2 + O(\Omega^3))N) \\
&= f(Q) + 2\operatorname{tr}(\Omega^T Q^T A Q N) \\
&\quad + \operatorname{tr}(\Omega^T Q^T A Q \Omega N - \Omega^T \Omega Q^T A Q N) + O(\Omega^3),
\end{aligned}$$

from which it follows that

$$\mathrm{D}\widehat{f}_Q(0)[Q\Omega] = 2\operatorname{tr}(Q^T AQ\Omega N)$$

$$\tfrac{1}{2}\mathrm{D}^2\widehat{f}_Q(0)[Q\Omega_1, Q\Omega_2] = \operatorname{tr}(\Omega_1^T Q^T AQ\Omega_2 N - \tfrac{1}{2}(\Omega_1^T\Omega_2 + \Omega_2^T\Omega_1)Q^T AQN)$$

$$\operatorname{grad}\widehat{f}_Q(0) = \operatorname{grad} f(Q) = Q[Q^T AQ, N]$$

$$\operatorname{Hess}\widehat{f}_Q(0)[Q\Omega] = \operatorname{Hess} f(Q)[Q\Omega] = \tfrac{1}{2}Q[[Q^T AQ, \Omega], N] + \tfrac{1}{2}Q[[N, \Omega], Q^T AQ],$$

where $[A, B] := AB - BA$. It is now straightforward to replace these expressions in the general formulation of Algorithm 10-11 and obtain a practical matrix algorithm.

An alternative way to obtain $\operatorname{Hess}\widehat{f}_Q(0)$ is to exploit Proposition 5.5.5, which yields $\operatorname{Hess}\widehat{f}_Q(0) = \nabla\operatorname{grad} f(Q)$. Since the manifold $\mathcal{M}$ is an embedded Riemannian submanifold of $\mathbb{R}^{n\times p}$, the covariant derivative ∇ is obtained by projecting the derivative in $\mathbb{R}^{n\times p}$ onto the tangent space to $\mathcal{M}$; see Section 5.3.3. We obtain $\operatorname{Hess} f(Q)[Q\Omega] = Q\operatorname{skew}(\Omega[Q^T AQ, N] + [\Omega^T Q^T AQ + Q^T AQ\Omega, N])$, which yields the same result as above.

All these ingredients can now be used in Algorithm 10-11 to obtain an iteration that satisfies the convergence properties proven in Section 7.4. Convergence to the critical points of the cost function means convergence to the matrices whose column vectors are the eigenvectors of A. Only the matrices containing eigenvectors in decreasing order of eigenvalue can be stable fixed points for the algorithm. They are asymptotically stable, with superlinear convergence, when all the eigenvalues are simple.

7.5.3 Computing an extreme eigenspace

Following up on the geometric Newton method for the generalized eigenvalue problem obtained in Section 6.4.3, we assume again that A and B are $n \times n$ symmetric matrices with B positive-definite, and we consider the generalized eigenvalue problem

$$Av = \lambda Bv.$$

We want to compute the leftmost p-dimensional invariant subspace of the pencil (A, B).

We consider the Rayleigh quotient function f defined by

$$f(\operatorname{span}(Y)) = \operatorname{tr}((Y^T AY)(Y^T BY)^{-1}), \tag{7.53}$$

where Y belongs to the set of full-rank $n \times p$ matrices and $\operatorname{span}(Y)$ denotes the column space of Y. The critical points of f are the invariant subspaces of the pencil (A, B), and the minimizers of f correspond to the leftmost invariant subspaces of (A, B).

In Section 6.4.3, we chose a noncanonical Riemannian metric (6.29) that yields a relatively short formula (6.34) for the Riemannian Hessian of f, obtained using the theory of Riemannian submersions (Proposition 5.3.4). In this section, we again use the definition

$$\mathcal{H}_Y = \{Z \in \mathbb{R}^{n\times p} : Y^T BZ = 0\}$$

for the horizontal spaces, and we take

$$R_{\mathcal{Y}}(\xi) = \mathrm{span}(Y + \bar{\xi}_Y), \tag{7.54}$$

where $\mathcal{Y} = \mathrm{span}(Y)$ and $\bar{\xi}_Y$ stands for the horizontal lift of the tangent vector $\xi \in T_{\mathcal{Y}} \mathrm{Grass}(p, n)$. But now we define the Riemannian metric as

$$\langle \xi, \zeta \rangle_{\mathcal{Y}} = \mathrm{tr}\left((Y^T BY)^{-1} \bar{\xi}_Y^{\ T} \bar{\zeta}_Y \right). \tag{7.55}$$

Notice that the horizontal space is not orthogonal to the vertical space with respect to the new Riemannian metric (7.55), so we have to renounce using the Riemannian submersion theory. However, we will see that with these choices for the horizontal space, the retraction, and the Riemannian metric, the second-order Taylor development of $\hat{f}_x := f \circ R_x$ admits quite a simple form.

For the Rayleigh cost function (7.53), using the notation

$$P_{U,V} = I - U(V^T U)^{-1} V^T \tag{7.56}$$

for the projector parallel to the span of U onto the orthogonal complement of the span of V, we obtain

$$
\begin{aligned}
&\hat{f}_{\mathcal{Y}}(\xi)\\
&= f(R_{\mathcal{Y}}(\xi)) = \mathrm{tr}\left(\left((Y+\bar{\xi}_Y)^T B(Y+\bar{\xi}_Y)\right)^{-1} \left((Y+\bar{\xi}_Y)^T A(Y+\bar{\xi}_Y)\right)\right)\\
&= \mathrm{tr}\left((Y^T BY)^{-1} Y^T AY\right) + 2\,\mathrm{tr}\left((Y^T BY)^{-1}\bar{\xi}_Y^{\ T} AY\right)\\
&\quad + \mathrm{tr}\left((Y^T BY)^{-1}\bar{\xi}_Y^{\ T}\left(A\bar{\xi}_Y - B\bar{\xi}_Y(Y^T BY)^{-1}(Y^T AY)\right)\right) + O(\|\xi\|^3)\\
&= \mathrm{tr}\left((Y^T BY)^{-1}Y^T AY\right) + 2\,\mathrm{tr}\left((Y^T BY)^{-1}\bar{\xi}_Y^{\ T} P_{BY,BY} AY\right)\\
&\quad + \mathrm{tr}\left((Y^T BY)^{-1}\bar{\xi}_Y^{\ T} P_{BY,BY}\left(A\bar{\xi}_Y - B\bar{\xi}_Y(Y^T BY)^{-1}(Y^T AY)\right)\right)\\
&\quad + O(\|\xi\|^3),
\end{aligned}
\tag{7.57}
$$

where the introduction of the projectors does not modify the expression since $P_{BY,BY}\bar{\xi}_Y = \bar{\xi}_Y$. By identification, using the noncanonical metric (7.55), we obtain

$$\overline{\mathrm{grad}\, f(\mathcal{Y})}_Y = \overline{\mathrm{grad}\,\hat{f}_{\mathcal{Y}}(0)}_Y = 2P_{BY,BY} AY \tag{7.58}$$

and

$$\overline{\mathrm{Hess}\,\hat{f}_{\mathcal{Y}}(0_{\mathcal{Y}})[\xi]}_Y = 2P_{BY,BY}\left(A\bar{\xi}_Y - B\bar{\xi}_Y(Y^T BY)^{-1}(Y^T AY)\right). \tag{7.59}$$

Notice that $\mathrm{Hess}\,\hat{f}_{\mathcal{Y}}(0_{\mathcal{Y}})$ is symmetric with respect to the metric, as required. We choose to take

$$H_{\mathcal{Y}} := \mathrm{Hess}\,\hat{f}_{\mathcal{Y}}(0_{\mathcal{Y}}). \tag{7.60}$$

Consequently, the approximation condition (7.36) is trivially satisfied. The model (7.6) is thus

$$
\begin{aligned}
\hat{m}_{\mathcal{Y}}(\xi) &= f(\mathcal{Y}) + \langle \mathrm{grad}\, f(\mathcal{Y}), \xi\rangle_{\mathcal{Y}} + \tfrac{1}{2}\langle H_{\mathcal{Y}}\xi, \xi\rangle_{\mathcal{Y}}\\
&= \mathrm{tr}\left((Y^T BY)^{-1}Y^T AY\right) + 2\,\mathrm{tr}\left((Y^T BY)^{-1}\bar{\xi}_Y^{\ T} AY\right)\\
&\quad + \mathrm{tr}\left((Y^T BY)^{-1}\bar{\xi}_Y^{\ T}\left(A\bar{\xi}_Y - B\bar{\xi}_Y(Y^T BY)^{-1}Y^T AY\right)\right).
\end{aligned}
\tag{7.61}
$$

(Observe that for $B = I$ and $p = 1$, and choosing $y = Y$ of unit norm, the model (7.61) becomes $\widehat{m}_Y(\xi) = \mathrm{tr}(y^T A y) + 2\bar{\xi}_y^{\ T} A y + \bar{\xi}_y^{\ T}(A - y^T A y I)\bar{\xi}_y$, $y^T \bar{\xi}_y = 0$. Assuming that the Hessian is nonsingular, this model has a unique critical point that is the solution of the Newton equation (6.17) for the Rayleigh quotient on the sphere.)

Since the Rayleigh cost function (7.53) is smooth on $\mathrm{Grass}(p, n)$ and since $\mathrm{Grass}(p, n)$ is compact, it follows that all the assumptions involved in the convergence analysis of the general RTR-tCG algorithm (Section 7.4) are satisfied. The only complication is that we do not have a closed-form expression for the distance involved in the superlinear convergence result (7.38). But since B is fixed and positive-definite, the distances induced by the noncanonical metric (7.55) and by the canonical metric—(7.55) with $B := I$— are locally equivalent, and therefore for a given sequence both distances yield the same rate of convergence. (Saying that two distances dist_1 and dist_2 are *locally equivalent* means that given $x \in \mathcal{M}$, there is a neighborhood $\mathcal{U}$ of x and constants $c_1, c_2 > 0$ such that, for all y in $\mathcal{U}$, we have $c_1 \mathrm{dist}_1(x, y) \le \mathrm{dist}_2(x, y) \le c_2 \mathrm{dist}_1(x, y)$.)

We have now all the required information to use the RTR-tCG method (Algorithm 10-11) for minimizing the Rayleigh cost function (7.53) on the Grassmann manifold $\mathrm{Grass}(p, n)$ endowed with the noncanonical metric (7.55). A matrix version of the inner iteration is displayed in Algorithm 12, where we omit the horizontal lift notation for conciseness and define

$$H_Y[Z] := P_{BY,BY}(AZ - BZ(Y^T BY)^{-1}Y^T AY). \qquad (7.62)$$

Note that omission of the factor 2 in both the gradient and the Hessian does not affect the sequence $\{\eta\}$ generated by the truncated CG algorithm.

According to the retraction formula (7.54), the returned η_k yields a candidate new iterate

$$Y_{k+1} = (Y_k + \eta_k)M_k,$$

where M_k is chosen such that $Y_{k+1}^T BY_{k+1} = I$. The candidate is accepted or rejected, and the trust-region radius is updated as prescribed in the outer RTR method (Algorithm 10), where ρ is computed using $\widehat{m}$ as in (7.61) and $\widehat{f}$ as in (7.57).

The resulting algorithm converges to the set of invariant subspaces of (A, B)—which are the critical points of the cost function (7.53)—and convergence to the leftmost invariant subspace $\mathcal{V}$ is expected to occur in practice since the other invariant subspaces are numerically unstable. Moreover, since $\mathcal{V}$ is a nondegenerate local minimum (under our assumption that $\lambda_p < \lambda_{p+1}$), it follows that the rate of convergence is $\min\{\theta + 1, 2\}$, where θ is the parameter appearing in the stopping criterion (7.10) of the inner (truncated CG) iteration.

Numerical experiments illustrating the convergence of the Riemannian trust-region algorithm for extreme invariant subspace computation are presented in Figures 7.1 and 7.2. The θ parameter of the inner stopping criterion (7.10) was chosen equal to 1 to obtain quadratic convergence (see Theorem 7.4.11). The right-hand plot shows a case with a small eigenvalue gap,

---

Algorithm 12 Truncated CG method for the generalized eigenvalue problem

---

Require: Symmetric $n \times n$ matrices A and B with B positive-definite; a B-orthonormal full-rank $n \times p$ matrix Y (i.e., $Y^T BY = I$); operator H_Y defined in (7.62); $\Delta > 0$.

1: Set $\eta^0 = 0 \in \mathbb{R}^{n \times p}$, $r_0 = P_{BY,BY} AY$, $\delta_0 = -r_0$; $j = 0$;
2: **loop**
3:     **if** $\operatorname{tr}\left(\delta_j^T H_Y[\delta_j]\right) \leq 0$ **then**
4:         Compute $\tau > 0$ such that $\eta = \eta^j + \tau \delta_j$ satisfies $\operatorname{tr}\left(\eta^T \eta\right) = \Delta$;
5:         **return** $\eta_k := \eta$;
6:     **end if**
7:     Set $\alpha_j = \operatorname{tr}\left(r_j^T r_j\right) / \operatorname{tr}\left(\delta_j^T H_Y[\delta_j]\right)$;
8:     Set $\eta^{j+1} = \eta^j + \alpha_j \delta_j$;
9:     **if** $\operatorname{tr}\left(\left(\eta^{j+1}\right)^T \eta^{j+1}\right) \geq \Delta$ **then**
10:        Compute $\tau \geq 0$ such that $\eta = \eta^j + \tau \delta_j$ satisfies $\operatorname{tr}\left(\eta^T \eta\right) = \Delta$;
11:        **return** $\eta_k := \eta$;
12:     **end if**
13:     Set $r_{j+1} = r_j + \alpha H_Y[\delta_j]$;
14:     Set $\beta_{j+1} = \operatorname{tr}\left(r_{j+1}^T r_{j+1}\right) / \operatorname{tr}\left(r_j^T r_j\right)$;
15:     Set $\delta_{j+1} = -r_{j+1} + \beta_{j+1} \delta_j$;
16:     **if** a stopping criterion is satisfied **then**
17:        **return** $\eta_k := \eta^j$;
18:     **end if**
19: **end loop**

---

which implies that the smallest eigenvalue of the Hessian of the cost function at the solution is much smaller than its largest eigenvalue. This suggests that the multiplicative constant in the superlinear convergence bound (7.38) is large, which explains why superlinear convergence sets in less clearly than on the left-hand plot featuring a large eigenvalue gap. An experiment with a smaller machine epsilon would reveal a quadratic convergence pattern; i.e., the number of digits of accuracy eventually approximately doubles at each new iterate. (All the experiments described in this book were performed with a machine epsilon of approximately $2 \cdot 10^{-16}$.)

The eigenvalue algorithm resulting from the proposed approach is surprisingly competitive with other eigenvalue methods in spite of the fact that it is just a brute-force application of a general optimization scheme that does not make any attempt to exploit the specific form of the Rayleigh quotient cost function. The efficiency of the algorithm is supported by its similarity to the Jacobi-Davidson eigenvalue method, in particular to the JDCG method of Notay. Nevertheless, the algorithm admits several enhancements, including subspace acceleration techniques and the possible monitoring of the ρ ratio within the inner iteration at a low computational cost. These enhancements, as well as comparisons with state-of-the-art eigenvalue methods, are

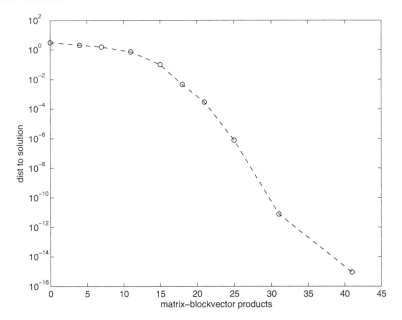

Figure 7.1 Numerical experiments on a trust-region algorithm for minimizing the Rayleigh cost function (7.53) on the Grassmann manifold $\mathrm{Grass}(p, n)$, with $n = 100$ and $p = 5$. $B = I$, and A is chosen with p eigenvalues evenly spaced on the interval $[1, 2]$ and the other $(n - p)$ eigenvalues evenly spaced on the interval $[10, 11]$; this is a problem with a large eigenvalue gap. The horizontal axis gives the number of multiplications of A times a block of p vectors. The vertical axis gives the distance to the solution, defined as the square root of the sum of the canonical angles between the current subspace and the leftmost p-dimensional invariant subspace of A. (This distance corresponds to the geodesic distance on the Grassmann manifold endowed with its canonical metric (3.44).)

presented in articles mentioned in Notes and References.

7.6 NOTES AND REFERENCES

The Riemannian trust-region approach was first proposed in [ABG04]. Most of the material in this section comes from [ABG07].

For more information on trust-region methods in $\mathbb{R}^n$, we refer the reader to Conn *et al.* [CGT00]. Trust-region methods are also discussed in textbooks on numerical optimization such as Nocedal and Wright [NW99]; see also Hei [Hei03], Gould *et al.* [GOST05], and Walmag and Delhez [WD05] for recent developments. Algorithm 10 reduces to [NW99, Alg. 4.1] in the classical $\mathbb{R}^n$ case; variants can be found in Conn *et al.* [CGT00, Ch. 10].

The method for computing an accurate solution of the trust-region sub-

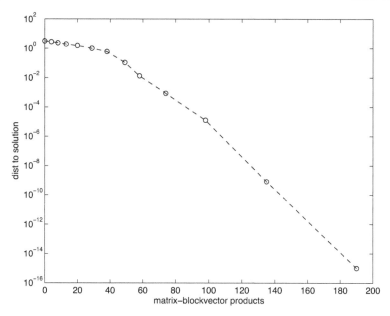

Figure 7.2 Same situation as in Figure 7.1 but now with $B = I$ and $A = \text{diag}(1, \ldots, n)$.

problem is due to Moré and Sorensen [MS83]. Proposition 7.3.1 is a straight-forward transcription of [CGT00, Th. 7.4.1], which itself generalizes results from [MS83] (or see [NW99, Th. 4.3]) to general norms.

The truncated CG method presented in Algorithm 11 closely follows the algorithm proposed by Steihaug [Ste83]; see also the work of Toint [Toi81]. Proposition 7.3.2 is due to Steihaug [Ste83, Th. 2.1]. The reader interested in the underlying principles of the Steihaug-Toint truncated CG method should refer to [Ste83], [NW99], or [CGT00].

Besides the truncated CG method, available algorithms for (approximately) solving trust-region subproblems include the dogleg method of Powell [Pow70], the double-dogleg method of Dennis and Mei [DM79], the method of Moré and Sorensen [MS83], the two-dimensional subspace minimization strategy of Byrd et al. [BSS88], the method based on the difference of convex functions proposed by Pham Dinh Tao and Le Thi Hoai An [TA98], the truncated Lanczos approach of Gould et al. [GLRT99], the matrix-free eigenproblem-based algorithm of Rojas et al. [RSS00], and the sequential subspace method of Hager [Hag01, HP05]. These and other methods are discussed in Conn et al. [CGT00, §7.5.4].

The classical global convergence results for trust-region methods in $\mathbb{R}^n$ can be found in Nocedal and Wright [NW99] (see in particular Theorem 4.8) and Conn et al. [CGT00]. A Taylor development similar to Lemma 7.4.7 can be found in Smith [Smi94]. The principle of the argument of Theorem 7.4.10 is closely related to the capture theorem, see Bertsekas [Ber95, Th 1.2.5]. A discussion on cubic versus quadratic convergence can be found in Dehaene and Vandewalle [DV00]. A proof of Corollary 7.4.6 is given in [ABG06a].

Experiments, comparisons, and further developments are presented in [ABG06b, ABGS05, BAG06] for the Riemannian trust-region approach to extreme invariant subspace computation (Section 7.5.3). Reference [ABG06b] works out a brute-force application of the Riemannian trust-region method to the optimization of the Rayleigh quotient cost function on the sphere: comparisons are made with other eigenvalue methods, in particular the JDCG algorithm of Notay [Not02] and the Tracemin algorithm of Sameh, Wisniewski, and Tong [SW82, ST00], and numerical results are presented. Reference [ABGS05] proposes a two-phase method that combines the advantages of the (unshifted) Tracemin method and of the RTR-tCG method with and order-2 model; it allows one to make efficient use of a preconditioner in the first iterations by relaxing the trust-region constraint. The implicit RTR method proposed in Baker et $al.$ [BAG06] makes use of the particular structure of the eigenvalue problem to monitor the value of the ratio ρ in the course of the inner iteration with little computational overhead, thereby avoiding the rejection of iterates because of poor model quality; for some problems, this technique considerably speeds up the iteration while the iterates are still far away from the solution, especially when a good preconditioner is available.

Chapter Eight

A Constellation of Superlinear Algorithms

The Newton method (Algorithm 5 in Chapter 6) applied to the gradient of a real-valued cost is the archetypal superlinear optimization method. The Newton method, however, suffers from a lack of global convergence and the prohibitive numerical cost of solving the Newton equation (6.2) necessary for each iteration. The trust-region approach, presented in Chapter 7, provides a sound framework for addressing these shortcomings and is a good choice for a generic optimization algorithm. Trust-region methods, however, are algorithmically complex and may not perform ideally on all problems. A host of other algorithms have been developed that provide lower-cost numerical iterations and stronger global convergence properties than the Newton iteration while still approximating the second-order properties of the Newton algorithm sufficiently well to obtain superlinear local convergence. The purpose of this chapter is to briefly review some of these techniques and show how they can be generalized to manifolds. These techniques admit so many variations that we have no pretention of being exhaustive. Most available optimization schemes in $\mathbb{R}^n$ have never been formulated on abstract manifolds. Considering each algorithm in detail is beyond the scope of this book. We will instead focus on resolving a common issue underlying most of these algorithms—approximating derivatives by finite differences on manifolds. To this end, we introduce the concept of vector transport, which relaxes the computational requirements of parallel translation in very much the same way as the concept of retraction relaxes the computational requirements of exponential mapping. Vector transport is a basic ingredient in generalizing the class of finite-difference and conjugate-gradient algorithms on manifolds.

We conclude the chapter by considering the problem of determining a solution, or more generally a least-squares solution, of a system of equations $F(x) = 0$, where F is a function on a manifold into $\mathbb{R}^n$. Although this problem is readily rewritten as the minimization of the squared norm of F, its particular structure lends itself to specific developments.

8.1 VECTOR TRANSPORT

In Chapter 4, on first-order algorithms, the notion of retraction was introduced as a general way to take a step in the direction of a tangent vector. (The tangent vector was, typically, the steepest-descent direction for the cost function.) In second-order algorithms, when the second-order information is not readily available through a closed-form Jacobian or Hessian, it will be necessary to approximate second derivatives by "comparing" first-order information (tangent vectors) at distinct points on the manifold. The notion of *vector transport* $\mathcal{T}$ on a manifold $\mathcal{M}$, roughly speaking, specifies how to transport a tangent vector ξ from a point $x \in \mathcal{M}$ to a point $R_x(\eta) \in \mathcal{M}$.

Vector transport, as defined below, is not a standard concept of differential geometry. (Neither is the notion of retraction.) However, as we will see, it is closely related to the classical concept of parallel translation. The reason for considering the more general notion of vector transport is similar to the reason for considering general retractions rather than the specific exponential mapping. Parallel translation along geodesics is a vector transport that is associated with any affine connection in a natural way. Conceptually appealing (like the exponential mapping), it can, however, be computationally demanding or cumbersome in numerical algorithms. Another vector transport may reduce (in some cases dramatically) the computational effort while retaining the convergence properties of the algorithm.

Let $T\mathcal{M} \oplus T\mathcal{M}$ denote the set

$$T\mathcal{M} \oplus T\mathcal{M} = \{(\eta_x, \xi_x) : \eta_x, \xi_x \in T_x\mathcal{M}, \ x \in \mathcal{M}\}.$$

This set admits a natural manifold structure for which the mappings

$$(\eta_x, \xi_x) \in T\mathcal{M} \oplus T\mathcal{M} \mapsto (\varphi_1(x), \ldots, \varphi_d(x), \eta_x\varphi_1, \ldots, \eta_x\varphi_d, \xi_x\varphi_1, \ldots, \xi_x\varphi_d)$$

are charts whenever φ is a chart of the manifold $\mathcal{M}$. The operation $\oplus$ is called the *Whitney sum*.

We refer to Figure 8.1 for an illustation of the following definition.

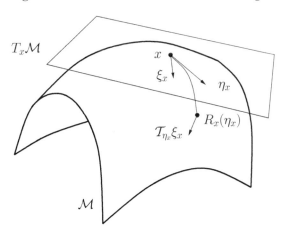

Figure 8.1 Vector transport.

Definition 8.1.1 (vector transport) *A* vector transport *on a manifold $\mathcal{M}$ is a smooth mapping*

$$T\mathcal{M} \oplus T\mathcal{M} \to T\mathcal{M} : (\eta_x, \xi_x) \mapsto \mathcal{T}_{\eta_x}(\xi_x) \in T\mathcal{M}$$

satisfying the following properties for all $x \in \mathcal{M}$:

(i) (Associated retraction) There exists a retraction R, called the retraction associated with $\mathcal{T}$, such that the following diagram commutes

$$
\begin{array}{ccc}
(\eta_x, \xi_x) & \xrightarrow{\;\mathcal{T}\;} & \mathcal{T}_{\eta_x}(\xi_x) \\
\downarrow & & \downarrow{\scriptstyle \pi} \\
\eta_x & \xrightarrow[\;R\;]{} & \pi\left(\mathcal{T}_{\eta_x}(\xi_x)\right)
\end{array}
$$

where $\pi\left(\mathcal{T}_{\eta_x}(\xi_x)\right)$ denotes the foot of the tangent vector $\mathcal{T}_{\eta_x}(\xi_x)$.
(ii) (Consistency) $\mathcal{T}_{0_x}\xi_x = \xi_x$ for all $\xi_x \in T_x\mathcal{M}$;
(iii) (Linearity) $\mathcal{T}_{\eta_x}(a\xi_x + b\zeta_x) = a\mathcal{T}_{\eta_x}(\xi_x) + b\mathcal{T}_{\eta_x}(\zeta_x)$.

The first point in Definition 8.1.1 means that $\mathcal{T}_{\eta_x}\xi_x$ is a tangent vector in $T_{R_x(\eta_x)}\mathcal{M}$, where R is the retraction associated with $\mathcal{T}$. When it exists, $(\mathcal{T}_{\eta_x})^{-1}(\xi_{R_x(\eta_x)})$ belongs to $T_x\mathcal{M}$. If η and ξ are two vector fields on $\mathcal{M}$, then $(\mathcal{T}_\eta)^{-1}\xi$ is naturally defined as the vector field satisfying

$$
\left((\mathcal{T}_\eta)^{-1}\xi\right)_x = (\mathcal{T}_{\eta_x})^{-1}\left(\xi_{R_x(\eta_x)}\right).
$$

8.1.1 Vector transport and affine connections

There is a close relationship between vector transport and affine connections.
If $\mathcal{T}$ is a vector transport and R is the associated retraction, then

$$
\nabla_{\eta_x}\xi := \left.\frac{\mathrm{d}}{\mathrm{d}t}\, \mathcal{T}_{t\eta_x}^{-1}\xi_{R(t\eta_x)}\right|_{t=0} \tag{8.1}
$$

defines an affine connection. The properties are readily checked from the definition.

Conversely, parallel translation is a particular vector transport that can be associated with any affine connection. Let $\mathcal{M}$ be a manifold endowed with an affine connection ∇ and recall from Section 5.4 the notation $t \mapsto P_\gamma^{t \leftarrow a}\xi(a)$ for the parallel vector field on the curve γ that satisfies $P_\gamma^{a \leftarrow a} = \gamma(a)$ and

$$
\frac{\mathrm{D}}{\mathrm{d}t}\left(P_\gamma^{t \leftarrow a}\xi(a)\right) = 0.
$$

Proposition 8.1.2 *If ∇ is an affine connection and R is a retraction on a manifold $\mathcal{M}$, then*

$$
\mathcal{T}_{\eta_x}(\xi_x) := P_\gamma^{1 \leftarrow 0}\xi_x \tag{8.2}
$$

is a vector transport with associated retraction R, where P_γ denotes the parallel translation induced by ∇ along the curve $t \mapsto \gamma(t) = R_x(t\eta_x)$. Moreover, $\mathcal{T}$ and ∇ satisfy (8.1).

Proof. It is readily checked that (8.2) defines a vector transport. For the second claim, let R be a retraction and let $\mathcal{T}$ be defined by the parallel translation induced by ∇, i.e.,

$$
\frac{\mathrm{D}}{\mathrm{d}t}\left(\mathcal{T}_{t\eta_x}\xi_x\right) = 0 \tag{8.3}
$$

with $\pi(\mathcal{T}_{t\eta_x}\xi_x) = R(t\eta_x)$ and $\mathcal{T}_{0_x}\xi_x = \xi_x$. Let $\hat{\nabla}$ be defined by

$$\hat{\nabla}_{\eta_x}\xi := \frac{\mathrm{d}}{\mathrm{d}t}\,\mathcal{T}_{t\eta_x}^{-1}\xi_{R(t\eta_x)}\Big|_{t=0}.$$

We want to show that $\nabla_{\eta_x}\xi = \hat{\nabla}_{\eta_x}\xi$ for all η_x,ξ. Let $\tilde{\xi}$ denote the vector field defined by $\tilde{\xi}_y = \mathcal{T}_{R_x^{-1}y}\xi_x$ for all y sufficiently close to x. We have

$$\hat{\nabla}_{\eta_x}\xi = \hat{\nabla}_{\eta_x}(\xi - \tilde{\xi}) + \hat{\nabla}_{\eta_x}\tilde{\xi} = \hat{\nabla}_{\eta_x}(\xi - \tilde{\xi}) = \nabla_{\eta_x}(\xi - \tilde{\xi}) = \nabla_{\eta_x}\xi,$$

where we have used the identities $\hat{\nabla}_{\eta_x}\tilde{\xi} = 0$ (which holds in view of the definitions of $\hat{\nabla}$ and $\tilde{\xi}$), $\hat{\nabla}_{\eta_x}(\xi - \tilde{\xi}) = \nabla_{\eta_x}(\xi - \tilde{\xi})$ (in view of $\xi_x - \tilde{\xi}_x = 0$), and $\nabla_{\eta_x}\tilde{\xi} = 0$ (since $\nabla_{\eta_x}\tilde{\xi} = \frac{\mathrm{D}}{\mathrm{d}t}\tilde{\xi}_{R(t\eta_x)}\big|_{t=0} = \frac{\mathrm{D}}{\mathrm{d}t}\mathcal{T}_{t\eta_x}\xi_x\big|_{t=0} = 0$). □

We also point out that if $\mathcal{M}$ is a Riemannian manifold, then the parallel translation defined by the Riemannian connection is an isometry, i.e.,

$$\langle P_\gamma^{t\leftarrow a}\xi(a), P_\gamma^{t\leftarrow a}\zeta(a)\rangle = \langle \xi(a), \zeta(a)\rangle.$$

Example 8.1.1   Sphere
We consider the sphere S^{n-1} with its structure of Riemannian submanifold of $\mathbb{R}^n$. Let $t \mapsto x(t)$ be a geodesic for the Riemannian connection (5.16) on S^{n-1}; see (5.25). Let u denote $\frac{1}{\|\dot{x}(0)\|}\dot{x}(0)$. The parallel translation (associated with the Riemannian connection) of a vector $\xi(0) \in T_{x(0)}$ along the geodesic is given by

$$\xi(t) = -x(0)\sin(\|\dot{x}(0)\|t)u^T\xi(0) + u\cos(\|\dot{x}(0)\|t)x^T(0)\xi(0) + (I - uu^T)\xi(0). \tag{8.4}$$

Example 8.1.2   Stiefel manifold
There is no known closed form for the parallel translation along geodesics for the Stiefel manifold $\mathrm{St}(p,n)$ endowed with the Riemannian connection inherited from the embedding in $\mathbb{R}^{n\times p}$.

Example 8.1.3   Grassmann manifold
Consider the Grassmann manifold viewed as a Riemannian quotient manifold of $\mathbb{R}^{n\times p}$ with the inherited Riemannian connection. Let $t \mapsto \mathcal{Y}(t)$ be a geodesic for this connection, with $\mathcal{Y}(0) = \mathrm{span}(Y_0)$ and $\overline{\dot{\mathcal{Y}}(0)}_{Y_0} = U\Sigma V^T$, a thin singular value decomposition (i.e., U is $n \times p$ orthonormal, V is $p \times p$ orthonormal, and Σ is $p \times p$ diagonal with nonnegative entries). We assume for simplicity that Y_0 is chosen orthonormal. Let $\xi(0)$ be a tangent vector at $\mathcal{Y}(0)$. Then the parallel translation of $\xi(0)$ along the geodesic is given by

$$\overline{\xi(t)}_{Y(t)} = -Y_0 V \sin(\Sigma t)U^T\overline{\xi(0)}_{Y_0} + U\cos(\Sigma t)U^T\overline{\xi(0)}_{Y_0} + (I - UU^T)\overline{\xi(0)}_{Y_0}. \tag{8.5}$$

Parallel translation is not the only way to achieve vector transport. As was the case with the choice of retraction, there is considerable flexibility in how a vector translation is chosen for a given problem. The approach taken will depend on the problem considered and the resourcefulness of the scientist designing the algorithm. In the next three subsections we present three approaches that can be used to generate computationally tractable vector translation mappings for the manifolds associated with the class of applications considered in this book.

8.1.2 Vector transport by differentiated retraction

Let $\mathcal{M}$ be a manifold endowed with a retraction R. Then a vector transport on $\mathcal{M}$ is defined by

$$\mathcal{T}_{\eta_x}\xi_x := \mathrm{D}R_x\left(\eta_x\right)[\xi_x];\qquad(8.6)$$

i.e.,

$$\mathcal{T}_{\eta_x}\xi_x = \left.\frac{\mathrm{d}}{\mathrm{d}t}R_x(\eta_x+t\xi_x)\right|_{t=0};$$

see Figure 8.2. Notice in particular that, in view of the local rigidity condition $\mathrm{D}R_x(0_x) = \mathrm{id}$, the condition $\mathcal{T}_{0_x}\xi = \xi$ for all $\xi \in T_x\mathcal{M}$ is satisfied.

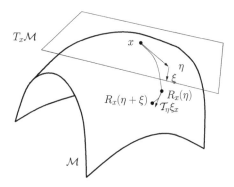

Figure 8.2 The vector transport $\mathcal{T}_\eta(\xi) := \mathrm{D}R_x\left(\eta\right)[\xi]$.

The definition (8.6) also provides a way to associate an affine connection with a retraction using (8.6) and (8.1).

We also point out that the vector transport (8.6) of a tangent vector along itself is given by

$$\mathcal{T}_{\eta_x}\eta_x = \left.\frac{\mathrm{d}}{\mathrm{d}t}(R_x(t\eta_x))\right|_{t=1}.$$

Example 8.1.4 *Sphere*
On the sphere S^{n-1} with the projection retraction

$$R_x(\xi_x) = (x+\xi_x)/\|x+\xi_x\|,$$

the vector transport (8.6) yields

$$\mathcal{T}_{\eta_x}\xi_x = \frac{1}{\|x+\eta_x\|}\mathrm{P}_{x+\eta_x}\xi_x$$

$$= \frac{1}{\|x+\eta_x\|}\left(I - \frac{1}{\|x+\eta_x\|^2}(x+\eta_x)(x+\eta_x)^T\right)\xi_x,$$

where, as usual, we implicitly use the natural inclusion of $T_x S^{n-1}$ in $\mathbb{R}^n$.

Example 8.1.5 *Stiefel manifold*

Consider the QR-based retraction (4.8) on the Stiefel manifold:

$$R_X(Z) = \mathrm{qf}(X + Z).$$

We need a formula for $\mathrm{Dqf}\,(Y)\,[U]$ *with* $Y \in \mathbb{R}_*^{n \times p}$ *and* $U \in T_Y \mathbb{R}_*^{n \times p} = \mathbb{R}^{n \times p}$. *Let* $t \mapsto W(t)$ *be a curve on* $\mathbb{R}_*^{n \times p}$ *with* $W(0) = Y$ *and* $\dot{W}(0) = U$ *and let* $W(t) = X(t)R(t)$ *denote the QR decomposition of* $W(t)$. *We have*

$$\dot{W} = \dot{X}R + X\dot{R}. \tag{8.7}$$

Since $XX^T + (I - XX^T) = I$, *we have the decomposition*

$$\dot{X} = XX^T \dot{X} + (I - XX^T)\dot{X}. \tag{8.8}$$

Multiplying (8.7) by $I - XX^T$ *on the left and by* R^{-1} *on the right yields the expression* $(I - XX^T)\dot{X} = (I - XX^T)\dot{W}R^{-1}$ *for the second term of (8.8). It remains to obtain an expression for* $X^T \dot{X}$. *Multiplying (8.7) on the left by* X^T *and on the right by* R^{-1} *yields*

$$X^T \dot{W} R^{-1} = X^T \dot{X} + \dot{R}R^{-1}. \tag{8.9}$$

In view of the form

$$T_X \mathrm{St}(p,n) = \{X\Omega + X_\perp K : \Omega^T = -\Omega, \; K \in \mathbb{R}^{(n-p) \times p}\}$$

for the tangent space to the Stiefel manifold at a point X, *it follows that the term* $X^T \dot{X}$ *in (8.9) belongs to the set of skew-symmetric* $p \times p$ *matrices, while the term* $\dot{R}R^{-1}$ *belongs to the set of upper triangular matrices. Let* $\rho_{\mathrm{skew}}(B)$ *denote the the skew-symmetric term of the decomposition of a square matrix* B *into the sum of a skew-symmetric term and an upper triangular term, i.e,*

$$(\rho_{\mathrm{skew}}(B))_{i,j} = \begin{cases} B_{i,j} & \text{if } i > j, \\ 0 & \text{if } i = j, \\ -B_{j,i} & \text{if } i < j. \end{cases}$$

From (8.9), we have $X^T \dot{X} = \rho_{\mathrm{skew}}(X^T \dot{W} R^{-1})$. *Replacing these results in (8.8) gives*

$$\dot{X} = XX^T \dot{X} + (I - XX^T)\dot{X} = X\rho_{\mathrm{skew}}(X^T \dot{W} R^{-1}) + (I - XX^T)\dot{W}R^{-1},$$

hence

$$\mathrm{Dqf}\,(Y)\,[U] = \mathrm{qf}(Y)\rho_{\mathrm{skew}}(\mathrm{qf}(Y)^T U (\mathrm{qf}(Y)^T Y)^{-1})$$
$$+ (I - \mathrm{qf}(Y)\mathrm{qf}(Y)^T)U(\mathrm{qf}(Y)^T Y)^{-1}.$$

Finally, we have, for $Z, U \in T_X \mathrm{St}(p,n)$,

$$\begin{aligned} \mathcal{T}_Z U &= \mathrm{D}R_X\,(Z)\,[U] \\ &= \mathrm{Dqf}\,(X + Z)\,[U] \\ &= R_X(Z)\rho_{\mathrm{skew}}(R_X(Z)^T U(R_X(Z)^T(X+Z))^{-1}) \\ &\quad + (I - R_X(Z)R_X(Z)^T)U(R_X(Z)^T(X+Z))^{-1}. \end{aligned}$$

Example 8.1.6 *Grassmann manifold*

   As previously, we view the Grassmann manifold $\mathrm{Grass}(p, n)$ *as a Riemannian quotient manifold of* $\mathbb{R}_*^{n \times p}$. *We consider the retraction*

$$R_{\mathcal{Y}}(\eta) = \mathrm{span}(Y + \overline{\eta}_Y).$$

We obtain

$$\overline{\mathrm{D} R_{\mathcal{Y}}(\eta)[\xi]}_{R_{\mathcal{Y}}(\eta)} = P_{Y + \overline{\eta}_Y}^h \overline{\xi}_Y,$$

where P_Y^h *denotes the orthogonal projection onto the orthogonal complement of the span of* Y; *see* (3.41).

8.1.3 Vector transport on Riemannian submanifolds

If $\mathcal{M}$ is an embedded submanifold of a Euclidean space $\mathcal{E}$ and $\mathcal{M}$ is endowed with a retraction R, then we can rely on the natural inclusion $T_y \mathcal{M} \subset \mathcal{E}$ for all $y \in \mathcal{N}$ to simply define the vector transport by

$$\mathcal{T}_{\eta_x} \xi_x := \mathrm{P}_{R_x(\eta_x)} \xi_x, \tag{8.10}$$

where P_x denotes the orthogonal projector onto $T_x \mathcal{N}$.

Example 8.1.7 *Sphere*

   On the sphere S^{n-1} *endowed with the retraction* $R(\eta_x) = (x + \eta_x)/\|x + \eta_x\|$, (8.10) *yields*

$$\mathcal{T}_{\eta_x} \xi_x = \left(I - \frac{(x + \eta_x)(x + \eta_x)^T}{\|x + \eta_x\|^2} \right) \xi_x \quad \in T_{R(\eta_x)} S^{n-1}.$$

Example 8.1.8 *Orthogonal Stiefel manifold*

   Let R *be a retraction on the Stiefel manifold* $\mathrm{St}(p, n)$. *(Possible choices of* R *are given in Section 4.1.1.) Formula* (8.10) *yields*

$$\mathcal{T}_{\eta_X} \xi_X = (I - Y Y^T) \xi_X + Y \mathrm{skew}(Y^T \xi_X) \quad \in T_Y \mathrm{St}(p, n),$$

where $Y := R_X(\eta_X)$.

8.1.4 Vector transport on quotient manifolds

Let $\mathcal{M} = \overline{\mathcal{M}}/\sim$ be a quotient manifold, where $\overline{\mathcal{M}}$ is an open subset of a Euclidean space $\mathcal{E}$ (this includes the case where $\overline{\mathcal{M}}$ itself is a Euclidean space). Let $\mathcal{H}$ be a horizontal distribution on $\overline{\mathcal{M}}$ and let $P_{\overline{x}}^h : T_{\overline{x}} \overline{\mathcal{M}} \to \mathcal{H}_{\overline{x}}$ denote the projection parallel to the vertical space $\mathcal{V}_{\overline{x}}$ onto the horizontal space $\mathcal{H}_{\overline{x}}$. Then (using the natural identification $T_{\overline{y}} \overline{\mathcal{M}} \simeq \mathcal{E}$ for all $\overline{y} \in \overline{\mathcal{M}}$),

$$\overline{(\mathcal{T}_{\eta_x} \xi_x)}_{\overline{x} + \overline{\eta}_{\overline{x}}} := \mathrm{P}_{\overline{x} + \overline{\eta}_{\overline{x}}}^h \overline{\xi}_{\overline{x}} \tag{8.11}$$

defines a vector transport on $\mathcal{M}$.

Example 8.1.9 *Projective space*

As in Section 3.6.2, we view the projective space $\mathbb{RP}^{n-1}$ as a Riemannian quotient manifold of $\mathbb{R}^n_*$. Equation (8.11) yields

$$\overline{(\mathcal{T}_{\eta_{x\mathbb{R}}}\xi_{x\mathbb{R}})}_{x+\overline{\eta}_x} = \mathrm{P}^h_{x+\overline{\eta}_x}\overline{\xi}_x,$$

where $\mathrm{P}^h_y z = z - yy^T z$ denotes the projection onto the horizontal space at y.

Example 8.1.10 *Grassmann manifold*

Again as in Section 3.6.2, we view the Grassmann manifold $\mathrm{Grass}(p,n)$ as the Riemannian quotient manifold $\mathbb{R}^{n\times p}_*/\mathrm{GL}_p$. Equation (8.11) leads to

$$\overline{(\mathcal{T}_{\eta_Y}\xi_Y)}_{Y+\overline{\eta}_Y} = \mathrm{P}^h_{Y+\overline{\eta}_Y}\overline{\xi}_Y, \tag{8.12}$$

where $\mathrm{P}^h_Y Z = Z - Y(Y^T Y)^{-1}Y^T Z$ denotes the projection onto the horizontal space at Y.

8.2 APPROXIMATE NEWTON METHODS

Let $\mathcal{M}$ be a manifold equipped with a retraction R and an affine connection ∇. Let ξ be a vector field on $\mathcal{M}$ and consider the problem of seeking a zero of ξ. The Newton equation (6.1) reads

$$\nabla_{\eta_x}\xi = -\xi_x$$

for the unknown $\eta_x \in T_x\mathcal{M}$. In Chapter 6, it was assumed that a procedure for computing $\nabla_{\eta_x}\xi$ is available at all $x \in \mathcal{M}$. In contrast, approximate Newton methods seek to relax the solution of Newton's equation in a way that retains the superlinear convergence of the algorithm. The kth iteration of the algorithm thus replaces (6.1) with the solution $\eta_k \in T_{x_k}\mathcal{M}$ of a relaxed equation

$$(J(x_k) + E_k)\eta_k = -\xi_{x_k} + \varepsilon_k, \tag{8.13}$$

where $J(x_k)$ is the Jacobian of ξ defined by

$$J(x_k) : T_{x_k}\mathcal{M} \to T_{x_k}\mathcal{M} : \eta_k \mapsto \nabla_{\eta_k}\xi.$$

The operator E_k denotes the approximation error on the Jacobian, while the tangent vector ε_k denotes the residual error in solving the (inexact) Newton equation.

The next result gives sufficiently small bounds on E_k and ε_k to preserve the fast local convergence of the exact Newton method.

Theorem 8.2.1 (local convergence of inexact Newton) *Suppose that at each step of Newton's method (Algorithm 4), the Newton equation (6.1) is replaced by the inexact equation (8.13). Assume that there exists $x_* \in \mathcal{M}$ such that $\xi_{x_*} = 0$ and $J(x_*)$ is invertible. Let $(\mathcal{U}', \varphi)$, $x_* \in \mathcal{U}'$, be a chart of* *the manifold $\mathcal{M}$ and let the coordinate expressions be denoted by $\hat{\cdot}$. Assume that there exist constants β_J and β_η such that*

$$\|\hat{E}_k\| \leq \beta_J \|\hat{\xi}_k\| \tag{8.14}$$

and

$$\|\hat{\varepsilon}_k\| \leq \min\{\|\hat{\xi}_k\|^\theta, \kappa\}\|\hat{\xi}_k\| \tag{8.15}$$

for all k, with $\theta > 0$. Then there exists a neighborhood $\mathcal{U}$ of x_* in $\mathcal{M}$ such that, for all $x_0 \in \mathcal{U}$, the inexact algorithm generates an infinite sequence $\{x_k\}$ converging superlinearly to x_*.

Proof. (Sketch.) The assumptions and notation are those of the proof of Theorem 6.3.2, and we sketch how that proof can be adapted to handle Theorem 8.2.1. By a smoothness argument,

$$\|\hat{\xi}_{\hat{x}}\| \leq \gamma_\xi \|\hat{x} - \hat{x}_*\|.$$

It follows from Lemma 6.3.1 that

$$\|(\hat{J}(\hat{x}_k) + \hat{E}_k)^{-1}\| \leq \|\hat{J}(\hat{x}_k)^{-1}\| \|(I - (\hat{J}(\hat{x}_k))^{-1}\hat{E}_k)^{-1}\|$$

$$\leq 2\beta \frac{1}{1 - \|(\hat{J}(\hat{x}_k))^{-1}\hat{E}_k\|} \leq 2\beta \frac{1}{1 - 2\beta\|\hat{E}_k\|} \leq 2\beta \frac{1}{1 - 2\beta\beta_J\gamma_\xi\|\hat{x}_k - \hat{x}_*\|}.$$

Consequently, by choosing $\mathcal{U}$ sufficiently small, $\|(\hat{J}(\hat{x}_k) + \hat{E}_k)^{-1}\|$ is bounded by a constant, say $2\beta'$, for all $x \in \mathcal{U}$. From there, it is direct to update the end of the proof of Theorem 6.3.2 to obtain again a bound

$$\|\hat{x}_{k+1} - \hat{x}_*\| \leq (\beta'(\gamma_J + \gamma_\Gamma) + 2\beta'\gamma_\Gamma + 2\beta'\beta_J\gamma_\xi + \gamma_R)\|\hat{x}_k - \hat{x}_*\|^2$$
$$+ 2\beta'\gamma_\xi^{\theta+1}\|\hat{x}_k - \hat{x}_*\|^{\theta+1}$$

for all x_k in some neighborhood of x_*. □

Condition (8.15) on the residual in the Newton equation is easily enforced by using an iterative solver that keeps track of the residual of the linear system of equations; the inner iteration is merely stopped as soon as the required precision is reached. Pointers to the literature on iterative solvers for linear equations can be found in Notes and References. Enforcing condition (8.14), on the other hand, involves differential geometric issues; this is the topic of the next section.

8.2.1 Finite difference approximations

A standard way to approximate the Jacobian $J(x_k)$ without having to compute second-order derivatives is to evaluate finite differences of the vector field ξ. On manifolds, the idea of evaluating finite differences on ξ is hindered by the fact that when $y \neq z$, the quantity $\xi_y - \xi_z$ is ill-defined, as the two tangent vectors belong to two different abstract Euclidean spaces $T_y\mathcal{M}$ and $T_z\mathcal{M}$. In practice, we will encounter only the case where a tangent vector η_y is known such that $z = R(\eta_y)$. We can then compare ξ_y and $\xi_{R(\eta_y)}$ using a

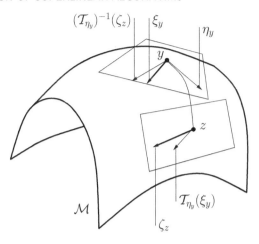

Figure 8.3 To compare a tangent vector $\xi_y \in T_y\mathcal{M}$ with a tangent vector $\zeta_z \in T_z\mathcal{M}$, $z = R(\eta_y)$, it is possible to transport ξ_y to $T_z\mathcal{M}$ through the mapping $\mathcal{T}_{\eta_y}$ or to transport ζ_z to $T_y\mathcal{M}$ through the mapping $(\mathcal{T}_{\eta_y})^{-1}$.

vector transport, as introduced in Section 8.1. Depending on the situation, we may want to compare the vectors in any of the two tangent spaces; see Figure 8.3.

To define finite differences in a neighborhood of a point x_* on a manifold $\mathcal{M}$ endowed with a vector transport $\mathcal{T}$, pick (smooth) vector fields E_i, $i = 1, \ldots, d$, such that $((E_1)_x, \ldots, (E_d)_x)$ forms a basis of $T_x\mathcal{M}$ for all x in a neighborhood $\mathcal{U}$ of x_*. Let R denote the retraction associated with the vector transport $\mathcal{T}$. Given a smooth vector field ξ and a real constant $h > 0$, let $A(x) : T_x\mathcal{M} \to T_x\mathcal{M}$ be the linear operator that satisfies, for $i = 1, \ldots, d$,

$$A(x)[E_i] = \frac{(\mathcal{T}_{h(E_i)_x})^{-1}\xi_{R(h(E_i)_x)} - \xi_x}{h}. \tag{8.16}$$

We thus have $A(x)[\eta_x] = \sum_{i=1}^{d} \eta^i|_x A(x)[E_i]$, where $\eta_x = \sum_{i=1}^{d} \eta^i|_x (E_i)_x$ is the decomposition of η_x in the basis $((E_1)_x, \ldots, (E_d)_x)$.

The next lemma gives a bound on how well $A(x)$ approximates the Jacobian $J(x) : \eta_x \mapsto \nabla_{\eta_x}\xi$ in a neighborhood of a zero of ξ. This result is instrumental in the local convergence analysis of the finite-difference quasi-Newton method introduced below.

Lemma 8.2.2 (finite differences) *Let ξ be a smooth vector field on a manifold $\mathcal{M}$ endowed with a vector transport $\mathcal{T}$ (Definition 8.1.1). Let x_* be a nondegenerate zero of ξ and let $(E_1, \ldots, E_d)$ be a basis of $\mathfrak{X}(\mathcal{U})$, where $\mathcal{U}$ is a neighborhood of x_*. Let A be defined by finite differences as in (8.16). Then there is $c > 0$ such that, for all x sufficiently close to x_* and all h sufficiently small, it holds that*

$$\|A(x)[E_i] - J(x)[E_i]\| \leq c(h + \|\xi_x\|). \tag{8.17}$$

Proof. This proof uses notation and conventions from the proof of Theorem 6.3.2. We work in local coordinates and denote coordinate expressions with a hat. (For example, $\hat{J}(\hat{x})$ denotes the coordinate expression of the operator $J(x)$.) There is a neighborhood $\mathcal{U}$ of x_* and constants $c_1, \ldots, c_6$ such that, for all $x \in \mathcal{U}$ and all $h > 0$ sufficiently small, the following bounds hold:

$$\|hA(x)[E_i] - J(x)[hE_i]\|$$
$$\leq c_1 \|\widehat{hA(x)[E_i]} - \widehat{J(x)[hE_i]}\|$$
$$= \|(\widehat{\mathcal{T}_{hE_i}})^{-1}\hat{\xi}_{\hat{R}_{\hat{x}}(h\hat{E}_i)} - \hat{\xi}_{\hat{x}} - \mathrm{D}\hat{\xi}(\hat{x})\left[h\hat{E}_i\right] - \hat{\Gamma}_{\hat{x},\hat{\xi}}[h\hat{E}_i]\|$$
$$\leq \|\hat{\xi}_{\hat{x}+h\hat{E}_i} - \hat{\xi}_{\hat{x}} - \mathrm{D}\hat{\xi}(\hat{x})\left[h\hat{E}_i\right]\| + \|(\widehat{\mathcal{T}_{hE_i}})^{-1}\hat{\xi}_{\hat{R}_{\hat{x}}(h\hat{E}_i)} - \hat{\xi}_{\hat{R}_{\hat{x}}(h\hat{E}_i)}\|$$
$$\quad + \|\hat{\xi}_{\hat{R}_{\hat{x}}(h\hat{E}_i)} - \hat{\xi}_{\hat{x}+h\hat{E}_i}\| + \|\hat{\Gamma}_{\hat{x},\hat{\xi}}[h\hat{E}_i]\|$$
$$\leq c_2 h^2 + c_3 h(\|\hat{x} - \hat{x}_*\| + h) + c_4 h^2 + c_5 \|\hat{x} - \hat{x}_*\| h$$
$$\leq c_6 h(h + \|\xi_x\|).$$

(A bound of the form $\|\hat{x} - \hat{x}_*\| \leq c\|\xi_x\|$ comes from the fact that x_* is a nondegenerate zero of ξ.) The claim follows. □

In the classical case, where $\mathcal{M}$ is a Euclidean space and the term

$$(\mathcal{T}_{h(E_i)_x})^{-1}\xi_{R(h(E_i)_x)}$$

in (8.16) reduces to ξ_{x+hE_i}, the bound (8.17) can be replaced by

$$\|A(x)[E_i] - J(x)[E_i]\| \leq ch, \tag{8.18}$$

i.e., $\|\xi_x\|$ no longer appears. The presence of $\|\xi_x\|$ is the counterpart to the fact that our definition of vector transport is particularly lenient. Fortunately, the perturbation $\|\xi_x\|$ goes to zero sufficiently fast as x goes to a zero of ξ. Indeed, using Lemma 8.2.2 and Theorem 8.2.1, we obtain the following result.

Proposition 8.2.3 *Consider the geometric Newton method (Algorithm 4) where the exact Jacobian $J(x_k)$ is replaced by the operator $A(x_k)$ defined in (8.16) with $h := h_k$. If*

$$\lim_{k \to \infty} h_k = 0,$$

then the convergence to nondegenerate zeros of ξ is superlinear. If, moreover, there exists some constant c such that

$$h_k \leq c\|\xi_{x_k}\|$$

for all k, then the convergence is (at least) quadratic.

8.2.2 Secant methods

An approximate Jacobian at $x \in \mathcal{M}$ is a linear operator in the d-dimensional tangent space $T_x\mathcal{M}$. Secant methods in $\mathbb{R}^n$ construct an approximate Jacobian A_{k+1} by imposing the secant equation

$$\xi_{x_{k+1}} - \xi_{x_k} = A_{k+1}\eta_k, \tag{8.19}$$

which can be seen as an underdetermined system of equations with d^2 unknowns. The remaining degrees of freedom in A_{k+1} are specified according to some algorithm that uses prior information where possible and also preserves or even improves the convergence properties of the underlying Newton method.

The generalization of the secant condition (8.19) on a manifold $\mathcal{M}$ endowed with a vector transport $\mathcal{T}$ is

$$\xi_{x_{k+1}} - \mathcal{T}_{\eta_k} \xi_{x_k} = A_{k+1}[\mathcal{T}_{\eta_k} \eta_k], \tag{8.20}$$

where η_k is the update vector at the iterate x_k, i.e., $R_{x_k}(\eta_k) = x_{k+1}$.

In the case where the manifold is Riemannian and ξ is the gradient of a real-valued function f of which a minimizer is sought, it is customary to require the following additional properties. Since the Hessian $J(x) = \operatorname{Hess} f(x)$ is symmetric (with respect to the Riemannian metric), one requires that the operator A_k be symmetric for all k. Further, in order to guarantee that η_k remains a descent direction for f, the updating formula should generate a positive-definite operator A_{k+1} whenever A_k is positive-definite. A well-known updating formula in $\mathbb{R}^n$ that aims at satisfying these properties is the Broyden-Fletcher-Goldfarb-Shanno (BFGS) scheme. On a manifold $\mathcal{M}$ endowed with a vector transport $\mathcal{T}$, the BFGS scheme generalizes as follows. With the notation

$$s_k := \mathcal{T}_{\eta_k} \eta_k \in T_{x_{k+1}} \mathcal{M},$$
$$y_k := \operatorname{grad} f(x_{k+1}) - \mathcal{T}_{\eta_k}(\operatorname{grad} f(x_k)) \in T_{x_{k+1}} \mathcal{M},$$

we define the operator $A_{k+1} : T_{x_{k+1}} \mathcal{M} \mapsto T_{x_{k+1}} \mathcal{M}$ by

$$A_{k+1}\eta = \tilde{A}_k \eta - \frac{\langle s_k, \tilde{A}_k \eta \rangle}{\langle s_k, \tilde{A}_k s_k \rangle} \tilde{A}_k s_k + \frac{\langle y_k, \eta \rangle}{\langle y_k, s_k \rangle} y_k \quad \text{for all } p \in T_{x_{k+1}} \mathcal{M},$$

with

$$\tilde{A}_k = \mathcal{T}_{\eta_k} \circ A_k \circ (\mathcal{T}_{\eta_k})^{-1}.$$

Note that the inner products are taken with respect to the Riemannian metric. Assume that A_k is symmetric positive-definite on $T_{x_k} \mathcal{M}$ (with respect to the inner product defined by the Riemannian metric) and that $\mathcal{T}_{\eta_k}$ is an isometry (i.e., the inverse of $\mathcal{T}_{\eta_k}$ is equal to its adjoint). Then $\tilde{A}_k$ is symmetric positive-definite, and it follows from the classical BFGS theory that A_{k+1} is symmetric positive-definite on $T_{x_{k+1}} \mathcal{M}$ if and only if $\langle y_k, s_k \rangle > 0$. The advantage of A_k is that it requires only first-order information that has to be computed anyway to provide the right-hand side of the Newton equation.

The local and global convergence analysis of the BFGS method in $\mathbb{R}^n$ is not straightforward. A careful generalization to manifolds, in the vein of the work done in Chapter 7 for trust-region methods, is beyond the scope of the present treatise.

8.3 CONJUGATE GRADIENTS

In this section we depart the realm of quasi-Newton methods to briefly consider conjugate gradient algorithms. We first summarize the principles of CG in $\mathbb{R}^n$.

The *linear* CG algorithm can be presented as a method for minimizing the function

$$\phi(x) = \tfrac{1}{2}x^T A x - x^T b, \qquad (8.21)$$

where $b \in \mathbb{R}^n$ and A is an $n \times n$ symmetric positive-definite matrix. One of the simplest ways to search for the minimizer of ϕ is to use a steepest-descent method, i.e., search along

$$-\mathrm{grad}\,\phi(x_k) = b - A x_k := r_k,$$

where r_k is called the *residual* of the iterate x_k. Unfortunately, if the matrix A is ill-conditioned, then the steepest-descent method may be very slow. (Recall that the convergence factor r in Theorem 4.5.6 goes to 1 as the ratio between the smallest and the largest eigenvalues of A—which are the eigenvalues of the constant Hessian of ϕ—goes to zero.) Conjugate gradients provide a remedy to this drawback by modifying the search direction at each step. Let x_0 denote the initial iterate and let $p_0, \ldots, p_k$ denote the successive search directions that can be used to generate x_{k+1}. A key observation is that, writing x_{k+1} as

$$x_{k+1} = x_0 + P_{k-1}y + \alpha p_k,$$

where $P_{k-1} = [p_1 | \ldots | p_{k-1}]$, $y \in \mathbb{R}^{k-1}$, and $\alpha \in \mathbb{R}$, we have

$$\phi(x_{k+1}) = \phi(x_0 + P_{k-1}y) + \alpha y^T P_{k-1}^T A p_k + \frac{\alpha^2}{2} p_k^T A p_k - \alpha p_k^T r_0.$$

Hence the minization of $\phi(x_{k+1})$ splits into two independent minimizations—one for y and one for α—when the search direction p_k is chosen to be A-orthogonal to the previous search directions, i.e.,

$$P_{k-1}^T A p_k = 0.$$

It follows that if the search directions $p_0, \ldots, p_k$ are *conjugate* with respect to A, i.e.,

$$p_i^T A p_j = 0 \qquad \text{for all } i \neq j,$$

then an algorithm, starting from x_0 and performing successive exact line-search minimizations of ϕ along $p_0, \ldots, p_k$, returns a point x_{k+1} that is the minimizer of ϕ over the set $x_0 + \mathrm{span}\{p_0, \ldots, p_k\}$.

Thus far we have only required that the search directions be conjugate with respect to A. The linear CG method further relates the search directions to the gradients by selecting each p_k to be in the direction of the minimizer of $\|p - r_k\|_2$ over all vectors p satisfying the A-orthogonality condition $[p_1 | \ldots | p_{k-1}]^T A p = 0$. It can be shown that this requirement is satisfied by

$$p_k = r_k + \beta_k p_{k-1}, \qquad (8.22)$$

where

$$\beta_k = -\frac{r_k^T A p_{k-1}}{p_{k-1}^T A p_{k-1}}. \tag{8.23}$$

Summarizing, the linear CG iteration is

$$x_{k+1} = x_k + \alpha_k p_k,$$

where α_k is chosen as

$$\alpha_k = -\frac{r_k^T p_k}{p_k^T A p_k}$$

to achieve exact minimization of ϕ along the line $x_k + \alpha p_k$ and where p_k is selected according to (8.22), (8.23). The first search direction p_0 is simply chosen as the steepest-descent direction at x_0. This algorithm is usually presented in a mathematically equivalent but numerically more efficient formulation, which is referred to as *the* (linear) CG algorithm. Notice that, since the minimizer of ϕ is $x = A^{-1}b$, the linear CG algorithm can also be used to solve systems of equations whose matrices are symmetric positive-definite.

Several generalizations of the linear CG algorithm have been proposed for cost functions f that are not necessarily of the quadratic form (8.21) with $A = A^T$ positive-definite. These algorithms are termed *nonlinear CG* methods. Modifications with respect to the linear CG algorithm occur at three places: (i) the residual r_k becomes the negative gradient $-\operatorname{grad} f(x_k)$, which no longer satisfies the simple recursive formula $r_{k+1} = r_k + \alpha_k A p_k$; (ii) computation of the line-search step α_k becomes more complicated and can be achieved approximately using various line-search procedures; (iii) several alternatives are possible for β_k that yield different nonlinear CG methods but nevertheless reduce to the linear CG method when f is strictly convex-quadratic and α_k is computed using exact line-search minimization. Popular choices for β_k in the formula

$$p_k = -\operatorname{grad} f(x_k) + \beta_k p_{k-1} \tag{8.24}$$

are

$$\beta_k = \frac{(\operatorname{grad} f(x_k))^T \operatorname{grad} f(x_k)}{(\operatorname{grad} f(x_{k-1}))^T \operatorname{grad} f(x_{k-1})} \quad \text{(Fletcher-Reeves)}$$

and

$$\beta_k = \frac{(\operatorname{grad} f(x_k))^T (\operatorname{grad} f(x_k) - \operatorname{grad} f(x_{k-1}))}{(\operatorname{grad} f(x_{k-1}))^T \operatorname{grad} f(x_{k-1})} \quad \text{(Polak-Ribière)}.$$

When generalizing nonlinear CG methods to manifolds, we encounter a familiar difficulty: in (8.24), the right-hand side involves the sum of an element $\operatorname{grad} f(x_k)$ of $T_{x_k}\mathcal{M}$ and an element p_{k-1} of $T_{x_{k-1}}\mathcal{M}$. Here again, the concept of vector transport provides an adequate and flexible solution. We are led to propose a "meta-algorithm" (Algorithm 13) for the conjugate gradient.

Algorithm 13 Geometric CG method

Require: Riemannian manifold $\mathcal{M}$; vector transport $\mathcal{T}$ on $\mathcal{M}$ with associated retraction R; real-valued function f on $\mathcal{M}$.
Goal: Find a local minimizer of f.
Input: Initial iterate $x_0 \in \mathcal{M}$.
Output: Sequence of iterates $\{x_k\}$.
 1: Set $\eta_0 = -\operatorname{grad} f(x_0)$.
 2: **for** $k = 0, 1, 2, \ldots$ **do**
 3:    Compute a step size α_k and set

$$x_{k+1} = R_{x_k}(\alpha_k \eta_k). \tag{8.25}$$

 4:    Compute β_{k+1} and set

$$\eta_{k+1} = -\operatorname{grad} f(x_{k+1}) + \beta_{k+1} \mathcal{T}_{\alpha_k \eta_k}(\eta_k). \tag{8.26}$$

 5: **end for**

In Step 3 of Algorithm 13, the computation of α_k can be done, for example, using a line-search backtracking procedure as described in Algorithm 1. If the numerical cost of computing the exact line-search solution is not prohibitive, then the minimizing value of α_k should be used. Exact line-search minimization yields $0 = \frac{\mathrm{d}}{\mathrm{d}t} f(R_{x_k}(t\eta_k))\big|_{t=\alpha_k} = \mathrm{D}f(x_{k+1})\left[\frac{\mathrm{d}}{\mathrm{d}t}R_{x_k}(t\eta_k)\big|_{t=\alpha_k}\right]$. Assuming that $\mathcal{T}_{\alpha_k \eta_k}(\eta_k)$ is collinear with $\frac{\mathrm{d}}{\mathrm{d}t}R_{x_k}(t\eta_k)\big|_{t=\alpha_k}$ (see Section 8.1.2), this leads to $\langle \operatorname{grad} f(x_{k+1}), \mathcal{T}_{\alpha_k \eta_k}(\eta_k)\rangle = \mathrm{D}f(x_{k+1})\left[\mathcal{T}_{\alpha_k \eta_k}(\eta_k)\right] = 0$. In view of (8.26), one finds that

$$\langle \operatorname{grad} f(x_{k+1}), \eta_{k+1}\rangle = -\langle \operatorname{grad} f(x_{k+1}), \operatorname{grad} f(x_{k+1})\rangle < 0,$$

i.e., η_{k+1} is a descent direction for f.

Several choices are possible for β_{k+1} in Step 4 of Algorithm 13. Imposing the condition that η_{k+1} and $\mathcal{T}_{\alpha_k \eta_k}(\eta_k)$ be conjugate with respect to Hess $f(x_{k+1})$ yields

$$\beta_{k+1} = \frac{\langle \mathcal{T}_{\alpha_k \eta_k}(\eta_k), \operatorname{Hess} f(x_{k+1})[\operatorname{grad} f(x_{k+1})]\rangle}{\langle \mathcal{T}_{\alpha_k \eta_k}(\eta_k), \operatorname{Hess} f(x_{k+1})[\mathcal{T}_{\alpha_k \eta_k}(\eta_k)]\rangle}. \tag{8.27}$$

The β of Fletcher-Reeves becomes

$$\beta_{k+1} = \frac{\langle \operatorname{grad} f(x_{k+1}), \operatorname{grad} f(x_{k+1})\rangle}{\langle \operatorname{grad} f(x_k), \operatorname{grad} f(x_k)\rangle}, \tag{8.28}$$

whereas the β of Polak-Ribière naturally generalizes to

$$\beta_{k+1} = \frac{\langle \operatorname{grad} f(x_{k+1}), \operatorname{grad} f(x_{k+1}) - \mathcal{T}_{\alpha_k \eta_k}(\operatorname{grad} f(x_k))\rangle}{\langle \operatorname{grad} f(x_k), \operatorname{grad} f(x_k)\rangle}. \tag{8.29}$$

Whereas the convergence theory of linear CG is well understood, nonlinear CG methods have convergence properties that depend on the choice of α_k and β_k, even in the case of $\mathbb{R}^n$. We do not further discuss such convergence issues in the present framework.

8.3.1 Application: Rayleigh quotient minimization

As an illustration of the geometric CG algorithm, we apply Algorithm 13 to the problem of minimizing the Rayleigh quotient function (2.1) on the Grassmann manifold. For simplicity, we consider the standard eigenvalue problem (namely, $B := I$), which leads to the cost function

$$f : \mathrm{Grass}(p, n) \to \mathbb{R} : \mathrm{span}(Y) \mapsto \mathrm{tr}((Y^T Y)^{-1} Y^T A Y),$$

where A is an arbitrary $n \times n$ symmetric matrix. As usual, we view $\mathrm{Grass}(p, n)$ as a Riemannian quotient manifold of $\mathbb{R}_*^{n \times p}$ (see Section 3.6.2). Formulas for the gradient and the Hessian of f can be found in Section 6.4.2. For Step 3 of Algorithm 13 (the line-search step), we select x_{k+1} as the Armijo point (Definition 4.2.2) with $\bar{\alpha} = 1$, $\sigma = 0.5$, and $\beta = 0.5$. For Step 4 (selection of the next search direction), we use the Polak-Ribière formula (8.29). The retraction is chosen as in (4.11), and the vector transport is chosen according to (8.12). The algorithm further uses a restart strategy that consists of choosing $\beta_{k+1} := 0$ when k is a multiple of the dimension $d = p(n - p)$ of the manifold. Numerical results are presented in Figures 8.4 and 8.5.

The resulting algorithm appears to be an efficient method for computing an extreme invariant subspace of a symmetric matrix. One should bear in mind, however, that this is only a brute-force application of a very general optimization scheme to a very specific problem. As such, the algorithm admits several enhancements that exploit the simple structure of the Rayleigh quotient cost function. A key observation is that it is computationally inexpensive to optimize the Rayleigh quotient over a low-dimensional subspace since this corresponds to a small-dimensional eigenvalue problem. This suggests a modification of the nonlinear CG scheme where the next iterate x_{k+1} is obtained by minimizing the Rayleigh quotient over the space spanned by the columns of x_k, η_{k-1} and $\mathrm{grad}\, f(x_k)$. The algorithm obtained using this modification, barring implementation issues, is equivalent to the locally optimal CG method proposed by Knyazev (see Notes and References in Chapter 4).

An interesting point of comparison between the numerical results displayed in Figures 7.1 and 7.2 for the trust-region approach and in Figures 8.4 and 8.5 is that the trust-region algorithm reaches twice the precision of the CG algorithm. The reason is that, around a minimizer v of a smooth cost function f, one has $f(R_v(\eta)) = f(v) + O(\|\eta\|^2)$, whereas $\|\mathrm{grad}\, f(R_v(\eta))\| = O(\|\eta\|)$. Consequently, the numerical evaluation of $f(x_k)$ returns exactly $f(v)$ as soon as the distance between x_k and v is of the order of the square root of the machine epsilon, and the line-search process in Step 3 of Algorithm 13 just returns $x_{k+1} = x_k$. In contrast, the linear CG method used in the inner iteration of the trust-region method, with its exact minimization formula for α_k, makes it possible to obtain accuracies of the order of the machine epsilon. Another potential advantage of the trust-region approach over nonlinear CG methods is that it requires significantly fewer evaluations of the cost function f since it relies only on its local model m_{x_k} to carry out the inner iteration process. This is important when the cost function is expensive to compute.

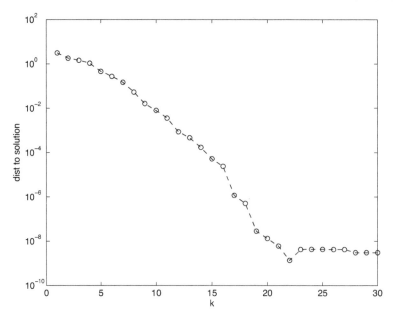

Figure 8.4 Minimization of the Rayleigh quotient (2.1) on Grass(p, n), with $n =$ 100 and $p = 5$. $B = I$ and A is chosen with p eigenvalues evenly spaced on the interval $[1, 2]$ and the other $(n - p)$ eigenvalues evenly spaced on the interval $[10, 11]$; this is a problem with a large eigenvalue gap. The distance to the solution is defined as the square root of the sum of the canonical angles between the current subspace and the leftmost p-dimensional invariant subspace of A. (This distance corresponds to the geodesic distance on the Grassmann manifold endowed with its canonical metric (3.44).)

8.4 LEAST-SQUARE METHODS

The problem addressed by the geometric Newton method presented in Algorithm 4 is to compute a zero of a vector field on a manifold $\mathcal{M}$ endowed with a retraction R and an affine connection ∇. A particular instance of this method is Algorithm 5, which seeks a critical point of a real-valued function f by looking for a zero of the gradient vector field of f. This method itself admits enhancements in the form of line-search and trust-region methods that ensure that f decreases at each iteration and thus favor convergence to local minimizers.

In this section, we consider more particularly the case where the real-valued function f takes the form

$$f : \mathcal{M} \to \mathbb{R} : x \mapsto \tfrac{1}{2}\|F(x)\|^2, \tag{8.30}$$

where

$$F : \mathcal{M} \to \mathcal{E} : x \mapsto F(x)$$

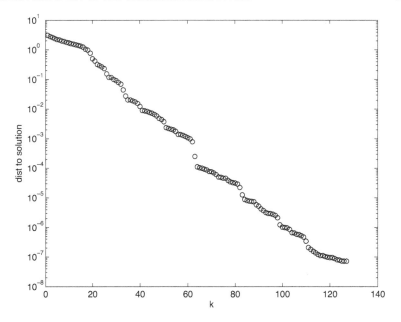

Figure 8.5 Same situation as in Figure 8.4 but now with $B = I$ and $A = \text{diag}(1, \ldots, n)$.

is a function on a Riemannian manifold $(\mathcal{M}, g)$ into a Euclidean space $\mathcal{E}$. The goal is to minimize $f(x)$. This is a *least-squares problem* associated with the least-squares cost $\sum_i (F_i(x))^2$, where $F_i(x)$ denotes the ith component of $F(x)$ in some orthonormal basis of $\mathcal{E}$. We assume throughout that $\dim(\mathcal{E}) \geq \dim(\mathcal{M})$, in other words, there are at least as many equations as "unknowns". Minimizing f is clearly equivalent to minimizing $\|F(x)\|$. Using the squared cost is important for regularity purposes, whereas the $\frac{1}{2}$ factor is chosen to simplify the equations.

Recall that $\|F(x)\|^2 := \langle F(x), F(x) \rangle$, where $\langle \cdot, \cdot \rangle$ denotes the inner product on $\mathcal{E}$. We have, for all $\xi \in T_x \mathcal{M}$,

$$\mathrm{D}f(x)[\xi] = \langle \mathrm{D}F(x)[\xi], F(x) \rangle = \langle \xi, (\mathrm{D}F(x))^*[F(x)] \rangle,$$

where $(\mathrm{D}F(x))^*$ denotes the adjoint of the operator $\mathrm{D}F(x) : T_x \mathcal{M} \to \mathcal{E}$, i.e.,

$$\langle y, \mathrm{D}F(x)[\xi] \rangle = g((\mathrm{D}F(x))^*[y], \xi)$$

for all $y \in T_{F(x)}\mathcal{E} \simeq \mathcal{E}$ and all $\xi \in T_x \mathcal{M}$. Hence

$$\mathrm{grad}\, f(x) = (\mathrm{D}F(x))^*[F(x)].$$

Further, we have, for all $\xi, \eta \in T_x \mathcal{M}$,

$$\nabla^2 f(x)[\xi, \eta] = \langle \mathrm{D}F(x)[\xi], \mathrm{D}F(x)[\eta] \rangle + \langle F(x), \nabla^2 F(x)[\xi, \eta] \rangle, \qquad (8.31)$$

where $\nabla^2 f(x)$ is the $(0, 2)$-tensor defined in Section 5.6.

8.4.1 Gauss-Newton methods

Recall that the geometric Newton method (Algorithm 5) computes an update vector $\eta \in T_x \mathcal{M}$ by solving the equation

$$\operatorname{grad} f(x) + \operatorname{Hess} f(x)[\eta] = 0,$$

or equivalently,

$$\mathrm{D} f(x)[\xi] + \nabla^2 f(x)[\xi, \eta] = 0 \quad \text{for all } \xi \in T_x \mathcal{M}.$$

The *Gauss-Newton* method is an approximation of this geometric Newton method for the case where $f(x) = \|F(x)\|^2$ as in (8.30). It consists of approximating $\nabla^2 f(x)[\xi, \eta]$ by the term $\langle \mathrm{D}F(x)[\xi], \mathrm{D}F(x)[\eta] \rangle$; see (8.31). This yields the Gauss-Newton equation

$$\langle \mathrm{D}F(x)[\xi], F(x) \rangle + \langle \mathrm{D}F(x)[\xi], \mathrm{D}F(x)[\eta] \rangle = 0 \quad \text{for all } \xi \in T_x \mathcal{M},$$

or equivalently,

$$(\mathrm{D}F(x))^*[F(x)] + ((\mathrm{D}F(x))^* \circ \mathrm{D}F(x))[\eta] = 0.$$

The geometric Gauss-Newton method is given in Algorithm 14. (Note that the affine connection ∇ is not required to state the algorithm.)

Algorithm 14 Riemannian Gauss-Newton method

Require: Riemannian manifold $\mathcal{M}$; retraction R on $\mathcal{M}$; function $F : \mathcal{M} \to \mathcal{E}$ where $\mathcal{E}$ is a Euclidean space.
Goal: Find a (local) least-squares solution of $F(x) = 0$.
Input: Initial iterate $x_0 \in \mathcal{M}$.
Output: Sequence of iterates $\{x_k\}$.
  1: **for** $k = 0, 1, 2, \ldots$ **do**
  2:     Solve the Gauss-Newton equation

$$((\mathrm{D}F(x_k))^* \circ \mathrm{D}F(x_k))[\eta_k] = -(\mathrm{D}F(x_k))^*[F(x_k)] \qquad (8.32)$$

     for the unknown $\eta_k \in T_{x_k} \mathcal{M}$.
  3:     Set

$$x_{k+1} := R_{x_k}(\eta_k).$$

  4: **end for**

In the following discussion, we assume that the operator $\mathrm{D}F(x_k)$ is injective (i.e., full rank, since we have assumed $n \geq d$). The Gauss-Newton equation (8.32) then reads

$$\eta_k = ((\mathrm{D}F(x_k))^* \circ (\mathrm{D}F(x_k)))^{-1} [(\mathrm{D}F(x_k))^*[F(x_k)]];$$

i.e.,

$$\eta_k = (\mathrm{D}F(x_k))^\dagger [F(x_k)], \qquad (8.33)$$

where $(\mathrm{D}F(x_k))^\dagger$ denotes the *Moore-Penrose inverse* or *pseudo-inverse* of the operator $\mathrm{D}F(x_k)$.

Key advantages of the Gauss-Newton method over the plain Newton method applied to $f(x) := \|F(x)\|^2$ are the lower computational complexity of producing the iterates and the property that, as long as $DF(x_k)$ has full rank, the Gauss-Newton direction is a descent direction for f. Note also that the update vector η_k turns out to be the least-squares solution

$$\arg\min_{\eta \in T_{x_k}\mathcal{M}} \|DF(x_k)[\eta] + F(x_k)\|^2.$$

In fact, instead of finding the critical point of the quadratic model of f, the Gauss-Newton method computes the minimizer of the norm of the "model" $F(x_k) + DF(x_k)[\eta]$ of F.

Usually, Algorithm 14 is used in combination with a line-search scheme that ensures a sufficient decrease in f. If the sequence $\{\eta_k\}$ generated by the method is gradient-related, then global convergence follows from Theorem 4.3.1.

The Gauss-Newton method is in general not superlinearly convergent. In view of Theorem 8.2.1, on the convergence of inexact Newton methods, it *is* superlinearly convergent to a nondegenerate minimizer x_* of f when the neglected term $\langle F(x), \nabla^2 F(x)[\xi, \eta]\rangle$ in (8.31) vanishes at x_*. In particular, this is the case when $F(x_*) = 0$, i.e., the (local) least-squares solution x_* turns out to be a zero of F.

8.4.2 Levenberg-Marquardt methods

An alternative to the line-search enhancement of Algorithm 14 (Gauss-Newton) is to use a trust-region approach. The model is chosen as

$$m_{x_k}(\eta) = \tfrac{1}{2}\|F(x_k)\|^2 + g(\eta, (DF(x)_k)^*[F(x_k)])$$
$$+ \tfrac{1}{2}g(\eta, ((DF(x))^* \circ DF(x))[\eta]])$$

so that the critical point of the model is the solution η_k of the Gauss-Newton equation (8.32). (We assume that $DF(x_k)$ is full rank for simplicity of the discussion.) All the convergence analyses of Riemannian trust-region methods apply.

In view of the characterization of the solutions of the trust-region subproblems in Proposition 7.3.1, the minimizer of $m_{x_k}(\eta)$ within the trust region $\|\eta\| \leq \Delta_k$ is either the solution of the Gauss-Newton equation (8.32) when it falls within the trust region, or the solution of

$$((DF(x_k))^* \circ DF(x_k) + \mu_k\, \mathrm{id})\eta = -(DF(x_k))^*F(x_k), \qquad (8.34)$$

where μ_k is such that the solution η_k satisfies $\|\eta_k\| = \Delta_k$. Equation (8.34) is known as the *Levenberg-Marquard* equation.

Notice that the presence of $\mu\,\mathrm{id}$ as a modification of the approximate Hessian $(DF(x))^* \circ DF(x)$ of f is analogous to the idea in (6.6) of making the modified Hessian positive-definite by adding a sufficiently positive-definite perturbation to the Hessian.

8.5 NOTES AND REFERENCES

On the Stiefel manifold, it is possible to obtain a closed form for the parallel translation along geodesics associated with the Riemannian connection obtained when viewing the manifold as a Riemannian quotient manifold of the orthogonal group; see Edelman *et al.* [EAS98]. We refer the reader to Edelman *et al.* [EAS98] for more information on the geodesics and parallel translations on the Stiefel manifold. Proof that the Riemannian parallel translation is an isometry can be found in [O'N83, Lemma 3.20].

More information on iterative methods for linear systems of equations can be found in, e.g., Axelsson [Axe94], Saad [Saa96], van der Vorst [vdV03], and Meurant [Meu06].

The proof of Lemma 8.2.2 is a generalization of the proof of [DS83, Lemma 4.2.1].

For more information on quasi-Newton methods in $\mathbb{R}^n$, see, e.g., Dennis and Schnabel [DS83] or Nocedal and Wright [NW99]. An early reference on quasi-Newton methods on manifolds (more precisely, on submanifolds of $\mathbb{R}^n$) is Gabay [Gab82]. The material on BFGS on manifolds comes from [Gab82], where we merely replaced the usual parallel translation by the more general notion of vector transport. Hints for the convergence analysis of BFGS on manifolds can also be found in [Gab82].

The linear CG method is due to Hestenes and Stiefel [HS52]. Major results for nonlinear CG algorithms are due to Fletcher and Reeves [FR64] and Polak and Ribiere [PR69]. More information can be found in, e.g., [NW99]. A counterexample showing lack of convergence of the Polak-Ribière method can be found in Powell [Pow84].

Smith [Smi93, Smi94] proposes a nonlinear CG algorithm on Riemannian manifolds that corresponds to Algorithm 13 with the retraction R chosen as the Riemannian exponential map and the vector transport $\mathcal{T}$ defined by the parallel translation induced by the Riemannian connection. Smith points out that the Polak-Ribière version of the algorithm has n-step quadratic convergence towards nondegenerate local minimizers of the cost function.

The Gauss-Newton method on Riemannian manifolds can be found in Adler *et al.* [ADM⁺02] in a formulation similar to Algorithm 14.

The original Levenberg-Marquardt algorithm [Lev44, Mar63] did not make the connection with the trust-region approach; it proposed heuristics to adapt μ directly.

More information on the classical version of the methods presented in this chapter can be found in textbooks on numerical optimization such as [Fle01, DS83, NS96, NW99, BGLS03].

Appendix A

Elements of Linear Algebra, Topology, and Calculus

A.1 LINEAR ALGEBRA

We follow the usual conventions of matrix computations. $\mathbb{R}^{n \times p}$ is the set of all $n \times p$ real matrices (m rows and p columns). $\mathbb{R}^n$ is the set $\mathbb{R}^{n \times 1}$ of column vectors with n real entries. $A(i, j)$ denotes the i, j entry (ith row, jth column) of the matrix A. Given $A \in \mathbb{R}^{m \times n}$ and $B \in \mathbb{R}^{n \times p}$, the matrix product $AB \in \mathbb{R}^{m \times p}$ is defined by $(AB)(i, j) = \sum_{k=1}^{n} A(i, k)B(k, j)$, $i = 1, \ldots, m$, $j = 1, \ldots, p$. A^T is the *transpose* of the matrix A: $(A^T)(i, j) = A(j, i)$. The entries $A(i, i)$ form the *diagonal* of A. A matrix is *square* if is has the same number of rows and columns. When A and B are square matrices of the same dimension, $[A, B] = AB - BA$ is termed the *commutator* of A and B. A matrix A is *symmetric* if $A^T = A$ and *skew-symmetric* if $A^T = -A$. The commutator of two symmetric matrices or two skew-symmetric matrices is symmetric, and the commutator of a symmetric and a skew-symmetric matrix is skew-symmetric. The *trace* of A is the sum of the diagonal elements of A,

$$\text{tr}(A) = \sum_{i=1}^{\min(n,p)} A(i, i).$$

We have the following properties (assuming that A and B have adequate dimensions)

$$\text{tr}(A) = \text{tr}(A^T), \tag{A.1a}$$

$$\text{tr}(AB) = \text{tr}(BA), \tag{A.1b}$$

$$\text{tr}([A, B]) = 0, \tag{A.1c}$$

$$\text{tr}(B) = 0 \quad \text{if } B^T = -B, \tag{A.1d}$$

$$\text{tr}(AB) = 0 \quad \text{if } A^T = A \text{ and } B^T = -B. \tag{A.1e}$$

An $n \times n$ matrix A is *invertible* (or *nonsingular*) if there exists an $n \times n$ matrix B such that $AB = BA = I_n$, where I_n denotes the $n \times n$ *identity matrix* with ones on the diagonal and zeros everywhere else. If this is the case, then B is uniquely determined by A and is called the *inverse* of A, denoted by A^{-1}. A matrix that is not invertible is called *singular*. A matrix Q is *orthonormal* if $Q^T Q = I$. A square orthonormal matrix is termed *orthogonal* and satisfies $Q^{-1} = Q^T$.

The notion of an n-dimensional *vector space* over $\mathbb{R}$ is an abstraction of $\mathbb{R}^n$ endowed with its operations of addition and multiplication by a scalar. Any

n-dimensional real vector space $\mathcal{E}$ is isomorphic to $\mathbb{R}^n$. However, producing a diffeomorphism involves generating a basis of $\mathcal{E}$, which may be computationally intractable; this is why most of the following material is presented on abstract vector spaces. We consider only finite-dimensional vector spaces over $\mathbb{R}$.

A *normed vector space* $\mathcal{E}$ is a vector space endowed with a *norm*, i.e., a mapping $x \in \mathcal{E} \mapsto \|x\| \in \mathbb{R}$ with the following properties. For all $a \in \mathbb{R}$ and all $x, y \in \mathcal{E}$,

1. $\|x\| \geq 0$, and $\|x\| = 0$ if and only if x is the zero vector;
2. $\|ax\| = |a| \, \|x\|$;
3. $\|x + y\| \leq \|x\| + \|y\|$.

Given two normed vector spaces $\mathcal{E}$ and $\mathcal{F}$, a mapping $A : \mathcal{E} \mapsto \mathcal{F}$ is a *(linear) operator* if $A[\alpha x + \beta y] = \alpha A[x] + \beta A[y]$ for all $x, y \in \mathcal{E}$ and all $\alpha, \beta \in \mathbb{R}$. The set $\mathfrak{L}(\mathcal{E}; \mathcal{F})$ of all operators from $\mathcal{E}$ to $\mathcal{F}$ is a vector space. An operator $A \in \mathfrak{L}(\mathbb{R}^n; \mathbb{R}^m)$ can be represented by an $m \times n$ matrix (also denoted by A) such that $A[x] = Ax$ for all $x \in \mathbb{R}^n$. This representation is an isomorphism that matches the composition of operators with the multiplication of matrices. Let $\mathcal{E}$, $\mathcal{F}$, and $\mathcal{G}$ be normed vector spaces, let $\| \cdot \|_{\mathfrak{L}(\mathcal{E};\mathcal{F})}$ be a norm on $\mathfrak{L}(\mathcal{E}; \mathcal{F})$, $\| \cdot \|_{\mathfrak{L}(\mathcal{F};\mathcal{G})}$ be a norm on $\mathfrak{L}(\mathcal{F}; \mathcal{G})$, and $\| \cdot \|_{\mathfrak{L}(\mathcal{E};\mathcal{G})}$ be a norm on $\mathfrak{L}(\mathcal{E}; \mathcal{G})$. These norms are called *mutually consistent* if $\|B \circ A\|_{\mathfrak{L}(\mathcal{E};\mathcal{G})} \leq \|A\|_{\mathfrak{L}(\mathcal{E};\mathcal{F})} \|B\|_{\mathfrak{L}(\mathcal{F};\mathcal{G})}$ for all $A \in \mathfrak{L}(\mathcal{E}; \mathcal{F})$ and all $B \in \mathfrak{L}(\mathcal{F}; \mathcal{G})$. A *consistent* or *submultiplicative norm* is a norm that is mutually consistent with itself. The *operator norm* or *induced norm* of $A \in \mathfrak{L}(\mathcal{E}; \mathcal{F})$ is

$$\|A\| := \max_{x \in \mathcal{E}, x \neq 0} \frac{\|A[x]\|}{\|x\|}.$$

Operator norms are mutually consistent.

Given normed vector spaces $\mathcal{E}_1$, $\mathcal{E}_2$, and $\mathcal{F}$, a mapping A from $\mathcal{E}_1 \times \mathcal{E}_2$ to $\mathcal{F}$ is called a *bilinear operator* if for any $x_2 \in \mathcal{E}_2$ the linear mapping $x_1 \mapsto A[x_1, x_2]$ is a linear operator from $\mathcal{E}_1$ to $\mathcal{F}$, and for any $x_1 \in \mathcal{E}_1$ the linear mapping $x_2 \mapsto A[x_1, x_2]$ is a linear operator from $\mathcal{E}_2$ to $\mathcal{F}$. The set of bilinear operators from $\mathcal{E}_1 \times \mathcal{E}_2$ to $\mathcal{F}$ is denoted by $\mathfrak{L}(\mathcal{E}_1, \mathcal{E}_2; \mathcal{F})$, and we use the notation $\mathfrak{L}_2(\mathcal{E}; \mathcal{F})$ for $\mathfrak{L}(\mathcal{E}, \mathcal{E}; \mathcal{F})$. These definitions are readily extended to multilinear operators. A bilinear operator $A \in \mathfrak{L}_2(\mathcal{E}; \mathcal{F})$ is *symmetric* if $A[x, y] = A[y, x]$ for all $x, y \in \mathcal{E}$. A symmetric bilinear operator $A \in \mathfrak{L}_2(\mathcal{E}, \mathbb{R})$ is *positive-definite* if $A[x, x] > 0$ for all $x \in \mathcal{E}$, $x \neq 0$.

By *Euclidean space* we mean a finite-dimensional vector space endowed with an inner product, i.e., a bilinear, symmetric positive-definite form $\langle \cdot, \cdot \rangle$. The canonical example is $\mathbb{R}^n$, endowed with the inner product

$$\langle x, y \rangle := x^T y.$$

An *orthonormal basis* of an n-dimensional Euclidean space $\mathcal{E}$ is a sequence $(e_1, \dots, e_n)$ of elements of $\mathcal{E}$ such that

$$\langle e_i, e_j \rangle = \begin{cases} 1 & \text{if } i = j, \\ 0 & \text{if } i \neq j. \end{cases}$$

Given an orthonormal basis of $\mathcal{E}$, the mapping that sends the elements of $\mathcal{E}$ to their vectors of coordinates in $\mathbb{R}^d$ is an isomorphism (an invertible mapping that preserves the vector space structure and the inner product). An operator $T : \mathcal{E} \to \mathcal{E}$ is termed *symmetric* if $\langle T[x], y \rangle = \langle x, T[y] \rangle$ for all $x, y \in \mathcal{E}$. Given an operator $T : \mathcal{E} \to \mathcal{F}$ between two Euclidean spaces $\mathcal{E}$ and $\mathcal{F}$, the *adjoint* of T is the operator $T^* : \mathcal{F} \to \mathcal{E}$ satisfying $\langle T[x], y \rangle = \langle x, T^*[y] \rangle$ for all $x \in \mathcal{E}$ and all $y \in \mathcal{F}$. The *kernel* of the operator T is the linear subspace $\ker(T) = \{x \in \mathcal{E} : T[x] = 0\}$. The *range* (or *image*) of T is the set $\mathrm{range}(T) = \{T[x] : x \in \mathcal{E}\}$. Given a linear subspace $\mathcal{S}$ of $\mathcal{E}$, the *orthogonal complement* of $\mathcal{S}$ is $\mathcal{S}^\perp = \{x \in \mathcal{E} : \langle x, y \rangle = 0 \text{ for all } y \in \mathcal{S}\}$. Given $x \in \mathcal{E}$, there is a unique decomposition $x = x_1 + x_2$ with $x_1 \in \mathcal{S}$ and $x_2 \in \mathcal{S}^\perp$; x_1 is the *orthogonal projection* of x onto $\mathcal{S}$ and is denoted by $\Pi_{\mathcal{S}}(x)$. The *Moore-Penrose inverse* or *pseudo-inverse* of an operator T is the operator

$$T^\dagger : \mathcal{F} \to \mathcal{E} : y \mapsto (T|_{(\ker(T))^\perp})^{-1}[\Pi_{\mathrm{range}(T)}y],$$

where the restriction $T|_{(\ker(T))^\perp} : (\ker(T))^\perp \to \mathrm{range}(T)$ is invertible by construction and $\Pi_{\mathrm{range}(T)}$ is the orthogonal projector in $\mathcal{F}$ onto $\mathrm{range}(T)$. The *Euclidean norm* on a Euclidean space $\mathcal{E}$ is

$$\|x\| := \sqrt{\langle x, x \rangle}.$$

The Euclidean norm on $\mathbb{R}^n$ is

$$\|x\| := \sqrt{x^T x}.$$

The Euclidean norm on $\mathbb{R}^{n \times p}$ endowed with the inner product $\langle X, Y \rangle = \mathrm{tr}(X^T Y)$ is the *Frobenius norm* given by $\|A\|_F = (\sum_{i,j}(A(i,j))^2)^{1/2}$. The operator norm on $\mathbb{R}^{n \times n} \simeq \mathcal{L}(\mathbb{R}^n; \mathbb{R}^n)$, where $\mathbb{R}^n$ is endowed with its Euclidean norm, is the *spectral norm* given by

$$\|A\|_2 = \sqrt{\lambda_{\max}(A^T A)},$$

where $\lambda_{\max}(A^T A)$ is the largest eigenvalue of the positive-semidefinite matrix $A^T A$.

An operator T in $\mathcal{L}(\mathcal{E}; \mathcal{E})$ is *invertible* if for all $y \in \mathcal{E}$ there exists $x \in \mathcal{E}$ such that $y = T[x]$. An operator that is not invertible is termed *singular*. Let $\mathrm{id}_\mathcal{E}$ denote the identity operator on $\mathcal{E}$: $\mathrm{id}_\mathcal{E}[x] = x$ for all $x \in \mathcal{E}$. Let $T \in \mathcal{L}(\mathcal{E}; \mathcal{E})$ be a symmetric operator. A real number λ is an *eigenvalue* of T if the operator $T - \lambda \mathrm{id}_\mathcal{E}$ is singular; any vector $x \neq 0$ in the kernel of $T - \lambda \mathrm{id}_\mathcal{E}$ is an *eigenvector* of T corresponding to the eigenvalue λ.

References: [GVL96], [Hal74], [Die69].

A.2 TOPOLOGY

A topology on a set X is an abstraction of the notion of open sets in $\mathbb{R}^n$. Defining a topology on X amounts to saying which subsets of X are open while retaining certain properties satisfied by open sets in $\mathbb{R}^n$. Specifically, a *topology* on a set X is a collection $\mathcal{T}$ of subsets of X, called *open sets*, such that

1. X and $\emptyset$ belong to $\mathcal{T}$;
2. the union of the elements of any subcollection of $\mathcal{T}$ is in $\mathcal{T}$;
3. the intersection of the elements of any finite subcollection of $\mathcal{T}$ is in $\mathcal{T}$.

A *topological space* is a couple $(X, \mathcal{T})$ where X is a set and $\mathcal{T}$ is a topology on X. When the topology is clear from the context or is irrelevant, we simply refer to the topological space X.

Let X be a topological space. A subset A of X is said to be *closed* if the set $X - A := \{x \in X : x \notin A\}$ is open. A *neighborhood* of a point $x \in X$ is a subset of X that includes an open set containing x. A *limit point* (or *accumulation point*) of a subset A of X is a point x of X such that every neighborhood of x intersects A in some point other than x itself. A subset of X is closed if and only if it contains all its limit points. A sequence $\{x_k\}_{i=1,2,\ldots}$ of points of X *converges* to the point $x \in X$ if, for every neighborhood U of x, there is a positive integer K such that x_k belongs to U for all $k \geq K$.

In view of the limited number of axioms that a topology has to satisfy, it is not suprising that certain properties that hold in $\mathbb{R}^n$ do not hold for an arbitrary topology. For example, singletons (subsets containing only one element) may not be closed; this is the case for the overlapping interval topology, a topology of $[-1, 1]$ whose open sets are intervals of the form $[-1, b)$ for $b > 0$, $(a, 1]$ for $a < 0$, and (a, b) for $a < 0$, $b > 0$. Another example is that sequences may converge to more than one point; this is the case with the cofinite topology of an infinite set, whose open sets are all the subsets whose complements are finite (i.e., have finitely many elements). To avoid these strange situations, the following *separation axioms* have been introduced.

Let X be a topological space. X is T_1, or *accessible* or *Fréchet*, if for any distinct points x and y of X, there is an open set that contains x and not y. Equivalently, every singleton is closed. X is T_2, or *Hausdorff*, if any two distinct points of X have disjoint neighborhoods. If X is Hausdorff, then every sequence of points of X converges to at most one point of X.

Let $\mathcal{T}_1$ and $\mathcal{T}_2$ be two topologies on the same set X. If $\mathcal{T}_1 \subseteq \mathcal{T}_2$, we say that $\mathcal{T}_2$ is *finer* than $\mathcal{T}_1$.

A *basis* for a topology on a set X is a collection $\mathcal{B}$ of subsets of X such that

1. each $x \in X$ belongs to at least one element of $\mathcal{B}$;
2. if $x \in (B_1 \cap B_2)$ with $B_1, B_2 \in \mathcal{B}$, then there exists $B_3 \in \mathcal{B}$ such that $x \in B_3 \subseteq B_1 \cap B_2$.

If $\mathcal{B}$ is a basis for a topology $\mathcal{T}$ on X, then $\mathcal{T}$ equals the collection of all unions of elements of $\mathcal{B}$. A topological space X is called *second-countable* if it has a countable basis (i.e., a basis with countably many elements) for its topology.

Let X and Y be topological spaces. The *product topology* on $X \times Y$ is the topology having as a basis the collection $\mathcal{B}$ of all sets of the form $U \times V$, where U is an open subset of X and V is an open subset of Y.

If Y is a subset of a topological space $(X, \mathcal{T})$, then the collection $\mathcal{T}_Y = \{Y \cap U : U \in \mathcal{T}\}$ is a topology on Y called the *subspace topology*.

If $\sim$ is an equivalence relation on a topological space X, then the collection of all subsets U of the quotient set $X/\sim$ such that $\pi^{-1}(U)$ is open in X is called the *quotient topology* of $X/\sim$. (We refer the reader to Section 3.4 for a discussion of the notions of equivalence relation and quotient set.)

Subspaces and products of Hausdorff spaces are Hausdorff, but quotient spaces of Hausdorff spaces need not be Hausdorff. Subspaces and countable products of second-countable spaces are second-countable, but quotients of second-countable spaces need not be second-countable.

Let X be a topological space. A collection $\mathcal{A}$ of subsets of X is said to *cover* X, or to be a *covering* of X, if the union of the elements of $\mathcal{A}$ is equal to X. It is called an open covering of X if its elements are open subsets of X. The space X is said to be *compact* if every open covering $\mathcal{A}$ of X contains a finite subcollection that also covers X. The *Heine-Borel theorem* states that a subset of $\mathbb{R}^n$ (with the subspace topology) is compact if and only if it is closed and bounded.

Let $\mathbb{F}$ denote either $\mathbb{R}$ or $\mathbb{C}$. The set $\mathbb{F}^n$ has a standard topology; the collection of "open balls" $\{y \in \mathbb{F}^n : \sum_i |y_i - x_i|^2 < \epsilon\}$, $x \in \mathbb{F}^n$, $\epsilon > 0$, is a basis of that topology. A finite-dimensional vector space $\mathcal{E}$ over a field $\mathbb{F}$ ($\mathbb{R}$ or $\mathbb{C}$) inherits a natural topology: let $F : \mathcal{E} \to \mathbb{F}^n$ be an isomorphism of $\mathcal{E}$ with $\mathbb{F}^n$ and endow $\mathcal{E}$ with the topology where a subset X of $\mathcal{E}$ is open if and only if $F(X)$ is open in $\mathbb{F}^n$. Hence, $\mathbb{R}^{n \times p}$ has a natural topology as a finite-dimensional vector space, and the noncompact Stiefel manifold $\mathbb{R}^{n \times p}_*$ has a natural topology as a subset of $\mathbb{R}^{n \times p}$.

Reference: [Mun00].

A.3 FUNCTIONS

There is no general agreement on the way to define a function, its range, and its domain, so we find it useful to state our conventions. A *function* (or *map*, or *mapping*)

$$f : A \to B$$

is a set of ordered pairs (a, b), $a \in A$, $b \in B$, with the property that, if (a, b) and (a, c) are in the set, then $b = c$. If $(a, b) \in f$, we write b as $f(a)$. Note that we do not require that $f(a)$ be defined for all $a \in A$. This is convenient, as it allows us to simply say, for example, that the tangent is a function from $\mathbb{R}$ to $\mathbb{R}$. The *domain* of f is $\mathrm{dom}(f) := \{a \in A : \exists b \in B : (a, b) \in f\}$, and the *range* (or *image*) of f is $\mathrm{range}(f) := \{b \in B : \exists a \in A : (a, b) \in f\}$. If $\mathrm{dom}(f) = A$, then f is said to be *on* A. If $\mathrm{range}(f) = B$, then f is *onto* B. An onto function is also called a *surjection*. An *injection* is a function f with the property that if $x \neq y$, then $f(x) \neq f(y)$. A function from a set A to a set B is a *bijection* or a *one-to-one correspondence* if it is both an injection and a surjection from A to B. The *preimage* $f^{-1}(Y)$ of a set $Y \subseteq B$ under

f is the subset of A defined by

$$f^{-1}(Y) = \{x \in A : f(x) \in Y\}.$$

Given $y \in B$, the set $f^{-1}(y) := f^{-1}(\{y\})$ is called a *fiber* or *level set* of f.

A function $f : A \to B$ between two topological spaces is said to be *continuous* if for each open subset V of B, the set $f^{-1}(V)$ is an open subset of A. By the *extreme value theorem*, if a real-valued function f is continuous on a compact set X, then there exist points c and d in X such that $f(c) \leq f(x) \leq f(d)$ for every $x \in X$.

A.4 ASYMPTOTIC NOTATION

Let $\mathcal{E}, \mathcal{F}, \mathcal{G}$ be normed vector spaces and let $F : \mathcal{E} \to \mathcal{F}$ and $G : \mathcal{E} \to \mathcal{G}$ be defined on a neighborhood of $x_* \in \mathcal{E} \cup \{\infty\}$. The notation

$$F(x) = O(G(x)) \text{ as } x \to x_*$$

(or simply $F(x) = O(G(x))$ when x_* is clear from the context) means that

$$\limsup_{\substack{x \to x_* \\ x \neq x_*}} \frac{\|F(x)\|}{\|G(x)\|} < \infty.$$

In other words, $F(x) = O(G(x))$ as $x \to x_*$, $x_* \in \mathcal{E}$, means that there is $C \geq 0$ and $\delta > 0$ such that

$$\|F(x)\| \leq C\|G(x)\| \tag{A.2}$$

for all x with $\|x - x_*\| < \delta$, and $F(x) = O(G(x))$ as $x \to \infty$ means that there is $C \geq 0$ and $\delta > 0$ such that (A.2) holds for all x with $\|x\| > \delta$. The notation

$$F(x) = o(G(x)) \text{ as } x \to x_*$$

means that

$$\lim_{\substack{x \to x_* \\ x \neq x_*}} \frac{\|F(x)\|}{\|G(x)\|} = 0.$$

Finally, the notation

$$F(x) = \Omega(G(x)) \text{ as } x \to x_*$$

means that there exist $C > 0$, $c > 0$, and a neighborhood $\mathcal{N}$ of x_* such that

$$c\|G(x)\| \leq \|F(x)\| \leq C\|G(x)\|$$

for all $x \in \mathcal{N}$.

We use similar notation to compare two sequences $\{x_k\}$ and $\{y_k\}$ in two normed spaces. The notation $y_k = O(x_k)$ means that there is $C \geq 0$ such that

$$\|y_k\| \leq C\|x_k\|$$

for all k sufficiently large. The notation $y_k = o(x_k)$ means that

$$\lim_{k \to \infty} \frac{\|y_k\|}{\|x_k\|} = 0.$$

The notation $y_k = \Omega(x_k)$ means that there exist $C > 0$ and $c > 0$ such that

$$c\|x_k\| \leq \|y_k\| \leq C\|x_k\|$$

for all k sufficiently large.

The loose notation $y_k = O(x_k^p)$ is used to denote that $\|y_k\| = O(\|x_k\|^p)$, and likewise for o and Ω.

A.5 DERIVATIVES

We present the concept of a derivative for functions between two finite-dimensional normed vector spaces. The extension to manifolds can be found in Chapter 3.

Let $\mathcal{E}$ and $\mathcal{F}$ be two finite-dimensional vector spaces over $\mathbb{R}$. (A particular case is $\mathcal{E} = \mathbb{R}^m$ and $\mathcal{F} = \mathbb{R}^n$.) A function $F : \mathcal{E} \to \mathcal{F}$ is *(Fréchet)-differentiable* at a point $x \in \mathcal{E}$ if there exists a linear operator

$$\mathrm{D}F(x) : \mathcal{E} \to \mathcal{F} : h \mapsto \mathrm{D}F(x)[h],$$

called the *(Fréchet) differential* (or the *Fréchet derivative*) of F at x, such that

$$F(x + h) = F(x) + \mathrm{D}F(x)[h] + o(\|h\|);$$

in other words,

$$\lim_{y \to x} \frac{\|F(y) - F(x) - \mathrm{D}F(x)[y - x]\|}{\|y - x\|} = 0.$$

The element $\mathrm{D}F(x)[h] \in \mathcal{F}$ is called the *directional derivative* of F at x along h. (We use the same notation $\mathrm{D}F(x)$ for the differential of a function F between two manifolds $\mathcal{M}_1$ and $\mathcal{M}_2$; then $\mathrm{D}F(x)$ is a linear operator from the vector space $T_x\mathcal{M}_1$ to the vector space $T_{F(x)}\mathcal{M}_2$; see Section 3.5.)

By convention, the notation "D" applies to the expression that follows. Hence $\mathrm{D}(f \circ g)(x)$ and $\mathrm{D}f(g(x))$ are two different things: the derivative of $f \circ g$ at x for the former, the derivative of f at $g(x)$ for the latter. We have the *chain rule*

$$\mathrm{D}(f \circ g)(x) = \mathrm{D}f(g(x)) \circ \mathrm{D}g(x);$$

i.e.,

$$\mathrm{D}(f \circ g)(x)[h] = \mathrm{D}f(g(x))\,[\mathrm{D}g(x)[h]]$$

for all h.

The function $F : \mathcal{E} \to \mathcal{F}$ is said to be *differentiable on an open domain* $\Omega \subseteq \mathcal{E}$ if F is differentiable at every point $x \in \Omega$. Note that, for all $x \in \Omega$,

$DF(x)$ belongs to the vector space $\mathfrak{L}(\mathcal{E}; \mathcal{F})$ of all linear operators from $\mathcal{E}$ to $\mathcal{F}$, which is itself a normed vector space with the induced norm

$$\|u\| = \max_{\|x\|=1} \|u(x)\|, \quad u \in \mathfrak{L}(\mathcal{E}; \mathcal{F}).$$

The function F is termed *continuously differentiable* (or C^1) on the open domain Ω if $DF : \mathcal{E} \to \mathfrak{L}(\mathcal{E}; \mathcal{F})$ is continuous on Ω.

Assume that bases $(e_1, \ldots, e_m)$ and $(e'_1, \ldots, e'_n)$ are given for $\mathcal{E}$ and $\mathcal{F}$ and let $\hat{F} : \mathbb{R}^m \to \mathbb{R}^n$ be the expression of F in these bases; i.e., $\sum_j \hat{F}^j(\hat{x})e'_j = F(\sum_i \hat{x}^i e_i)$. Then F is continuously differentiable if and only if the partial derivatives of $\hat{F}$ exist and are continuous, and we have

$$DF(x)[h] = \sum_j \sum_i \partial_i \hat{F}^j(\hat{x})\hat{h}^i e'_j.$$

It can be shown that this expression does not depend on the chosen bases.

If f is a real-valued function on a Euclidean space $\mathcal{E}$, then, given $x \in \mathcal{E}$, we define $\operatorname{grad} f(x)$, the *gradient* of f at x, as the unique element of $\mathcal{E}$ that satisfies

$$\langle \operatorname{grad} f(x), h \rangle = Df(x)[h] \quad \text{for all } h \in \mathcal{E}.$$

Given an orthonormal basis $(e_1, \ldots, e_d)$ of $\mathcal{E}$, we have

$$\operatorname{grad} f(x) = \sum_i \partial_i \hat{f}(x^1, \ldots, x^d)e_i,$$

where $x^1 e_1 + \cdots + x^d e_d = x$ and $\hat{f}$ is the expression of f in the basis.

If $F : \mathcal{E} \to \mathcal{F}$ is a linear function, then $DF(x)[h] = F(h)$ for all $x, h \in \mathcal{E}$. In particular, the derivative of the function $\operatorname{tr} : \mathbb{R}^{n \times n} \to \mathbb{R}$ is given by

$$D\operatorname{tr}(X)[H] = \operatorname{tr}(H).$$

For the function $\operatorname{inv} : \mathbb{R}^{p \times p}_* \to \mathbb{R}^{p \times p}_* : M \mapsto M^{-1}$, we have

$$D\operatorname{inv}(X)[Z] = -X^{-1}ZX^{-1}.$$

In other words, if $t \mapsto X(t)$ is a smooth curve in the set of invertible matrices, then

$$\frac{dX^{-1}}{dt} = -X^{-1}\frac{dX}{dt}X^{-1}. \tag{A.3}$$

The derivative of the determinant is given by *Jacobi's formula*,

$$D\det(X)[Z] = \operatorname{tr}(\operatorname{adj}(X)Z),$$

where $\operatorname{adj}(X) := \det(X)X^{-1}$. For $X \in \mathbb{R}^{n \times p}$, let $\operatorname{qf}(X)$ denote the Q factor of the *thin QR decomposition* $X = QR$, where $Q \in \mathbb{R}^{n \times p}$ is orthonormal and $R \in \mathbb{R}^{p \times p}$ is upper triangular with strictly positive diagonal elements. We have

$$D\operatorname{qf}(X)[Z] = X\rho_{\text{skew}}(Q^T ZR^{-1}) + (I - XX^T)ZR^{-1},$$

where $X = QR$ is the thin QR decomposition of X and $\rho_{\text{skew}}(A)$ denotes the skew-symmetric part of the decomposition of A into the sum of a skew-symmetric matrix and an upper triangular matrix.

If the mapping $\mathrm{D}F : \mathcal{E} \to \mathcal{L}(\mathcal{E}; \mathcal{F})$ is differentiable at a point $x \in \mathcal{E}$, we say that F is *twice differentiable* at x. The differential of $\mathrm{D}F$ at x is called the *second derivative* of F at x and denoted by $\mathrm{D}^2 F(x)$. This is an element of $\mathcal{L}(\mathcal{E}; \mathcal{L}(\mathcal{E}; \mathcal{F}))$, but this space is naturally identified with the space $\mathcal{L}_2(\mathcal{E}; \mathcal{F})$ of bilinear mappings of $\mathcal{E} \times \mathcal{E}$ into $\mathcal{F}$, and we use the notation $\mathrm{D}^2 F(x)[h_1, h_2]$ for $(\mathrm{D}^2 F(x)[h_1])[h_2]$. The second derivative satisfies the symmetry property $\mathrm{D}^2 F(x)[h_1, h_2] = \mathrm{D}^2 F(x)[h_2, h_1]$ for all $x, h_1, h_2 \in \mathcal{E}$. If $g \in \mathcal{E}$ and h is a differentiable function on $\mathcal{E}$ into $\mathcal{E}$, then

$$\mathrm{D}\left(\mathrm{D}F\left(\cdot\right)[h(\cdot)]\right)(x)[g] = \mathrm{D}^2 F(x)[h(x), g] + \mathrm{D}F\left(x\right)\left[\mathrm{D}h\left(x\right)[g]\right].$$

If $\mathcal{E}$ is a Euclidean space and f is a twice-differentiable, real-valued function on $\mathcal{E}$, then the unique symmetric operator $\mathrm{Hess}\, f(x) : \mathcal{E} \to \mathcal{E}$ defined by

$$\langle \mathrm{Hess}\, f(x)[h_1], h_2 \rangle = \mathrm{D}^2 f(x)[h_1, h_2] \quad \text{for all } h_1, h_2 \in \mathcal{E},$$

is termed the *Hessian operator* of f at x. We have

$$\mathrm{Hess}\, f(x)[h] = \mathrm{D}(\mathrm{grad}\, f)(x)[h]$$

for all $h \in \mathcal{E}$. Given an orthonormal basis $(e_1, \ldots, e_d)$ of $\mathcal{E}$, we have

$$\mathrm{Hess}\, f(x)[e_i] = \sum_j \partial_i \partial_j \widehat{f}(x^1, \ldots, x^d) e_j$$

and

$$\mathrm{D}^2 f(x)[e_i, e_j] = \partial_i \partial_j \widehat{f}(x^1, \ldots, x^d),$$

where $\widehat{f}$ is the expression of f in the basis.

The definition of the second derivative is readily generalized to derivatives of higher order. By induction on p, we define a *p-times-differentiable mapping* $F : \mathcal{E} \to \mathcal{F}$ as a $(p-1)$-times-differentiable mapping whose $(p-1)$th derivative $\mathrm{D}^{p-1}F$ is differentiable, and we call the derivative $\mathrm{D}(\mathrm{D}^{p-1}F)$ the *pth derivative* of F, written $\mathrm{D}^p F$. The element $\mathrm{D}^p F(x)$ is identified with an element of the space $\mathcal{L}_p(\mathcal{E}; \mathcal{F})$ of the p-linear mappings of $\mathcal{E}$ into $\mathcal{F}$, and we write it

$$(h_1, \ldots, h_p) \mapsto \mathrm{D}^p F(x)[h_1, \ldots, h_p].$$

The function $F : \mathbb{R}^n \to \mathbb{R}^m$ is *smooth* if it is continuously differentiable to all orders. This happens if and only if the partial derivatives of $\widehat{F}$ exist and are continuous to all orders, and we have

$$\mathrm{D}^p F(x)[h_1, \ldots, h_p] = \sum_j \sum_{i_1, \ldots, i_p} \partial_{i_1} \cdots \partial_{i_p} \widehat{F}^j(\widehat{x}) \widehat{h}_1^{i_1} \cdots \widehat{h}_p^{i_p}\, e_j',$$

where the superscripts of $\widehat{h}_1, \ldots, \widehat{h}_p$ denote component numbers.

Reference: [Die69] and [Deh95] for the derivatives of several matrix functions.

A.6 TAYLOR'S FORMULA

Let $\mathcal{E}$ and $\mathcal{F}$ be two finite-dimensional normed vector spaces and let $F : \mathcal{E} \to \mathcal{F}$ be $(p+1)$-times continuously differentiable on an open convex domain $\Omega \subseteq \mathcal{E}$. (The set Ω is *convex* if it contains all the line segments connecting any pair of its points.) *Taylor's theorem* (with the remainder in *Cauchy form*) states that, for all x and $x + h$ in Ω,

$$F(x+h) = F(x) + \frac{1}{1!}\mathrm{D}F(x)[h] + \frac{1}{2!}\mathrm{D}^2F(x)[h,h]$$

$$+ \cdots + \frac{1}{p!}\mathrm{D}^pF(x)[h,\dots,h] + R_p(h;x), \quad \text{(A.4)}$$

where

$$R_p(h;x) = \int_0^1 \frac{(1-t)^p}{p!}\mathrm{D}^{p+1}F(x+th)[h,\dots,h]\,\mathrm{d}t = O(\|h\|^{p+1}).$$

If F is real-valued, then the remainder $R_p(h;x)$ can also be expressed in *Lagrange form*: for all x and $x + h$ in Ω, there exists $t \in (0,1)$ such that

$$F(x+h) = F(x) + \frac{1}{1!}\,\mathrm{D}F(x)[h] + \frac{1}{2!}\,\mathrm{D}^2F(x)[h,h]$$

$$+ \cdots + \frac{1}{p!}\,\mathrm{D}^pF(x)[h,\dots,h] + \frac{1}{(p+1)!}\,\mathrm{D}^{p+1}F(x+th)[h,\dots,h]. \quad \text{(A.5)}$$

The function

$$h \mapsto F(x) + \frac{1}{1!}\,\mathrm{D}F(x)[h] + \frac{1}{2!}\,\mathrm{D}^2F(x)[h,h] + \cdots + \frac{1}{p!}\,\mathrm{D}^pF(x)[h,\dots,h]$$

is called the pth-order *Taylor expansion* of F around x.

The result $R_p(h;x) = O(\|h\|^{p+1})$ can be obtained under a weaker differentiability assumption. A function G between two normed vector spaces $\mathcal{A}$ and $\mathcal{B}$ is said to be *Lipschitz-continuous* at $x \in \mathcal{A}$ if there exist an open set $\mathcal{U} \subseteq \mathcal{A}$, $x \in \mathcal{U}$, and a constant α such that for all $y \in \mathcal{U}$,

$$\|G(y) - G(x)\| \le \alpha\|y - x\|. \quad \text{(A.6)}$$

The constant α is called a *Lipschitz constant* for G at x. If this holds for a specific $\mathcal{U}$, then G is said to be Lipschitz-continuous at x in the neighborhood $\mathcal{U}$. If (A.6) holds for every $x \in \mathcal{U}$, then G is said to be Lipschitz-continuous in $\mathcal{U}$ with Lipschitz constant α. If G is continuously differentiable, then it is Lipschitz-continuous in any bounded domain $\mathcal{U}$.

Proposition A.6.1 *Let $\mathcal{E}$ and $\mathcal{F}$ be two finite-dimensional normed vector spaces, let $F : \mathcal{E} \to \mathcal{F}$ be p-times continuously differentiable in an open convex set $\mathcal{U} \subseteq \mathcal{E}$, $x \in \mathcal{U}$, and let the differential $\mathrm{D}^pF : \mathcal{E} \to \mathfrak{L}_p(\mathcal{E};\mathcal{F})$ be Lipschitz continuous at x in the neighborhood $\mathcal{U}$ with Lipschitz constant α (using the induced norm in $\mathfrak{L}_p(\mathcal{E};\mathcal{F})$). Then, for any $x + h \in \mathcal{U}$,*

$$\left\| F(x+h) - F(x) - \frac{1}{1!}\,\mathrm{D}F(x)[h] - \cdots - \frac{1}{p!}\,\mathrm{D}^pF(x)[h,\dots,h] \right\|$$

$$\le \frac{\alpha}{(p+1)!}\|h\|^{p+1}.$$

In particular, for $p = 1$, i.e., F continuously differentiable with a Lipschitz-continuous differential, we have

$$\|F(x + h) - F(x) - \mathrm{D}F(x)[h]\| \leq \frac{\alpha}{2}\|h\|^2.$$

Bibliography

[ABG04]   P.-A. Absil, C. G. Baker, and K. A. Gallivan. Trust-region methods on Riemannian manifolds with applications in numerical linear algebra. In *Proceedings of the 16th International Symposium on Mathematical Theory of Networks and Systems (MTNS2004), Leuven, Belgium, 5–9 July 2004*, 2004.

[ABG06a]   P.-A. Absil, C. G. Baker, and K. A. Gallivan. Convergence analysis of Riemannian trust-region methods. Technical Report FSU-SCS-2006-175, School of Computational Science, Florida State University, http://www.scs.fsu.edu/publications/, 2006.

[ABG06b]   P.-A. Absil, C. G. Baker, and K. A. Gallivan. A truncated-CG style method for symmetric generalized eigenvalue problems. *J. Comput. Appl. Math.*, 189(1–2):274–285, 2006.

[ABG07]   P.-A. Absil, C. G. Baker, and K. A. Gallivan. Trust-region methods on Riemannian manifolds. *Found. Comput. Math.*, 7(3):303–330, July 2007.

[ABGS05]   P.-A. Absil, C. G. Baker, K. A. Gallivan, and A. Sameh. Adaptive model trust region methods for generalized eigenvalue problems. In Vaidy S. Sunderam, Geert Dick van Albada, and Peter M. A. Sloot, editors, *International Conference on Computational Science*, volume 3514 of *Lecture Notes in Computer Science*, pages 33–41. Springer-Verlag, 2005.

[ABM06]   F. Alvarez, J. Bolte, and J. Munier. A unifying local convergence result for Newton's method in Riemannian manifolds. Found. Comput. Math., to appear. Published online, http://dx.doi.org/10.1007/s10208-006-0221-6, 2006.

[Abs03]   P.-A. Absil. *Invariant Subspace Computation: A Geometric Approach*. PhD thesis, Faculté des Sciences Appliquées, Université de Liège, Secrétariat de la FSA, Chemin des Chevreuils 1 (Bât. B52), 4000 Liège, Belgium, 2003.

[AC98]   Shun-ichi Amari and Andrzej Cichocki. Adaptive blind signal processing—neural network approaches. *Proc. IEEE*, 86(10):2026–2048, 1998.

[ACC00]   Shun-ichi Amari, Tian-Ping Chen, and Andrzej Cichocki. Non-holonomic orthogonal learning algorithms for blind source separation. *Neural Comput.*, 12:1463–1484, 2000.

[ADM+02]   Roy L. Adler, Jean-Pierre Dedieu, Joseph Y. Margulies, Marco Martens, and Mike Shub. Newton's method on Riemannian manifolds and a geometric model for the human spine. *IMA J. Numer. Anal.*, 22(3):359–390, July 2002.

[AG05]   P.-A. Absil and K. A. Gallivan. Accelerated line-search and trust-region methods. Technical Report FSU-SCS-2005-095, School of Computational Science, Florida State University, http://www.scs.fsu.edu/publications/, 2005.

[AG06]   P.-A. Absil and K. A. Gallivan. Joint diagonalization on the oblique manifold for independent component analysis. In *Proceedings of the IEEE International Conference on Acoustics, Speech, and Signal Processing (ICASSP)*, volume 5, pages V–945–V–948, 2006.

[AHLT05]   Peter Arbenz, Ulrich L. Hetmaniuk, Richard B. Lehoucq, and Raymond S. Tuminaro. A comparison of eigensolvers for large-scale 3D modal analysis using AMG-preconditioned iterative methods. *Int. J. Numer. Meth. Eng.*, 64(2):204–236, 2005.

[AK04]   Bijan Afsari and P. S. Krishnaprasad. Some gradient based joint diagonalization methods for ICA. In Springer LCNS Series, editor, *Proceedings of the 5th International Conference on Independent Component Analysis and Blind Source Separation*, 2004.

[AMA05]   P.-A. Absil, R. Mahony, and B. Andrews. Convergence of the iterates of descent methods for analytic cost functions. *SIAM J. Optim.*, 6(2):531–547, 2005.

[AMR88]   R. Abraham, J. E. Marsden, and T. Ratiu. *Manifolds, Tensor Analysis, and Applications*, volume 75 of *Applied Mathematical Sciences*. Springer-Verlag, New York, second edition, 1988.

[AMS04]   P.-A. Absil, R. Mahony, and R. Sepulchre. Riemannian geometry of Grassmann manifolds with a view on algorithmic computation. *Acta Appl. Math.*, 80(2):199–220, January 2004.

[AMSV02]   P.-A. Absil, R. Mahony, R. Sepulchre, and P. Van Dooren. A Grassmann-Rayleigh quotient iteration for computing invariant subspaces. *SIAM Rev.*, 44(1):57–73, 2002.

[Arm66]   Larry Armijo. Minimization of functions having Lipschitz continuous first partial derivatives. *Pacific J. Math.*, 16:1–3, 1966.

[AS04]   P.-A. Absil and R. Sepulchre. Continuous dynamical systems that realize discrete optimization on the hypercube. *Systems Control Lett.*, 52(3-4):297–304, 2004.

[ASVM04]   P.-A. Absil, R. Sepulchre, P. Van Dooren, and R. Mahony. Cubically convergent iterations for invariant subspace computation. *SIAM J. Matrix Anal. Appl.*, 26(1):70–96, 2004.

[Axe94]   Owe Axelsson. *Iterative Solution Methods*. Cambridge University Press, Cambridge, 1994.

[BAG06]   C. G. Baker, P.-A. Absil, and K. A. Gallivan. An implicit Riemannian trust-region method for the symmetric generalized eigenproblem. In Vassil N. Alexandrov, Geert Dick van Albada, Peter M.A. Sloot, and Jack Dongarra, editors, *Computational Science—ICCS 2006*, volume 3991 of *LNCS*, pages 210–217. Springer, New York, 2006.

[BC70]   F. Brickell and R. S. Clark. *Differentiable Manifolds*. Van Nostrand Reinhold, London, 1970.

[BCS00]   D. Bao, S.-S. Chern, and Z. Shen. *An Introduction to Riemann-Finsler Geometry*, volume 200 of *Graduate Texts in Mathematics*. Springer-Verlag, New York, 2000.

[BDDR00]   Zhaojun Bai, James Demmel, Jack Dongarra, and Axel Ruhe, editors. *Templates for the Solution of Algebraic Eigenvalue Problems*. Software, Environments, and Tools. Society for Industrial and Applied Mathematics (SIAM), Philadelphia, PA, 2000. A practical guide.

[BDJ99]   Michael W. Berry, Zlatko Drmač, and Elizabeth R. Jessup. Matrices, vector spaces, and information retrieval. *SIAM Rev.*, 41(2):335–362, 1999.

[BDR05]   Roland Badeau, Bertrand David, and Gaël Richard. Fast approximated power iteration subspace tracking. *IEEE Trans. Signal Process.*, 53(8, part 1):2931–2941, 2005.

[Ber95]   Dimitri P. Bertsekas. *Nonlinear Programming*. Athena Scientific, Belmont, MA, 1995.

[BGL05]   Michele Benzi, Gene H. Golub, and Jörg Liesen. Numerical solution of saddle point problems. *Acta Numer.*, 14:1–137, 2005.

[BGLS03]   J. Frédéric Bonnans, J. Charles Gilbert, Claude Lemaréchal, and Claudia A. Sagastizábal. *Numerical Optimization*. Universitext. Springer-Verlag, Berlin, 2003. Theoretical and practical aspects. Translated and revised from the 1997 French original.

[Bha87]   Rajendra Bhatia. *Perturbation bounds for matrix eigenvalues*, volume 162 of *Pitman Research Notes in Mathematics Series*. Longman Scientific & Technical, Harlow, 1987.

[BI04]   Anthony M. Bloch and Arieh Iserles. On the optimality of double-bracket flows. *Int. J. Math. Math. Sci.*, 2004(61-64):3301–3319, 2004.

[BL89a]   D. A. Bayer and J. C. Lagarias. The nonlinear geometry of linear programming. I. Affine and projective scaling trajectories. *Trans. Amer. Math. Soc.*, 314(2):499–526, 1989.

[BL89b]   D. A. Bayer and J. C. Lagarias. The nonlinear geometry of linear programming. II. Legendre transform coordinates and central trajectories. *Trans. Amer. Math. Soc.*, 314(2):527–581, 1989.

[Boo75]   William M. Boothby. *An Introduction to Differentiable Manifolds and Riemannian Geometry*. Academic Press [A subsidiary of Harcourt Brace Jovanovich, Publishers], New York-London, 1975. Pure and Applied Mathematics, No. 63.

[Bra03]   Jan Brandts. The Riccati algorithm for eigenvalues and invariant subspaces of matrices with inexpensive action. *Linear Algebra Appl.*, 358:335–365, 2003. Special issue on accurate solution of eigenvalue problems (Hagen, 2000).

[Bro91]   R. W. Brockett. Dynamical systems that sort lists, diagonalize matrices, and solve linear programming problems. *Linear Algebra Appl.*, 146:79–91, 1991.

[Bro93]   Roger W. Brockett. Differential geometry and the design of gradient algorithms. In *Differential geometry: partial differential equations on manifolds (Los Angeles, CA, 1990)*, volume 54 of *Proc. Sympos. Pure Math.*, pages 69–92. Amer. Math. Soc., Providence, RI, 1993.

[BS89]   Steve Batterson and John Smillie. The dynamics of Rayleigh quotient iteration. *SIAM J. Numer. Anal.*, 26(3):624–636, 1989.

[BSS88]   Richard H. Byrd, Robert B. Schnabel, and Gerald A. Shultz. Approximate solution of the trust region problem by minimization over two-dimensional subspaces. *Math. Programming*, 40(3, (Ser. A)):247–263, 1988.

[BX05]   Stephen Boyd and Lin Xiao. Least-squares covariance matrix adjustment. *SIAM J. Matrix Anal. Appl.*, 27(2):532–546, 2005.

[CA01]   T. P. Chen and S. Amari. Unified stabilization approach to principal and minor components extraction algorithms. *Neural Networks*, 14(10):1377–1387, 2001.

[CD00]   Jann-Long Chern and Luca Dieci. Smoothness and periodicity of some matrix decompositions. *SIAM J. Matrix Anal. Appl.*, 22(3):772–792, 2000.

[CDLP05]   M. Chu, N. Del Buono, L. Lopez, and T. Politi. On the low-rank approximation of data on the unit sphere. *SIAM J. Matrix Anal. Appl.*, 27(1):46–60, 2005.

[CE75]   Jeff Cheeger and David G. Ebin. *Comparison Theorems in Riemannian Geometry.* North-Holland Publishing Co., Amsterdam, 1975. North-Holland Mathematical Library, Vol. 9.

[CG90]   P. Comon and G. H. Golub. Tracking a few extreme singular values and vectors in signal processing. *Proc. IEEE*, 78(8):1327–1343, 1990.

[CG02]   Moody T. Chu and Gene H. Golub. Structured inverse eigenvalue problems. *Acta Numer.*, 11:1–71, 2002.

[CG03]   Andrzej Cichocki and Pando Georgiev. Blind source separation algorithms with matrix constraints. *IEICE Trans. Fundam.*, E86-A(1):1–9, 2003.

[CGT00]   Andrew R. Conn, Nicholas I. M. Gould, and Philippe L. Toint. *Trust-Region Methods.* MPS/SIAM Series on Optimization. Society for Industrial and Applied Mathematics (SIAM), Philadelphia, PA, 2000.

[Cha84]   Françoise Chatelin. Simultaneous Newton's iteration for the eigenproblem. In *Defect correction methods (Oberwolfach, 1983)*, volume 5 of *Comput. Suppl.*, pages 67–74. Springer, Vienna, 1984.

[Chu92]   Moody T. Chu. Numerical methods for inverse singular value problems. *SIAM J. Numer. Anal.*, 29(3):885–903, 1992.

[Chu94]   Moody T. Chu. A list of matrix flows with applications. In *Hamiltonian and gradient flows, algorithms and control*, volume 3 of *Fields Inst. Commun.*, pages 87–97. Amer. Math. Soc., Providence, RI, 1994.

[CI01]   E. Celledoni and A. Iserles. Methods for the approximation of the matrix exponential in a Lie-algebraic setting. *IMA J. Numer. Anal.*, 21(2):463–488, 2001.

[Cou20]   R. Courant. Über die Eigenwert bei den Differentialgleichungen
            der Mathematischen physik. *Math. Z.*, 7:1–57, 1920.

[Dar94]   R. W. R. Darling. *Differential Forms and Connections*. Cam-
            bridge University Press, Cambridge, 1994.

[dC76]   Manfredo P. do Carmo. *Differential Geometry of Curves and
            Surfaces*. Prentice-Hall Inc., Englewood Cliffs, NJ, 1976. Trans-
            lated from the Portuguese.

[dC92]   M. P. do Carmo. *Riemannian geometry*. Mathematics: The-
            ory & Applications. Birkhäuser Boston Inc., Boston, MA, 1992.
            Translated from the second Portuguese edition by Francis Fla-
            herty.

[DDL99]   R. D. DeGroat, E. M. Dowling, and A. D. Linebarger. Subspace
            tracking. In V. K. Madiesetti and D. B. Williams, editors,
            Digital Signal Processing Handbook. CRC, Boca Raton, FL,
            1999.

[DE99]   Luca Dieci and Timo Eirola. On smooth decompositions of
            matrices. *SIAM J. Matrix Anal. Appl.*, 20(3):800–819, 1999.

[Deh95]   Jeroen Dehaene.   *Continuous-time matrix algorithms,
            systolic algorithms and adaptive neural networks*. PhD the-
            sis, Katholieke Universiteit Leuven, Faculteit Toegepaste
            Wetenschappen,   Departement   elektrotechniek-ESAT,
            Kard. Mercierlaan 94, 3001 Leuven, Belgium, 1995.
            ftp://ftp.esat.kuleuven.ac.be/pub/SISTA/dehaene/phd/.

[Dem87]   J. W. Demmel. Three methods for refining estimates of invari-
            ant subspaces. *Computing*, 38(1):43–57, 1987.

[Den71]   J. E. Dennis, Jr. Toward a unified convergence theory for
            Newton-like methods. In *Nonlinear Functional Anal. and Appl.
            (Proc. Advanced Sem., Math. Res. Center, Univ. of Wiscon-
            sin, Madison, WI., 1970*, pages 425–472. Academic Press, New
            York, 1971.

[Die69]   J. Dieudonné. *Foundations of Modern Analysis*, volume 10-I
            of *Pure and Applied Mathematics*. Academic Press, New York,
            1969. Enlarged and corrected printing.

[DM79]   J. E. Dennis, Jr. and H. H. W. Mei. Two new unconstrained op-
            timization algorithms which use function and gradient values.
            J. Optim. Theory Appl., 28(4):453–482, 1979.

[DMV99]   Jeroen Dehaene, Marc Moonen, and Joos Vandewalle. Analysis
            of a class of continuous-time algorithms for principal compo-
            nent analysis and subspace tracking. *IEEE Trans. Circuits
            Systems I Fund. Theory Appl.*, 46(3):364–372, 1999.

[DN05]   Jean-Pierre Dedieu and Dmitry Nowicki. Symplectic methods for the approximation of the exponential map and the Newton iteration on Riemannian submanifolds. *J. Complexity*, 21(4):487–501, 2005.

[Dou00]   Scott C. Douglas. Self-stabilized gradient algorithms for blind source separation with orthogonality constraints. *IEEE Trans. Neural Networks*, 11(6):1490–1497, 2000.

[DPM03]   Jean-Pierre Dedieu, Pierre Priouret, and Gregorio Malajovich. Newton's method on Riemannian manifolds: Covariant alpha theory. *IMA J. Numer. Anal.*, 23(3):395–419, 2003.

[DS83]   John E. Dennis, Jr. and Robert B. Schnabel. *Numerical methods for unconstrained optimization and nonlinear equations*. Prentice Hall Series in Computational Mathematics. Prentice Hall Inc., Englewood Cliffs, NJ, 1983.

[DV00]   J. Dehaene and J. Vandewalle. New Lyapunov functions for the continuous-time QR algorithm. In *Proceedings CD of the 14th International Symposium on the Mathematical Theory of Networks and Systems (MTNS2000), Perpignan, France, July 2000*, 2000.

[DW92]   Marc De Wilde. Géométrie différentielle globale. course notes, Institut de Mathématique, Université de Liège, 1992.

[EAS98]   Alan Edelman, Tomás A. Arias, and Steven T. Smith. The geometry of algorithms with orthogonality constraints. *SIAM J. Matrix Anal. Appl.*, 20(2):303–353, 1998.

[EP99]   Lars Eldén and Haesun Park. A Procrustes problem on the Stiefel manifold. *Numer. Math.*, 82(4):599–619, 1999.

[EY36]   C. Ekcart and G. Young. The approximation of one matrix by another of lower rank. *Psychometrika*, 1:211–218, 1936.

[Fan49]   Ky Fan. On a theorem of Weyl concerning eigenvalues of linear transformations. I. *Proc. Nat. Acad. Sci. U.S.A.*, 35:652–655, 1949.

[Fat98]   Jean-Luc Fattebert. A block Rayleigh quotient iteration with local quadratic convergence. *Electron. Trans. Numer. Anal.*, 7:56–74, 1998. Large scale eigenvalue problems (Argonne, IL, 1997).

[Fay91a]   L. Faybusovich. Dynamical systems which solve optimization problems with linear constraints. *IMA J. Math. Control Inform.*, 8(2):135–149, 1991.

[Fay91b]   Leonid Faybusovich. Hamiltonian structure of dynamical sys-
             tems which solve linear programming problems. *Phys. D*, 53(2-
             4):217–232, 1991.

[FD95]   Zuqiang Fu and Eric M. Dowling. Conjugate gradient eigen-
             structure tracking for adaptive spectral estimation. *IEEE
             Trans. Signal Process.*, 43:1151–1160, 1995.

[FF63]   D. K. Faddeev and V. N. Faddeeva. *Computational Methods
             of Linear Algebra.* Translated by Robert C. Williams. W. H.
             Freeman and Co., San Francisco, 1963.

[FGP94]   J. Ferrer, Ma. I. García, and F. Puerta. Differentiable families
             of subspaces. *Linear Algebra Appl.*, 199:229–252, 1994.

[Fis05]   Ernst Fischer. Über quadratische Formen mit reelen Koeffizien-
             ten. *Monatsch Math. Phys.*, 16:234–249, 1905.

[Fle01]   R. Fletcher. *Practical Methods of Optimization.* Wiley-
             Interscience [John Wiley & Sons], New York, second edition,
             2001.

[FR64]   R. Fletcher and C. M. Reeves. Function minimization by con-
             jugate gradients. *Comput. J.*, 7:149–154, 1964.

[FS02]   O. P. Ferreira and B. F. Svaiter. Kantorovich's theorem on
             Newton's method in Riemannian manifolds. *J. Complexity*,
             18(1):304–329, 2002.

[Gab82]   D. Gabay. Minimizing a differentiable function over a differen-
             tial manifold. *J. Optim. Theory Appl.*, 37(2):177–219, 1982.

[GD04]   J. C. Gower and G. B. Dijksterhuis. *Procrustes Problems*, vol-
             ume 30 of *Oxford Statistical Science Series*. Oxford University
             Press, Oxford, 2004.

[GDS05]   L. M. Graña Drummond and B. F. Svaiter. A steepest de-
             scent method for vector optimization. *J. Comput. Appl. Math.*,
             175(2):395–414, 2005.

[GH83]   John Guckenheimer and Philip Holmes. *Nonlinear Oscilla-
             tions, Dynamical Systems, and Bifurcations of Vector Fields*,
             volume 42 of *Applied Mathematical Sciences*. Springer-Verlag,
             New York, 1983.

[GHL90]   Sylvestre Gallot, Dominique Hulin, and Jacques Lafontaine.
             Riemannian Geometry. Universitext. Springer-Verlag, Berlin,
             second edition, 1990.

[GL93]   Michel Gevers and Gang Li. *Parametrizations in Control, Estimation and Filtering Problems: Accuracy Aspects.* Communications and Control Engineering Series. Springer-Verlag London Ltd., London, 1993.

[GLR86]   I. Gohberg, P. Lancaster, and L. Rodman. *Invariant subspaces of matrices with applications.* Canadian Mathematical Society Series of Monographs and Advanced Texts. John Wiley & Sons Inc., New York, 1986. , A Wiley-Interscience Publication.

[GLRT99]   Nicholas I. M. Gould, Stefano Lucidi, Massimo Roma, and Philippe L. Toint. Solving the trust-region subproblem using the Lanczos method. *SIAM J. Optim.*, 9(2):504–525, 1999.

[GOST05]   Nicholas I. M. Gould, Dominique Orban, Annick Sartenaer, and Phillipe L. Toint. Sensitivity of trust-region algorithms to their parameters. *4OR*, 3(3):227–241, 2005.

[GP74]   Victor Guillemin and Alan Pollack. *Differential Topology.* Prentice-Hall Inc., Englewood Cliffs, NJ, 1974.

[GP07]   Igor Grubišić and Raoul Pietersz. Efficient rank reduction of correlation matrices. *Linear Algebra Appl.*, 422(2-3):629–653, 2007.

[GR97]   M. Géradin and D. Rixen. *Mechanical Vibrations: Theory and Applications to Structural Dynamics.* John Wiley & Sons, Chichester, U.K., 1997.

[GS01]   F. Grognard and R. Sepulchre. Global stability of a continuous-time flow which computes time-optimal switchings. In *Proceedings of the 16th IEEE Conference on Decision and Control*, pages 3826–3831, 2001.

[GvdV00]   Gene H. Golub and Henk A. van der Vorst. Eigenvalue computation in the 20th century. *J. Comput. Appl. Math.*, 123(1-2):35–65, 2000. Numerical analysis 2000, Vol. III. Linear algebra.

[GVL96]   Gene H. Golub and Charles F. Van Loan. *Matrix Computations.* Johns Hopkins Studies in the Mathematical Sciences. Johns Hopkins University Press, Baltimore, MD, third edition, 1996.

[Hag01]   William W. Hager. Minimizing a quadratic over a sphere. *SIAM J. Optim.*, 12(1):188–208, 2001.

[Hal74]   Paul R. Halmos. *Finite-Dimensional Vector Spaces.* Undergraduate Texts in Mathematics. Springer-Verlag, New York, second edition, 1974.

[Hei03]   Long Hei. A self-adaptive trust region algorithm. *J. Comput. Math.*, 21(2):229–236, 2003.

[Hel78]   Sigurdur Helgason. *Differential Geometry, Lie Groups, and Symmetric Spaces*, volume 80 of *Pure and Applied Mathematics*. Academic Press Inc. [Harcourt Brace Jovanovich Publishers], New York, 1978.

[Hel93a]   U. Helmke. Balanced realizations for linear systems: a variational approach. *SIAM J. Control Optim.*, 31(1):1–15, 1993.

[Hel93b]   U. Helmke. Isospectral flows and linear programming. *J. Austral. Math. Soc. Ser. B*, 34(4):495–510, 1993.

[HH00]   U. Helmke and K. Hüper. A Jacobi-type method for computing balanced realizations. *Systems Control Lett.*, 39(1):19–30, 2000.

[HHLM07]   Uwe Helmke, Knut Hüper, Pei Yean Lee, and John B. Moore. Essential matrix estimation using Gauss-Newton iterations on a manifold. *Int. J. Computer Vision*, 74(2), 2007.

[Hir76]   Morris W. Hirsch. *Differential Topology*, volume 33 of *Graduate Texts in Mathematics*. Springer-Verlag, New York, 1976.

[HJ85]   Roger A. Horn and Charles R. Johnson. *Matrix Analysis*. Cambridge University Press, Cambridge, 1985.

[HJ91]   Roger A. Horn and Charles R. Johnson. *Topics in Matrix Analysis*. Cambridge University Press, Cambridge, 1991.

[HK51]   Magnus R. Hestenes and William Karush. A method of gradients for the calculation of the characteristic roots and vectors of a real symmetric matrix. *J. Research Nat. Bur. Standards*, 47:45–61, 1951.

[HL06]   U. Hetmaniuk and R. Lehoucq. Basis selection in LOBPCG. *J. Comput. Phys.*, 218(1):324–332, 2006.

[HM94]   Uwe Helmke and John B. Moore. *Optimization and Dynamical Systems*. Communications and Control Engineering Series. Springer-Verlag London Ltd., London, 1994. With a foreword by R. Brockett.

[Hop84]   J. J. Hopfield. Neurons with graded response have collective computational cababilities like those of two-state neurons. *Proc. Natl. Acad. Sci. USA*, 81:3088–3092, 1984.

[HP05]   William W. Hager and Soonchul Park. Global convergence of SSM for minimizing a quadratic over a sphere. *Math. Comp.*, 74(251):1413–1423, 2005.

[HR57]   André Haefliger and Georges Reeb. Variétés (non séparées) à
           une dimension et structures feuilletées du plan. *Enseignement
           Math. (2)*, 3:107–125, 1957.

[HS52]   Magnus R. Hestenes and Eduard Stiefel. Methods of conjugate
           gradients for solving linear systems. *J. Research Nat. Bur.
           Standards*, 49:409–436 (1953), 1952.

[HS03]   Michiel E. Hochstenbach and Gerard L. G. Sleijpen. Two-
           sided and alternating Jacobi-Davidson. *Linear Algebra Appl.*,
           358:145–172, 2003. Special issue on accurate solution of eigen-
           value problems (Hagen, 2000).

[HSS06]   Knut Hüper, Hao Shen, and Abd-Krim Seghouane. Local con-
           vergence properties of FastICA and some generalisations. In
           *Proceedings of the IEEE International Conference on Acous-
           tics, Speech, and Signal Processing (ICASSP)*, volume 5, pages
           V–1009–V–1012, 2006.

[HT85]   J. J. Hopfield and D. W. Tank. "Neural" computation of deci-
           sion optimization problems. *Biol. Cybernet.*, 52:141–152, 1985.

[HT04]   Knut Hüper and Jochen Trumpf. Newton-like methods for nu-
           merical optimization on manifolds. In *Proceedings of the 38th
           IEEE Asilomar Conference on Signals, Systems, and Comput-
           ers, Pacific Grove, CA, November 7–10, 2004*, 2004.

[Hüp02]   Knut Hüper. A calculus approach to matrix eigenvalue algo-
           rithms. Habilitation Dissertation, July 2002. Mathematisches
           Institut, Universität Würzburg, Germany.

[HXC+99]   Y. Hua, Y. Xiang, T. Chen, K. Abed-Meraim, and Y. Miao.
           A new look at the power method for fast subspace tracking.
           Digital Signal Process., 9(4):297–314, Oct. 1999.

[HZ03]   Richard Hartley and Andrew Zisserman. *Multiple View Ge-
           ometry in Computer Vision*. Cambridge University Press,
           Cambridge, second edition, 2003. With a foreword by Olivier
           Faugeras.

[IMKNZ00] Arieh Iserles, Hans Z. Munthe-Kaas, Syvert P. Nørsett, and
           Antonella Zanna. Lie-group methods. *Acta Numer.*, 9:215–
           365, 2000.

[IZ05]   Arieh Iserles and Antonella Zanna. Efficient computation of
           the matrix exponential by generalized polar decompositions.
           SIAM J. Numer. Anal., 42(5):2218–2256, 2005.

[JH05]   Christopher J. James and Christian W. Hesse. Independent
            component analysis for biomedical signals. *Physiol. Meas.*,
            26:R15–R19, 2005.

[JM02]   Marcel Joho and Heinz Mathis. Joint diagonalization of corre-
            lation matrices by using gradient methods with application to
            blind signal separation. In *Proceedings of the IEEE Sensor Ar-
            ray and Multichannel Signal Processing Workshop SAM*, pages
            273–277, 2002.

[JR02]   Marcel Joho and Kamran Rahbar. Joint diagonalization of cor-
            relation matrices by using Newton methods with applications
            to blind signal separation. In *Proceedings of the IEEE Sen-
            sor Array and Multichannel Signal Processing Workshop SAM*,
            pages 403–407, 2002.

[JW92]   Richard A. Johnson and Dean W. Wichern. *Applied Multivari-
            ate Statistical Analysis*. Prentice Hall Inc., Englewood Cliffs,
            NJ, third edition, 1992.

[Kan52]   L. V. Kantorovich. *Functional analysis and applied mathemat-
            ics*. NBS Rep. 1509. U. S. Department of Commerce National
            Bureau of Standards, Los Angeles, CA, 1952. Translated by C.
            D. Benster.

[Kli82]   Wilhelm Klingenberg. *Riemannian Geometry*, volume 1 of *de
            Gruyter Studies in Mathematics*. Walter de Gruyter & Co.,
            Berlin, 1982.

[KN63]   Shoshichi Kobayashi and Katsumi Nomizu. *Foundations of
            Differential Geometry*. Interscience Publishers, a division of
            John Wiley & Sons, New York-London, 1963. Volumes 1 and
            2.

[Kny01]   Andrew V. Knyazev. Toward the optimal preconditioned eigen-
            solver: locally optimal block preconditioned conjugate gradient
            method. *SIAM J. Sci. Comput.*, 23(2):517–541, 2001. Copper
            Mountain Conference (2000).

[Lan99]   Serge Lang. *Fundamentals of Differential Geometry*, volume
            191 of *Graduate Texts in Mathematics*. Springer-Verlag, New
            York, 1999.

[LE02]   Eva Lundström and Lars Eldén. Adaptive eigenvalue compu-
            tations using Newton's method on the Grassmann manifold.
            SIAM J. Matrix Anal. Appl., 23(3):819–839, 2001/02.

[LE00]   R. Lippert and A. Edelman. Nonlinear eigenvalue problems
            with orthogonality constraints (Section 9.4). In Zhaojun Bai,

James Demmel, Jack Dongarra, Axel Ruhe, and Henk van der Vorst, editors, *Templates for the Solution of Algebraic Eigenvalue Problems*, pages 290–314. SIAM, Philadelphia, 2000.

[Lei61]   Kurt Leichtweiss. Zur Riemannschen Geometrie in Grassmannschen Mannigfaltigkeiten. *Math. Z.*, 76:334–366, 1961.

[Lev44]   Kenneth Levenberg. A method for the solution of certain nonlinear problems in least squares. *Quart. Appl. Math.*, 2:164–168, 1944.

[LM04]   Pei Yean Lee and John B. Moore. Pose estimation via a Gauss-Newton-on-manifold approach. In *Proceedings of the 16th International Symposium on Mathematical Theory of Network and System (MTNS), Leuven*, 2004.

[Loj93]   Stanislas Łojasiewicz. Sur la géométrie semi- et sous-analytique. *Ann. Inst. Fourier (Grenoble)*, 43(5):1575–1595, 1993.

[LSG04]   Xiuwen Liu, Anuj Srivastava, and Kyle Gallivan. Optimal linear representations of images for object recognition. *IEEE Pattern Anal. and Mach. Intell.*, 26(5):662–666, May 2004.

[LST98]   Ralf Lösche, Hubert Schwetlick, and Gisela Timmermann. A modified block Newton iteration for approximating an invariant subspace of a symmetric matrix. *Linear Algebra Appl.*, 275/276:381–400, 1998.

[Lue72]   David G. Luenberger. The gradient projection method along geodesics. *Management Sci.*, 18:620–631, 1972.

[Lue73]   David G. Luenberger. *Introduction to Linear and Nonlinear Programming*. Addison-Wesley, Reading, MA, 1973.

[LW00]   Xue-Bin Liang and Jun Wang. A recurrent neural network for nonlinear optimization with a continuously differentiable objective function and bound constraints. *IEEE Trans. Neural Networks*, 11(6):1251–1262, 2000.

[MA03]   R. Mahony and P.-A. Absil. The continuous-time Rayleigh quotient flow on the sphere. *Linear Algebra Appl.*, 368C:343–357, 2003.

[Mah94]   Robert Mahony. *Optimization Algorithms on Homogeneous Spaces: with Applications in Linear Systems Theory*. PhD thesis, Department of Systems Engineering, Australian National University, 77 Massachusetts Avenue, Cambridge, MA 02139-4307, 1994.

[Mah96]   R. E. Mahony. The constrained Newton method on a Lie group
           and the symmetric eigenvalue problem. *Linear Algebra Appl.*,
           248:67–89, 1996.

[Man02]   Jonathan H. Manton. Optimization algorithms exploiting uni-
           tary constraints. *IEEE Trans. Signal Process.*, 50(3):635–650,
           2002.

[Mar63]   Donald W. Marquardt. An algorithm for least-squares esti-
           mation of nonlinear parameters. *J. Soc. Indust. Appl. Math.*,
           11:431–441, 1963.

[Meu06]   Gérard Meurant. *The Lanczos and conjugate gradient algo-
           rithms*, volume 19 of *Software, Environments, and Tools*. Soci-
           ety for Industrial and Applied Mathematics (SIAM), Philadel-
           phia, PA, 2006. From theory to finite precision computations.

[MH98a]   R. E. Mahony and U. Helmke. System assignment and pole
           placement for symmetric realisations. *J. Math. Systems Estim.
           Control*, 8(3):321–352, 1998.

[MH98b]   Yongfeng Miao and Yingbo Hua. Fast subspace tracking and
           neural network learning by a novel information criterion. *IEEE
           Trans. Signal Process.*, 46(7):1967–1979, Jul. 1998.

[MHM96]   R. E. Mahony, U. Helmke, and J. B. Moore. Gradient algo-
           rithms for principal component analysis. *J. Austral. Math. Soc.
           Ser. B*, 37(4):430–450, 1996.

[MHM05]   Jonathan H. Manton, Uwe Helmke, and Iven M. Y. Mareels.
           A dual purpose principal and minor component flow. *Systems
           Control Lett.*, 54(8):759–769, 2005.

[MKS01]   Yi Ma, Jana Kosecka, and Shankar S. Sastry. Optimization
           criteria and geometric algorithms for motion and structure es-
           timation. *Int. J. Computer Vision*, 44(3):219–249, 2001.

[MM02]   Robert Mahony and Jonathan H. Manton. The geometry of the
           Newton method on non-compact Lie groups. *J. Global Optim.*,
           23(3-4):309–327, 2002. Nonconvex optimization in control.

[MMH94]   J. B. Moore, R. E. Mahony, and U. Helmke. Numerical gradient
           algorithms for eigenvalue and singular value calculations. *SIAM
           J. Matrix Anal. Appl.*, 15(3):881–902, 1994.

[MMH03]   J. H. Manton, R. Mahony, and Y. Hua. The geometry of
           weighted low-rank approximations. *IEEE Trans. Signal Pro-
           cess.*, 51(2):500–514, 2003.

[MS83]   Jorge J. Moré and D. C. Sorensen. Computing a trust region
            step. *SIAM J. Sci. Statist. Comput.*, 4(3):553–572, 1983.

[MS86]   Ronald B. Morgan and David S. Scott. Generalizations of
            Davidson's method for computing eigenvalues of sparse sym-
            metric matrices. *SIAM J. Sci. Statist. Comput.*, 7(3):817–825,
            1986.

[Mun00]   James R. Munkres. *Topology*. Prentice Hall, Upper Saddle
            River, NJ, second edition, 2000.

[MV91]   Jürgen Moser and Alexander P. Veselov. Discrete versions of
            some classical integrable systems and factorization of matrix
            polynomials. *Comm. Math. Phys.*, 139(2):217–243, 1991.

[NA05]   Yasunori Nishimori and Shotaro Akaho. Learning algorithms
            utilizing quasi-geodesic flows on the Stiefel manifold. *Neuro-
            computing*, 67:106–135, 2005.

[NMH02]   Maziar Nikpour, Jonathan H. Manton, and Gen Hori. Algo-
            rithms on the Stiefel manifold for joint diagonalization. In *Proc.
            ICASSP*, pages II–1481–1484, 2002.

[Not02]   Y. Notay. Combination of Jacobi-Davidson and conjugate gra-
            dients for the partial symmetric eigenproblem. *Numer. Linear
            Algebra Appl.*, 9(1):21–44, 2002.

[Not03]   Yvan Notay. Convergence analysis of inexact Rayleigh quotient
            iteration. *SIAM J. Matrix Anal. Appl.*, 24(3):627–644, 2003.

[Not05]   Yvan Notay. Is Jacobi-Davidson faster than Davidson? *SIAM
            J. Matrix Anal. Appl.*, 26(2):522–543, 2005.

[NS96]   Stephen G. Nash and Ariela Sofer. *Linear and Nonlinear Pro-
            gramming*. McGraw-Hill, New York, 1996.

[NW99]   J. Nocedal and S. J. Wright. *Numerical Optimization*. Springer
            Series in Operations Research. Springer-Verlag, New York,
            1999.

[NZ05]   Guy Narkiss and Michael Zibulevsky. Sequential subspace opti-
            mization method for large-scale unconstrained problems. Tech-
            nical Report CCIT No. 559, EE Dept., Technion, Haifa, Israel,
            September 2005.

[OH05]   Shan Ouyang and Yingbo Hua. Bi-iterative least-square
            method for subspace tracking. *IEEE Trans. Signal Process.*,
            53(8, part 2):2984–2996, 2005.

[Oja89]   Erkki Oja. Neural networks, principal components, and sub-
            spaces. *Int. J. Neural Syst.*, 1:61–68, 1989.

[OM01]   Brynjulf Owren and Arne Marthinsen. Integration methods based on canonical coordinates of the second kind. *Numer. Math.*, 87(4):763–790, 2001.

[O'N83]   Barrett O'Neill. *Semi-Riemannian Geometry*, volume 103 of *Pure and Applied Mathematics*. Academic Press Inc. [Harcourt Brace Jovanovich Publishers], New York, 1983.

[OR70]   J. M. Ortega and W. C. Rheinboldt. *Iterative Solution of Nonlinear Equations in Several Variables*. Academic Press, New York, 1970.

[OW00]   B. Owren and B. Welfert. The Newton iteration on Lie groups. *BIT*, 40(1):121–145, 2000.

[Par80]   Beresford N. Parlett. *The symmetric eigenvalue problem*. Prentice-Hall Inc., Englewood Cliffs, N.J., 1980. Prentice-Hall Series in Computational Mathematics.

[Pha01]   Dinh Tuan Pham. Joint approximate diagonalization of positive definite Hermitian matrices. *SIAM J. Matrix Anal. Appl.*, 22(4):1136–1152, 2001.

[Plu05]   M. D. Plumbley. Geometrical methods for non-negative ICA: Manifolds, Lie groups and toral subalgebras. *Neurocomputing*, 67:161–197, 2005.

[PLV94]   R. V. Patel, A. J. Laub, and P. M. Van Dooren. *Numerical Linear Algebra Techniques for Systems and Control*. IEEE Press, Piscataway, NJ, 1994.

[Pol71]   E. Polak. *Computational Methods in Optimization. A Unified Approach*. Mathematics in Science and Engineering, Vol. 77. Academic Press, New York, 1971.

[Pow70]   M. J. D. Powell. A new algorithm for unconstrained optimization. In *Nonlinear Programming (Proc. Sympos., Univ. of Wisconsin, Madison, Wis., 1970)*, pages 31–65. Academic Press, New York, 1970.

[Pow84]   M. J. D. Powell. Nonconvex minimization calculations and the conjugate gradient method. In *Numerical Analysis (Dundee, 1983)*, volume 1066 of *Lecture Notes in Math.*, pages 122–141. Springer, Berlin, 1984.

[PR69]   E. Polak and G. Ribière. Note sur la convergence de méthodes de directions conjuguées. *Rev. Française Informat. Recherche Opérationnelle*, 3(16):35–43, 1969.

[Prz03]   Maria Przybylska. Isospectral-like flows and eigenvalue prob-
            lem. *Future Generation Computer Syst.*, 19:1165–1175, 2003.

[PW79]   G. Peters and J. H. Wilkinson. Inverse iteration, ill-conditioned
            equations and Newton's method. *SIAM Rev.*, 21(3):339–360,
            1979.

[RR00]   Kamran Rahbar and James P. Reilly. Geometric optimiza-
            tion methods for blind source separation of signals. In *In-
            ternational Conference on Independent Component Analysis
            ICA2000, Helsinki, Finland*, June 2000.

[RR02]   André C. M. Ran and Leiba Rodman. A class of robustness
            problems in matrix analysis. In *Interpolation theory, systems
            theory and related topics (Tel Aviv/Rehovot, 1999)*, volume 134
            of *Oper. Theory Adv. Appl.*, pages 337–383. Birkhäuser, Basel,
            2002.

[RSS00]   Marielba Rojas, Sandra A. Santos, and Danny C. Sorensen.
            A new matrix-free algorithm for the large-scale trust-region
            subproblem. *SIAM J. Optim.*, 11(3):611–646, 2000.

[Saa92]   Youcef Saad. *Numerical Methods for Large Eigenvalue Prob-
            lems*. Algorithms and Architectures for Advanced Scientific
            Computing. Manchester University Press, Manchester, U.K.,
            1992.

[Saa96]   Yousef Saad. *Iterative methods for sparse linear systems*.
            http://www-users.cs.umn.edu/~saad/, 1996.

[Sak96]   Takashi Sakai. *Riemannian Geometry*, volume 149 of *Trans-
            lations of Mathematical Monographs*. American Mathemati-
            cal Society, Providence, RI, 1996. Translated from the 1992
            Japanese original by the author.

[SBFvdV96] Gerard L. G. Sleijpen, Albert G. L. Booten, Diederik R.
            Fokkema, and Henk A. van der Vorst. Jacobi-Davidson type
            methods for generalized eigenproblems and polynomial eigen-
            problems. *BIT*, 36(3):595–633, 1996. International Linear Al-
            gebra Year (Toulouse, 1995).

[SE02]   Valeria Simoncini and Lars Eldén. Inexact Rayleigh quotient-
            type methods for eigenvalue computations. *BIT*, 42(1):159–
            182, 2002.

[SHS06]   Hao Shen, Knut Hüper, and Alexander J. Smola. Newton-like
            methods for nonparametric independent component analysis.
            In Irwin King, Jun Wang, Laiwan Chan, and DeLiang Wang,
            editors, *Neural Information Processing*, volume 4232 of *LNCS*,
            pages 1068–1077. Springer, 2006.

[Shu86]   Michael Shub. Some remarks on dynamical systems and nu-
            merical analysis. In L. Lara-Carrero and J. Lewowicz, editors,
            Proc. VII ELAM., pages 69–92. Equinoccio, U. Simón Bolívar,
            Caracas, 1986.

[SK04]   Anuj Srivastava and Eric Klassen. Bayesian and geometric
            subspace tracking. *Adv. in Appl. Probab.*, 36(1):43–56, 2004.

[SM06]   Andreas Stathopoulos and James R. McCombs. Nearly opti-
            mal preconditioned methods for Hermitian eigenproblems un-
            der limited memory. Part II: Seeking many eigenvalues. Techni-
            cal Report WM-CS-2006-02, Department of Computer Science,
            College of William and Mary, Williamsburg, VA, June 2006.

[Smi93]   Steven Thomas Smith. *Geometric Optimization Methods for
            Adaptive Filtering.* PhD thesis, Division of Applied Sciences,
            Harvard University, Cambridge, MA, May 1993.

[Smi94]   Steven T. Smith. Optimization techniques on Riemannian
            manifolds. In *Hamiltonian and gradient flows, algorithms and
            control*, volume 3 of *Fields Inst. Commun.*, pages 113–136.
            Amer. Math. Soc., Providence, RI, 1994.

[Smi97]   Paul Smit. *Numerical Analysis of Eigenvalue Algorithms Based
            on Subspace Iterations.* PhD thesis, CentER, Tilburg Univer-
            sity, P.O. Box 90153, 5000 LE Tilburg, The Netherlands, 1997.

[Sor02]   Danny C. Sorensen. Numerical methods for large eigenvalue
            problems. *Acta Numer.*, 11:519–584, 2002.

[Spi70]   Michael Spivak. *A comprehensive introduction to differential
            geometry. Vol. One.* Published by M. Spivak, Brandeis Univ.,
            Waltham, MA, 1970.

[Sri00]   Anuj Srivastava. A Bayesian approach to geometric subspace
            estimation. *IEEE Trans. Signal Process.*, 48(5):1390–1400,
            2000.

[SS92]   J. M. Sanz-Serna. Symplectic integrators for Hamiltonian prob-
            lems: an overview. *Acta Numer.*, 1:243–286, 1992.

[SS98]   Andreas Stathopoulos and Yousef Saad. Restarting techniques
            for the (Jacobi-)Davidson symmetric eigenvalue methods. *Elec-
            tron. Trans. Numer. Anal.*, 7:163–181, 1998. Large scale eigen-
            value problems (Argonne, IL, 1997).

[ST00]   Ahmed Sameh and Zhanye Tong. The trace minimization
            method for the symmetric generalized eigenvalue problem. *J.
            Comput. Appl. Math.*, 123(1-2):155–175, 2000. Numerical anal-
            ysis 2000, Vol. III. Linear algebra.

[Sta05]   Andreas Stathopoulos. Nearly optimal preconditioned methods for Hermitian eigenproblems under limited memory. Part I: Seeking one eigenvalue. Technical Report WM-CS-2005-03, Department of Computer Science, College of William and Mary, Williamsburg, VA, July 2005.

[Ste83]   Trond Steihaug. The conjugate gradient method and trust regions in large scale optimization. *SIAM J. Numer. Anal.*, 20(3):626–637, 1983.

[Ste01]   G. W. Stewart. *Matrix Algorithms. Vol. II.* Society for Industrial and Applied Mathematics (SIAM), Philadelphia, PA, 2001. Eigensystems.

[Str97]   P. Strobach. Bi-iteration SVD subspace tracking algorithms. *IEEE Trans. Signal Process.*, 45(5):1222–1240, 1997.

[SVdV96]   Gerard L. G. Sleijpen and Henk A. Van der Vorst. A Jacobi-Davidson iteration method for linear eigenvalue problems. *SIAM J. Matrix Anal. Appl.*, 17(2):401–425, 1996.

[SvdVM98]   Gerard L. G. Sleijpen, Henk A. van der Vorst, and Ellen Meijerink. Efficient expansion of subspaces in the Jacobi-Davidson method for standard and generalized eigenproblems. *Electron. Trans. Numer. Anal.*, 7:75–89, 1998. Large scale eigenvalue problems (Argonne, IL, 1997).

[SW82]   Ahmed H. Sameh and John A. Wisniewski. A trace minimization algorithm for the generalized eigenvalue problem. *SIAM J. Numer. Anal.*, 19(6):1243–1259, 1982.

[TA98]   Pham Dinh Tao and Le Thi Hoai An. A d.c. optimization algorithm for solving the trust-region subproblem. *SIAM J. Optim.*, 8(2):476–505, 1998.

[TL02]   Nickolay T. Trendafilov and Ross A. Lippert. The multimode Procrustes problem. *Linear Algebra Appl.*, 349:245–264, 2002.

[Toi81]   Ph. L. Toint. Towards an efficient sparsity exploiting Newton method for minimization. In I. S. Duff, editor, *Sparse Matrices and Their Uses*, pages 57–88. Academic Press, London, 1981.

[Tre99]   Nickolay T. Trendafilov. A continuous-time approach to the oblique Procrustes problem. *Behaviormetrika*, 26:167–181, 1999.

[Udr94]   Constantin Udrişte. *Convex functions and optimization methods on Riemannian manifolds*, volume 297 of *Mathematics and its Applications*. Kluwer Academic Publishers Group, Dordrecht, 1994.

[vdE02]   Jasper van den Eshof. The convergence of Jacobi-Davidson iterations for Hermitian eigenproblems. *Numer. Linear Algebra Appl.*, 9(2):163–179, 2002.

[vdV03]   Henk A. van der Vorst. *Iterative Krylov Methods for Linear Systems*, volume 13 of *Cambridge Monographs on Applied and Computational Mathematics*. Cambridge University Press, Cambridge, 2003.

[Vid95]   M. Vidyasagar. Minimum-seeking properties of analog neural networks with multilinear objective functions. *IEEE Trans. Automat. Control*, 40(8):1359–1375, 1995.

[Vid02]   M. Vidyasagar. *Nonlinear Systems Analysis*, volume 42 of *Classics in Applied Mathematics*. Society for Industrial and Applied Mathematics (SIAM), Philadelphia, PA, 2002. Reprint of the second (1993) edition.

[War83]   Frank W. Warner. *Foundations of differentiable manifolds and Lie groups*, volume 94 of *Graduate Texts in Mathematics*. Springer-Verlag, New York, 1983. Corrected reprint of the 1971 edition.

[WD05]   Jérôme M. B. Walmag and Éric J. M. Delhez. A note on trust-region radius update. *SIAM J. Optim.*, 16(2):548–562, 2005.

[Wil65]   J. H. Wilkinson. *The Algebraic Eigenvalue Problem*. Clarendon Press, Oxford, 1965.

[Yan95]   Bin Yang. Projection approximation subspace tracking. *IEEE Trans. Signal Process.*, 43(1):95–107, Jan. 1995.

[Yan07]   Y. Yang. Globally convergent optimization algorithms on Riemannian manifolds: Uniform framework for unconstrained and constrained optimization. *J. Optim. Theory Appl.*, 132(2):245–265, 2007.

[Yer02]   Arie Yeredor. Non-orthogonal joint diagonalization in the least-squares sense with application in blind source separation. *IEEE Trans. Signal Process.*, 50(7):1545–1553, 2002.

[YL99]   Wei-Yong Yan and James Lam. An approximate approach to H^2 optimal model reduction. *IEEE Trans. Automat. Control*, 44(7):1341–1358, 1999.

[Zho06]   Yunkai Zhou. Studies on Jacobi-Davidson, Rayleigh quotient iteration, inverse iteration generalized Davidson and Newton updates. *Numer. Linear Algebra Appl.*, 13(8):621–642, 2006.

Index